Ecological Knowledge and Environmental Problem-Solving

CONCEPTS AND CASE STUDIES

Committee on the Applications of Ecological
Theory to Environmental Problems
Commission on Life Sciences
National Research Council

NATIONAL ACADEMY PRESS
Washington, D.C. 1986

National Academy Press   2101 Constitution Ave., NW   Washington, DC 20418

NOTICE: The project that is the subject of this report was approved by the Governing Board of the National Research Council, whose members are drawn from the councils of the National Academy of Sciences, the National Academy of Engineering, and the Institute of Medicine. The members of the committee responsible for the report were chosen for their special competences and with regard for appropriate balance.

This report has been reviewed by a group other than the authors according to procedures approved by a Report Review Committee consisting of members of the National Academy of Sciences, the National Academy of Engineering, and the Institute of Medicine

The Research Council was established by the National Academy of Sciences in 1916 to associate the broad community of science and technology with the Academy's purposes of furthering knowledge and of advising the federal government. The Research Council operates in accordance with general policies determined by the Academy under the authority of its congressional charter of 1863, which establishes the Academy as a private, nonprofit, self-governing membership corporation. The Research Council has become the principal operating agency of both the National Academy of Sciences and the National Academy of Engineering in the conduct of their services to the government, the public, and the scientific and engineering communities. It is administered jointly by both Academies and the Institute of Medicine. The National Academy of Engineering and the Institute of Medicine were established in 1964 and 1970, respectively, under the charter of the National Academy of Sciences.

This study has been supported by the National Research Council Fund, a pool of private, discretionary, nonfederal funds that is used to support a program of Academy-initiated studies of national issues in which science and technology figure significantly. The NRC Fund consists of contributions from a consortium of private foundations, including the Carnegie Corporation of New York, the Charles E. Culpeper Foundation, the William and Flora Hewlett Foundation, the John D. and Catherine T. MacArthur Foundation, the Andrew W. Mellon Foundation, the Rockefeller Foundation, and the Alfred P. Sloan Foundation; the Academy Industry Program, which seeks annual contributions from companies that are concerned with the health of U.S. science and technology and with public policy issues with technological content; and the National Academy of Sciences and the National Academy of Engineering endowments.

The study also was supported in part by the U.S. Environmental Protection Agency under Assistance Agreement No. CX-811943-01-0 to the National Academy of Sciences; however, the information in this document may not necessarily reflect the views of the agency and no official endorsement should be inferred.

Library of Congress Cataloging-in-Publication Data

Ecological knowledge and environmental problem-solving: Concepts and case studies

Bibliography: p.

1. Nature conservation—Case studies.   2. Environmental protection—Case studies.   3. Conservation of natural resources—Case studies.   4. Ecology. I. National Research Council (U.S.). Commission on Life Sciences. Committee on Applications of Ecological Theory to Environmental Problems.

QH75.U86   1986   363.7   86-817
ISBN 0-309-03645-3

Printed in the United States of America

Committee on the Applications of Ecological Theory to Environmental Problems

GORDON H. ORIANS, *Chairman*, Institute for Environmental Studies, University of Washington, Seattle, Washington

JOHN BUCKLEY, P.O. Box 263, Whitney Point, New York

WILLIAM CLARK, International Institute for Applied Systems Analysis, Laxenburg, Austria

MICHAEL E. GILPIN, Department of Biology, University of California at San Diego, La Jolla, California

CARL F. JORDAN, Institute of Ecology, University of Georgia, Athens, Georgia

JOHN T. LEHMAN, Division of Biological Science, University of Michigan, Ann Arbor, Michigan

ROBERT M. MAY, Department of Biology, Princeton University, Princeton, New Jersey

GORDON A. ROBILLIARD, Entrix Inc., Concord, California

DANIEL S. SIMBERLOFF, Department of Biological Science, The Florida State University, Tallahassee, Florida

Consultant

W. JAMES ERCKMANN, Institute for Environmental Studies, University of Washington, Seattle, Washington

National Research Council Staff

DAVID POLICANSKY, *Staff Officer*
NORMAN GROSSBLATT, *Editor*
AGNES GASKIN, *Secretary*

Preface

This report is the product of the recognition by the Board on Basic Biology of the National Research Council's Commission on Life Sciences that it should be concerned with the basic biology of ecosystems. The first aspect of ecology to receive attention by the Board was the current status of ecological theory and concepts and their applicability to specific environmental problems. A workshop on this topic had been held September 22-23, 1979, at the National Academy of Sciences in Washington, D.C. The 15 participants represented diverse approaches to ecology and had extensive experiences in marine, freshwater, and terrestrial environments. Discussion during the workshop made it clear that various approaches to ecology could profitably be evaluated from the point of view of the data needed to test them, the difficulty of obtaining those data, the domain over which models can be presumed to be applicable, and special precautions in the use of results. The group also recognized that further exploration of models and concepts would be more productive if it were done in the context of potential applications to environmental problem-solving.

The deliberations of this workshop laid the groundwork for the establishment of the Committee on Applications of Ecological Theory to Environmental Problems, which consists of persons with varied backgrounds in basic ecology, environmental management, and problem-solving. The chief basis of the activities of the committee was the perception that, whereas much about the functioning of ecological systems remains poorly understood, we commonly fail to use even available information when attempting to solve environmental problems. This failure has many causes, among which are the difficulty of determining which components of ecological knowledge might be most usefully applied to particular environmental problems and the difficulty of finding the relevant information. Those difficulties in turn are partly a result of the failure of academic ecologists to attempt to determine where and how their knowledge might be applied and of a lack of communication among generators and users of new knowledge.

The committee recognized that it could not attempt to cover all aspects of ecology without producing a report of unacceptable length. Therefore, we concentrated our attention on ecological knowledge—broadly conceived to include theories, models, data, and concepts—and left methods and techniques of data analysis to be considered in detail elsewhere. This concentration should not be construed as a belief that methods and techniques are unimportant. The committee fully recognizes that such methods as satellite imaging and the use of powerful computers permit analyses of ecological systems on scales of space and time that were previously impossible. These methods are more than just "tools." Nonetheless, the most powerful analytical systems are not substitutes for biological insights or imaginative questioning and hypothesizing. It is to these issues that the committee directed the bulk of its efforts.

The committee also recognized that a report dominated by abstract considerations of ecological knowledge would fail to convey vividly and convincingly how ecological knowledge might improve environmental problem-solving. Therefore, we decided to develop a set of case histories that would illustrate in concrete ways how ecological information has been used and how it has influenced the ways in which problems have been conceived and solutions approached. Because the manifold problems with the quality of assessments of environmental effects of projects have been well noted and thoroughly discussed by others, the committee decided to use examples of the creative use of ecological information, believing that a good example is more instructive than a bad one.

The committee carried out its tasks by means of a series of meetings in which the major format of the report was decided on. Once the plan of the report was determined, subgroups were formed to draft the various chapters of Part I. Another subgroup had the specific task of finding a series of acceptable cases for study, choosing the best ones from among a long list, and selecting persons to write the case studies. The cases were selected to illustrate the use of ecological information in dealing with a wide variety of environmental problems, from management of single species to issues involving complete ecosystems over broad areas. Some of the case studies were written by persons actively involved in the projects themselves. Others were written by members of the committee who had familiarity with the examples and who could therefore write about them with some depth of understanding. To make the case studies more informative, the committee prepared an evaluation of each one, pointing out how the use of ecological information had been particularly effective and how even better use of ecological knowledge could have been made. We hope that those comments will help to improve future studies of similar types.

The work of the committee was supported largely by the National Research Council Fund. Additional support was received from the Office of Research and Development of the Environmental Protection Agency for preparing material on biological monitoring (Chapter 7). The Office of Federal Activities of the Environmental Protection Agency and the Office of Environmental Analysis of the Department of Energy provided support for a joint U.S.-Canadian workshop on cumulative environmental effects. The workshop provided much of the basis of the chapter on cumulative environmental effects (Chapter 9).

A project of this magnitude depends on the cooperation and efforts of many persons in addition to those serving on the committee. In our attempts to find good case studies, we contacted many ecologists in North America and Europe. We give special thanks to the authors of the case studies, who labored so hard to provide us with documents that followed the basic format we had established for them and who were also able to convince us that it was sometimes inappropriate to follow those guidelines slavishly.

The report has been vastly improved by the devoted efforts of the National Research Council staff under the guidance of David Policansky. Dr. Policansky not only provided a valuable interface between the committee and the Research Council, but also contributed extensively to the conceptual development of the project and to the writing of the report, including a case study. He was the person who consistently and insistently reminded the committee members of their duties and responsibilities, thereby preventing the report from appearing even later than it has. Similarly, W. James Erckmann, who assisted in production of the report, played a key role in the writing of the chapters of Part I and in the development of the project and the case studies. We are deeply indebted to Norman Grossblatt of the Commission on Life Sciences for his thoughtful, careful editing; he has immeasurably improved the clarity and style of the report. The report has also benefited from the patient and insightful clerical support of Agnes Gaskin of the Board on Basic Biology and

Jeanette Pederson of the Institute for Environmental Studies, as well as from the comments of several reviewers. Without support of such quality and quantity, our report could not have achieved the quality we now believe it possesses.

GORDON H. ORIANS
Chairman
Committee on Applications of
    Ecological Theory to Environmental
    Problems

Contents

Introduction

Concern for the quality of the environment has been evident in the United States for over 100 years. One result of this concern has been the passage of laws, regulations, and treaties that require actions to preserve environmental quality or that inhibit actions detrimental to the environment. These legislative acts have caused the investment of large amounts of time, effort, and money in analyses of the possible environmental impacts of specific development projects or management decisions. For two reasons, the effectiveness of these analyses and the magnitude of resources devoted to them are controversial (Goodman, 1975; Schindler, 1976; Suter, 1981): available scientific information is often not used effectively in their preparation, and research conducted to provide additional information is often inadequate. It is also commonly perceived that massive studies yield predictions that are little better than those produced with much less effort, cost, and time. This report, by the Committee on Applications of Ecological Theory to Environmental Problems, in the Board on Basic Biology of the National Research Council's Commission on Life Sciences, deals with both problems. It explores how the scientific tools of ecology can be used more effectively in dealing with a variety of environmental problems.

"Ecological theory," as described in standard textbooks on ecology, is seldom applied directly to environmental problems. But ecological "knowledge"—including not only theory, but also facts, observations, research results, syntheses, models, and methods of investigation—has been extremely important in developing approaches to a wide range of

1

environmental problems. This report discusses knowledge that has proved useful and how it has been used. It deals with a wide range of environmental issues, including prediction and management of environmental impacts, management of renewable resources, protection and restoration of species and ecosystems, control of agricultural and silvicultural pests, and use of generic ecological studies to promote understanding of classes of environmental problems. Rather than focusing on what is wrong with the way ecology is often applied, we use successful applications to show how ecological knowledge can be valuable when used appropriately.

Environmental assessment and management are based on societal goals, which are not always clearly articulated. The social effects of environmental changes and the way goals are developed by society are important aspects of environmental problems. Nevertheless, we treat only the biological side of environmental problems, to avoid taking on an unmanageably large task.

The report does not attempt to cover all aspects of ecology. To have done so would have made our task unacceptably difficult and the report unacceptably long. We have focused on different kinds of ecological knowledge and appropriate arenas for their use, but have restricted the coverage of subjects and have not provided details of methods, except as examples or in the case studies.

The subjects covered were determined by the experiences of the committee members. Ecological subjects that have interfaces with chemistry, geology, and hydrology have been slighted, except in some of the case studies (e.g., Chapters 20, 21, 23, and 24). Thus, detailed discussions of such subjects as CO_2 buildup, acid deposition, ecotoxicology, and the effects of complex chemical mixtures in the environment are lacking. For treatment of these important topics, the reader is referred to several recent reviews (e.g., Harshbarger and Black, in press; Maki *et al.*, 1980; National Research Council, 1976, 1981, 1982, 1983a,b,c, 1984, in press a,b; Reif, 1984).

Among the most important methods—some of which have changed the scope and scale of ecological research—that do not receive explicit treatment in this report are equations for estimating environmental partitioning of chemicals and related phenomena, such as transformations and transport; hazard evaluation protocols and risk assessment; validation; decision analysis; remote sensing; interpretation of photographs and satellite photography; and bioassays. Some methods—such as modeling, statistical analysis, and use of computers—are not discussed specifically, but appear in many of the chapters of Part I and the case studies.

STRUCTURE AND AUDIENCE OF THE REPORT

Structure

The report is divided into "What we know" (Part I) and "How our knowledge is applied" (Part II). Chapters 1-5 treat a number of categories of ecological knowledge and Chapters 6-10 discuss procedures and kinds of knowledge that are common to most environmental problems. Part II presents studies of 13 specific cases of environmental problem-solving.

The first three chapters deal with populations of single species, interactions between two species, and community ecology. They focus on the kinds of information that are available and how that information might be useful in solving environmental problems. Chapter 4 discusses the flow of energy and materials through ecosystems; Chapter 5 deals with the effects of changes in time and space scales on ecological processes and products. Chapter 6 explores the benefits of treating environmental manipulations as large-scale experiments so that we can learn more from them. Chapters 7 and 8 discuss biological monitoring and the management of ecological systems in the face of the substantial uncertainty associated with the behavior of those systems.

Chapter 9 deals with cumulative environmental changes that result from repeated perturbations. There is increasing recognition that some of our most severe environmental problems involve the cumulative effects of many small local actions—individually insignificant, but collectively creating major regional and even global changes. The discussion points out the major issues associated with cumulative effects, which have only recently begun to receive serious consideration, but which present some of the most difficult challenges to decision-makers.

Chapter 10 deals with procedures for application of ecological knowledge. It begins with a discussion of how ecology can help in conceptualizing a problem and in devising appropriate approaches to its solution. We focus on the design of ecological studies for predicting and managing the effects of manipulations, emphasizing the importance of monitoring projects and decisions to determine their consequences.

Part II discusses 13 cases chosen as instructive examples of the application of ecological knowledge to various kinds of problems. The cases were selected because the committee members and consultants who were asked for suggestions were familiar with them and believed that they would reveal the challenges of applying ecological knowledge, show the difficulty of predicting accurately the behavior of complex systems, and point out the substantial value of monitoring the results of manipulations. The cases were chosen to exemplify scientifically sound uses of ecological knowledge. No attempt was made to achieve comprehensive coverage; of necessity, the cases are a small sample of a large universe of potentially informative examples.

Audience

This report is intended for a broad audience, including:

- Those who prepare, receive, and use environmental evaluations and management plans.
    - Legislators and regulators concerned with environmental issues.
    - Teachers and students in the natural and environmental sciences.

TYPES OF ECOLOGICAL KNOWLEDGE

The chapters in Part I discuss the ways in which ecological knowledge has been applied in a variety of environmental problems, from the management of renewable resources to the prediction of environmental impacts. A few common themes run through these chapters. The themes are related to the basic characteristics of ecological systems that are of fundamental importance to understanding their responses to perturbations.

Complex Linkages

Because of the complex linkages between species in ecosystems, the effects of changes are often indirect. Obvious and direct influences on the objects of study are sometimes not as important as less obvious indirect influences. Problem-solvers can find it helpful to develop diagrams that show both direct and indirect influences and show the pathways through which indirect agents exert their influences (Andrewartha and Birch, 1984). Such diagrams are accounting schemes that help to identify important environmental factors. The case studies in Part II contain many examples of indirect effects, such as the effect of DDT on birds far from sites of its application (Chapter 24) and the mobilization of mercury that results from changing the level of a lake (Chapter 21). These phenomena could not have been predicted by analyses that concentrated on direct effects. In addition, many ecological interactions are characterized by strong non-linear effects brought about by slight changes in key factors.

Density Dependence

Because environments are finite and resources might not be renewed as quickly as they are harvested, the success of individuals often depends on the number of individuals of the same and other species in their environments. Often, but not always, conditions deteriorate for individuals as the numbers of individuals of the same species, predators, and parasites increase. Conversely, conditions can improve when the numbers of individuals of those classes decrease. It is therefore often possible to extract much higher sustainable yields from populations than would be expected on the basis of an examination of birth, growth, and death rates in an unharvested population existing close to the limit set by environmental resources. Alternatively, high population densities in most areas could prevent individuals displaced by a perturbation from moving into other areas. Very low densities might influence the probability of finding mates, erode genetic diversity of a population, and increase the probability of its extinction. The existence of density-dependent effects does not necessarily imply that populations exist at steady densities. These effects can cause populations to fluctuate regularly or irregularly (May 1973), as did the blowflies studied by Nicholson and Bailey (1935) and the bean weevil *Callosubrochus chinensis* studied by Utida (1957). Moreover, density-independent effects can often mask density-dependent ones, because they are often much larger.

The Uniqueness of Individuals

Each individual in a sexually reproducing population is unique, as opposed to many nonliving entities (e.g., all CO_2 molecules can be treated as identical). A species is composed of different age and sex classes, and genetic variation occurs within each class. Plants with open growth systems often have substantial intraindividual variability that is important for consumers of the plants (Whitham *et al.*, 1984). Because of genetic variability, management practices can create selective pressures that cause genetic and thus phenotypic changes in a population that sometimes subvert the goals of the management practices (Chapter 1).

Keystone Species

Keystone species are those which exert influences over other members of their ecological communities out of proportion to their abundances. Keystone species can have various roles in ecological communities. For example, dominant plants are the major photosynthesizers in communities, and they also form the physical structure in which many interactions in the system take place. Keystone predators preferentially eat prey that would be competitive dominants in the absence of predation. As a result, keystone predators create conditions favorable for the existence of species that are crowded out of systems with less predation. Keystone predators have been most commonly identified among predators on sessile prey— such as plants and sessile marine invertebrates of rocky shores—that compete for space (Harper, 1969; Hurlbert and Mulla, 1981; Lubchenco, 1978; Morin, 1983; Paine, 1974; Paine and Vadas, 1969). Keystone mutualists are essential for survival of other members of their communities, even though the amount of resources they consume is sometimes small. Well-known examples are pollinators and frugivores, ants that patrol and protect plants, and microorganisms associated with vascular plants (Alexander, 1971; Gilbert, 1975; Howe, 1981; Janzen, 1966). Because changes in the populations of keystone species can influence community dynamics in major ways, astute environmental problem-solvers are alert to keystone relationships in the communities being altered (Chapter 3).

Biological Magnification

Organisms are united through food chains, so materials taken up by prey can accumulate and become concentrated in their predators. Passage of substances through several trophic levels in a community can result in concentrations hundreds of times those initially present in the environment. The greatly magnified indirect effects of pesticides like DDT, which are metabolically stable and fat-soluble, constitute a striking case of biological magnification (Chapter 24).

Population Fragmentation

Human activities often change large patches of, say, forest or prairie into small patches surrounded by land devoted to other purposes (such as agriculture, forest plantations, highways, and cities). Fragmentation of habitats sets in motion processes that can cause unexpected changes in the system. Among these processes is the loss of species unable to exist in small patches (Chapter 5). An important management implication is the need to provide corridors of suitable habitat through which individuals can travel between more extensive patches of appropriate habitat. Ignoring the larger, heterogeneous environment in which local populations of interest are set can lead to poor management of those populations (Chapters 17 and 19).

Stability Boundaries

"Stability," the tendency of ecological systems to remain in a relatively constant state, can result from several processes (Orians, 1975). Of concern to the manager and environmental problem-solver are the resistance of a system or a component of it to change, the speed with which it returns to its previous state when the perturbation ends, and its stability boundaries—the range over which it can be changed without leaving it unable to return to its previous state (Holling, 1973). The crossing of stability boundaries or thresholds by individual organisms can lead to large changes in their functioning. Photoperiods above a threshold might be necessary to induce a plant to flower or an animal to come into breeding condition (e.g., Farner, 1964; Salisbury and Ross, 1978; Sundararaj and Vasali, 1976). A critical minimal density of prey might be necessary for a particular kind of predator to survive in an area. A specific patch size of suitable habitat might be necessary for a particular species to survive (Chapter 17). An organism might be able to withstand temperatures down to or up to a critical point, but not beyond.

Stability boundaries are also important for more complex ecological systems. For example, environmental change and overharvesting can cause persistent changes in the composition of fish stocks (Daan, 1980; Gulland and Garcia, 1984; Chapters 1 and 8). Overgrazed plant communities might be resistant to invasion by new species, even if grazing ceases. These "alternate stable states" are usually less desirable from the human perspective, but some are seen as beneficial, as when a shrub stage resistant to invasion by trees is established as a low-maintenance community along roads and powerline rights of way (Niering and Egler, 1955). Managing systems to achieve greater constancy, a common human objective, can reduce their ability to withstand perturbations. For example, the suppression of spruce budworm populations in eastern Canada through the use of insecticides achieved the intended preservation of the pulp and paper industry in the short run, but left the forests and the economy more vulnerable to outbreaks of a size and intensity that would be impossible in undisturbed forests (Holling, 1973).

It is especially important for the environmental problem-solver to recognize that these thresholds are sometimes not apparent until after they have been crossed. An important role of ecological knowledge is to suggest what such thresholds are likely to be.

Aggregate Variables

Ecologists and other scientists use aggregate variables in several ways. Typically, entities are grouped on the basis of shared characteristics. These might be traits shared because of common ancestry, such as the dorsal, hollow nervous system of all vertebrates; they might result from independent adaptations, as in the aggregate entity "herbivores." Ecological guilds—groups of species that use common resources in similar ways— are examples of aggregates consisting of units connected by strong interactions. Aggregates can also be based on some composite process or product, such as productivity or biomass, that is deemed to be of theoretical or applied interest.

Analysis of complex systems requires the use of aggregate variables. Aggregate variables should be designed to capture the essence of traits or processes. It is the nature of aggregate variables that they lose substantial amounts of information; but *essential* information should not be sacrificed. A major challenge in using ecological information to help in solving environmental problems is to delineate aggregrate variables, their properties, and what will be gained and lost if they are adopted.

Complexity and Uncertainty

Ecological theory is most fully developed for relatively simple systems, which have relatively small numbers of interacting units. Ecology's greatest predictive success occurs in cases that involve only one or two species. That is why, for example, management of game and fish populations through regulation of hunting and fishing is often successful. Predicting the outcome of interactions among many species is much more difficult. Therefore, most environmental problem-solving entails considerable uncertainty. Wise problem-solvers learn to expect the unexpected and to develop approaches that allow them to learn as they proceed, altering their responses as they learn more about the system and avoiding irrevocable actions (Chapter 8).

Scales in Space and Time

Processes and products in ecological systems are strongly influenced by differences in scale (Addicott, 1978; Andrewartha and Birch, 1954, 1984; Levin, 1974; Wiens, 1976; Chapter 5). Large lakes are not simply smaller lakes made bigger, and a large expanse of forest is not the same as a collection of small woodlots. Patchy environments have more edges in proportion to their area than do homogeneous environments; they create dispersal problems for organisms inhabiting them, and they lead to high local extinction rates. Speciation, particularly of large organisms, requires large expanses of suitable habitat (Soulé and Wilcox, 1980)—larger, in fact, than most parks and reserves. Success in achieving some local goal can set in motion processes that cause regional results opposite to those desired (Chapter 19). Populations of some species are maintained by high rates of reproduction in a small fraction of the area that they occupy; populations in the rest of the area depend on a steady supply of immigrants from sites with high reproductive rates. Evolutionary changes can cause both interacting organisms and the results of their interactions to change, often rapidly enough to be important for the environmental problem-solver (Chapter 1).

APPLICATIONS OF ECOLOGICAL KNOWLEDGE

The chapters of Part I are organized in part by their ecological content. Chapters 1-5 deal with a variety of kinds of ecological knowledge, many having specific applicability to different kinds of problem-solving. Chapters 6-10 deal with knowledge and procedures that are more general in their applicability. We briefly summarize here the parts of the report that deal with particular kinds of application (see Table 1) so that users can more easily tap the sources of knowledge most important to them.

Renewable-Resource Management

Renewable-resource management refers to commercial and recreational harvesting of animals and plants. The important topics of population growth, life histories, habitat selection and other behaviors, genetics of populations, and evolution in response to management are dealt with in Chapter 1. Equally important for managers are interactions between populations (Chapter 2) and energetics—i.e., productivity and nutrient requirements (Chapter 4). Examples of renewable-resource management are discussed

TABLE 1   Kinds of Applications of Ecological Knowledge and Where They Are Discussed

Application	Chapter
Renewable-resource management	1, 2, 4, 12, 19
Conservation of species	1-3, 5, 8, 17
Control of pests and diseases	1-3, 13-15, 24
Impact assessment and prediction of effects	6-9, 16, 21-24
Preservation of communities	3-5, 8, 18, 20
Preservation of habitat	7-9, 17, 18, 20, 23
Contaminants and toxic substances	4, 7, 9, 20-22, 24
Mitigation of effects of construction	7, 10, 16, 18, 21
Restoration	3-5, 8, 18
General applications	6-10

in detail in Chapter 12 (managing a fishery) and Chapter 19 (managing a forest).

Conservation of Species

Closely related to the commercial and recreational harvest of plants and animals is the protection of species that lack commercial value, but that have aesthetic value or that play important roles in ecosystems valued for other reasons. Species conservation requires knowledge of the ecology of the species (Chapter 1), its interactions with other species (Chapter 2), and its relationships to its ecological community (Chapter 3). The size and shape of the available habitat and the population size of the species needed to ensure its survival for acceptably long periods are central to management decisions (Chapter 5), as is knowledge of environmental and demographic variability (Chapter 8). Application of such knowledge to the conservation of populations of spotted owls is discussed in Chapter 17.

Control of Pests and Diseases

The control of pests and diseases usually requires knowledge of the natural history and population characteristics of the pests and their victims (Chapter 1) and understanding of their interactions (Chapter 2) and of the ecological communities in which they occur (Chapter 3). Attempts to control pests and diseases often result in the evolution of resistance, which tends to reduce the effectiveness of the programs (Chapter 1). Cases of control are discussed in Chapter 13 (vampire bats), Chapter 14 (biological control of citrus pests), Chapter 15 (malaria vectors), and Chapter 24 (results of attempts to control insects with DDT).

Impact Assessment and Prediction of Effects

Predicting effects and assessing potential impacts of activities require knowledge of present conditions (Chapter 7) and of the variability inherent in nature and the uncertainties of science (Chapter 8); without such knowledge, effects cannot be reliably detected and the causes of change cannot be understood. Because the effects of a project can combine with effects of other projects elsewhere, and the effects of distant projects can add to the effects of the project under consideration, potential cumulative effects might require modifications in project design (Chapter 9). It is often useful to compare the situation under consideration with similar ones elsewhere and to carry out pilot studies (Chapter 6). Cases of impact assessment are discussed in Chapter 16 (predicting the effects of construction on a herd of caribou), Chapter 21 (predicting the effects of freshwater impoundment), Chapter 22 (studies of the effects of ionizing radiation), Chapter 23 (studies of the effects of clearcutting), and Chapter 24 (predicting and learning the effects of the use of DDT and application of that knowledge to other cases).

Preservation of Communities

When the management goal is the preservation of entire communities of organisms, an understanding of community ecology (Chapter 3), of productivity and nutrient cycling (Chapter 4), and of the required space and time scales (Chapter 5) is important. Natural variability in climate and community patterns will also affect preservation efforts (Chapter 8). Chapter 18 (restoration of communities of plants on derelict lands) and Chapter 20 (protecting a lake ecosystem from the effects of eutrophication) discuss examples of such applications.

Preservation of Habitat

The conservation of communities and species requires preservation of the habitat in which those organisms live. Monitoring both before and after efforts are initiated is necessary to determine the success of the efforts (Chapter 7). The variability of natural systems and the cumulative effects of various activities, often far from the site of concern, can strongly influence the outcome of preservation attempts (Chapters 8 and 9). Some aspects of habitat preservation are treated in Chapters 17, 18, 20, and 23.

Contaminants and Toxic Substances

The pervasive and crucial issues of pollution and exposure to toxic chemicals are relatively neglected in this report, but the transport of chemicals is discussed in Chapter 4, indicator species and biological monitoring to detect pollutants and toxic chemicals are discussed in Chapter 7, and the problem of cumulative effects (e.g., several sources of pollutants) is discussed in Chapter 9. Specific examples are presented in Chapter 20 (excess nutrients), Chapter 21 (chemistry and hydrology related to heavy metals), Chapter 22 (ionizing radiation), and Chapter 24 (persistent pesticides).

Mitigation of Effects of Construction

Mitigation of undesired side effects of projects is often economically, aesthetically, or legally motivated. If done after the project is completed, it might take the form of restoration (see below); but the best approach to mitigation is to make it unnecessary by appropriate design of the project. This requires careful planning and scoping (Chapter 10), as well as careful monitoring (Chapter 7). Chapter 16 (protecting caribou during hydro-electric development) and Chapter 21 (raising the water level of a lake) provide examples of including mitigation efforts in the design of a project; Chapter 18 (restoration of derelict land) describes mitigation of effects of projects completed many years earlier.

Restoration

To restore an ecosystem, one must have a goal to work toward. This requires some knowledge of previous conditions. Success in reaching the goal depends on knowledge of community structure (Chapter 3), of productivity and nutrient requirements (Chapter 4), of the space and time scales that will be needed for success (Chapter 5), and of natural variation in biotic and abiotic factors (Chapter 8). Chapter 18 describes the restoration of plant communities on derelict lands in Britain.

General Applications

Some issues, procedures, and kinds of knowledge common to most environmental problems are treated in Chapters 6-10. There is always uncertainty, both in the abiotic environment and in organisms' responses to perturbations (Chapter 8). And there is always a need for rigorous scientific procedures and careful scoping, even when time and money are short (Chapter 10). It is always helpful to compare the case at issue with similar ones; it is often useful to conduct pilot experiments, if feasible (Chapter 6). Potential cumulative effects are often a concern, and monitoring is almost always necessary to ascertain current conditions, to detect variability, and to measure the effects of actions (Chapter 7).

THE CASE STUDIES

The objective of Part II is to show how ecological knowledge has been used in planning and carrying out problem-solving efforts, including how the knowledge has been limited and how it has had to be adapted to the details of specific problems. The case studies are referred to repeatedly in Part I.

The existence and application of scientific knowledge do not guarantee correct predictions. Ecological systems are complex, situations are often unique, and predictive ecological theories are sometimes inadequate. Nonetheless, when scientific predictions turn out to be inaccurate, the results can reveal important aspects of the limitations of even the best available scientific approaches and show the need to increase the knowledge base on which decision-makers depend.

We selected examples to cover a wide array of types of environmental problem-solving (Table 2). In some of our cases—such as those of the North Pacific halibut, the spotted owl, and the Newfoundland caribou— protection of a single species or achievement of a sustained harvest was the goal. In the malaria, red scale, and vampire bat cases, the goal was to eliminate an organism or at least to reduce it enough to render the damage it caused acceptable. Problem-solving oriented toward manipulating a single species can draw on general ecological knowledge, such as population dynamics and predator-prey theory, and on the results of previous efforts. As the case studies show, proper targeting of management usually requires a great deal of knowledge about the natural history of the organisms in question.

Other case studies deal with problem-solving at the community level. The objective might have been the achievement of a sustained steady yield of useful products from a combination of species, as in the New Brunswick forest case, or the creation of a functioning ecological community similar to one that existed before the influence of human disturbance, as in the Lake Washington and derelict lands cases. In some cases, the purpose was to anticipate, and thereby to reduce, the undesirable impacts of a future project. For example, in the Southern Indian Lake case, the lake was altered for hydroelectric power generation, in the hope that it would not impair other uses of the lake.

Still other case studies deal with projects in which the objective was to determine the impacts of practices that take place in many locations over a broad area. Some of these practices are subject to legal regulation (use of pesticides, forestry practices, and uses of nuclear energy), and case studies can help to determine what kinds of legal restraints might be properly applied to reduce unwanted consequences in natural environments. As examples we have chosen the use of DDT in the United States, forest clearcutting, and the effects of nuclear radiation on ecological communities. In the latter two, a major objective was scientific.

To ensure consistent emphasis of particular points, we adopted a general format for the presentation of case studies. The preparer of each case

TABLE 2   Synopsis of Case Studies Presented in Part II

Short Title	Type of Case	Subject
Halibut (Chapter 12)	Renewable-resource management	Managing an international marine fishery
Vampire Bat (Chapter 13)	Pest control	Chemical control of a parasite on cattle in Central America
Red Scale (Chapter 14)	Pest control	Biological control of a pest on citrus crops in California
Malaria (Chapter 15)	Disease control	Experimental study of chemical control of a malaria vector
Caribou (Chapter 16)	Impact assessment	Protecting a caribou herd during hydroelectric development
Spotted Owl (Chapter 17)	Species conservation	Managing a regional population of a rare species
Derelict Lands (Chapter 18)	Ecosystem restoration	Restoring seminatural plant communities in mined areas
New Brunswick Forest (Chapter 19)	Renewable-resource management	Developing a model for long-term regional forestry management
Lake Washington (Chapter 20)	Ecosystem protection	Predicting and controlling cumulative impacts of sewage effluent in an urban lake
Southern Indian Lake (Chapter 21)	Impact assessment	Predicting impacts of raising the level of a subarctic lake
Nuclear Radiation (Chapter 22)	Generic ecological studies	Ecological effects of nuclear radiation
Forest Clearcutting (Chapter 23)	Generic ecological studies	Ecological effects of clearcutting forests
DDT (Chapter 24)	Pesticide use	Ecological effects of DDT

study, whether a committee member or another person intimately involved with the project, was asked to treat in sequence, if possible, the history of the study project, the major environmental problems that it posed, the types of ecological knowledge used in the project, and the role of the knowledge in determining how the problems were perceived, approached, and dealt with. Each case study is followed by a committee comment on the project's strengths and weaknesses and on the lessons that can be drawn from it for other studies of a similar nature. Our concentration on ecological knowledge is not meant to suggest that other sources of knowledge were not used or were necessarily less important in dealing with the environmental problems posed; it merely reflects our specific objective of showing how ecological knowledge can be used creatively in helping to solve environmental problems.

RECOMMENDATIONS

This report does not attempt to provide a "cookbook" for solving environmental problems; that would be impractical. Rather, it is a guide to the process of "cooking," indicating where useful approaches to particular problems can be found in the literature and suggesting how appropriate knowledge and skills might be integrated for dealing with complex environmental problems.

The following recommendations constitute general advice to field workers, managers, and regulators and are intended to form the basis of scientifically sound but flexible environmental problem-solving. The overall approach we recommend is described in Chapter 10. Here we highlight some major recommendations that apply to most environmental problems. If followed, they should increase the probability that solutions chosen for dealing with environmental problems are imaginative and appropriate. More details can be found in the chapters referred to, including the case studies, which constitute recommendations by example.

• *Involve scientists from the beginning.* Scientists should be involved from the beginning of a project in setting goals, in identifying valued ecosystem components, and in scoping the problem (Chapter 10). Scientists do not determine the values attached by society to ecosystem components, but they might know which organisms have important roles in the ecosystem that are not understood or appreciated by the general public. Scientists can help to assemble information about a project site and about similar sites and projects elsewhere. They are also helpful in defining goals, because of their knowledge of potential outcomes of manipulations that might be considered. They can advise on the implications of trying to achieve particular goals, on the measurement of values, and on why seemingly compatible values might conflict. The involvement of scientists does not guarantee success, but it should increase the probability that project plans are appropriate.

• *Treat projects as experiments.* Many projects are carried out on a larger scale than scientific experiments can be. To learn the scientific and management lessons that these projects have to offer, we need to treat them as experiments (Chapter 6). Treating projects as environmental experiments increases the understanding of basic ecological science and allows the testing of predictions made in impact statements. Careful monitoring is essential to anyone who would understand the effects of projects (Chapter 7), test the predictions made in impact statements (Chapter 6), detect changes in baseline conditions (Chapter 8), and detect cumulative effects (Chapter 9). To be most successful, monitoring should be incorporated into the project and study plans (Chapters 7 and 10).

 • *Publish information.* Many people associated with projects learn a great deal from them, but without making their knowledge available to others through publication. If the knowledge were made widely available, similar projects could be managed better and could serve as controls. The cases described in Chapters 12, 14, 15, 18-20, 22, and 23 are good examples.

 • *Set proper boundaries on projects.* Setting appropriate boundaries (Chapters 5 and 9) involves matching the scale of management to the scales of ecosystems. Environmental problems that cross political boundaries are especially difficult to deal with (e.g., Chapters 12 and 23), so the appropriate jurisdiction for management should be chosen carefully (Chapter 9).

 • *Use natural-history information.* Carefully collected natural-history information can help in solving environmental problems (Chapter 1). The success of the cases described in Chapters 12-21 depended on such information; for example, the control of such pests as vampire bats (Chapter 13) and red scale (Chapter 15) was based on identifying key natural-history features of the focal organisms.

 • *Be aware of interactions.* Interactions among populations and among species in communities are complex (Chapters 2-4), and they can influence the effects of perturbations. It is important to remember that the most obvious connections are not necessarily the most important.

 • *Be alert for possible cumulative effects.* Small perturbations, even if individually trivial, can have important cumulative effects when they are repeated in space and time. Therefore, many projects need to be evaluated not only in isolation, but also in the context of the overall frequency of their occurrence. Similarly, when a particular environment is being studied, investigators need to be alert for possible effects of activities and perturbations elsewhere (Chapter 9).

 • *Plan for heterogeneity in space and time.* Environmental patchiness, variation in time, and variation in the species composition of ecological communities conspire to complicate environmental management (Chapter 5). These considerations are particularly important in harvesting populations (for example, the maximal sustainable yield changes as the environment changes) and in conservation. Interpopulation variation, both genetic and behavioral, can also complicate management, because of rapid evolutionary changes (e.g., Chapters 1 and 15).

• *Prepare for uncertainty and think probabilistically.* A goal of management is usually to minimize uncertainty. But uncertainty is unavoidable (Chapter 8), and managers are advised to expect it and to plan for it. In addition, pretending that uncertainty is absent leads to bad planning and inflexibility. Therefore, both scientists and managers must be willing to think in terms of probabilities and to deal with them, as weather forecasters and farmers, sailors, fliers, and the general public do every day.

REFERENCES

Alexander, M. 1971. Microbial Ecology. John Wiley & Sons, New York.

Addicott, J. F. 1978. The population dynamics of aphids on fireweed: A comparison of local populations and metapopulations. Can. J. Zool. 56:2554-2564.

Andrewartha, H. G., and L. C. Birch. 1954. The Distribution and Abundance of Animals. University of Chicago Press, Chicago.

Andrewartha, H. G., and L. C. Birch. 1984. The Ecological Web: More on the Distribution and Abundance of Animals. University of Chicago Press, Chicago.

Daan, N. 1980. A review of replacement of depleted stocks by other species and the mechanisms underlying such replacement. Rapp. P-v. Reun. Cons. Int. Expl. Mer 177:405-421.

Farner, D. S. 1964. The photoperiodic control of reproductive cycles in birds. Am. Sci. 52:137-156.

Gilbert, L. E. 1975. Ecological consequences of a coevolved mutualism between butterflies and plants. Pp. 210-240 in L. E. Gilbert and P. H. Raven, eds. Coevolution of Animals and Plants. University of Texas Press, Austin.

Goodman, D. 1975. The theory of diversity-stability relationships in ecology. Q. Rev. Biol. 50:237-266.

Gulland, J. A., and S. Garcia. 1984. Observed patterns in multi-species fisheries. Pp. 155-190 in R. M. May, ed. Exploitation of Marine Communities. Dahlem Konferenzen. Springer, Berlin.

Harper, J. L. 1969. The role of predation in vegetational diversity. Pp. 48-62 in Diversity and Stability in Ecological Systems. Brookhaven Symposia in Biology 22. Brookhaven National Laboratory, Upton, N.Y.

Harshberger, J. C., and J. J. Black. In press. A strategy for using fish bioassays and surveys to identify and eliminate point source environmental carcinogens. In Towards a Transboundary Monitoring Network: Proceedings of a Workshop. International Joint Commission, Washington, D.C., and Ottawa, Ont.

Holling, C. S. 1973. Resilience and stability of ecological systems. Annu. Rev. Ecol. Syst. 4:1-23.

Howe, H. F. 1981. Removal of wild nutmeg (*Virola suranimensis*) crops by birds. Ecology 62:1093-1106.

Hurlbert, S. H., and M. S. Mulla. 1981. Impacts of mosquitofish (*Gambusia affinis*) predation on plankton communities. Hydrobiologia 83:125-151.

Janzen, D. H. 1966. Coevolution of mutualism between ants and acacias in Central America. Evolution 20:249-275.

Levin, S. A. 1974. Dispersion and population interactions. Am. Nat. 108:207-228.

Lubchenco, J. 1978. Plant species diversity in a marine intertidal community: Importance of hervibore food preference and algal competition abilities. Am. Nat. 112:23-39.

Maki, A. W., K. L. Dickson, and J. R. Cairns, Jr., eds. 1980. Biotransformation and Fate of Chemicals in the Aquatic Environment. American Society for Microbiology, Washington, D.C.

May, R. M. 1973. Stability and Complexity in Model Ecosystems. Princeton University Press, Princeton, N.J.

Morin, P. J. 1983. Predation, competition, and the composition of larval anuran guilds. Ecol. Monogr. 53:119-138.

National Research Council. 1976. Vapor-Phase Organic Pollutants: Volatile Hydrocarbons and Oxidation Products. National Academy of Sciences, Washington, D.C.

National Research Council, 1981. Testing for Effects of Chemicals on Ecosystems. National Academy Press, Washington, D.C.

National Research Council. 1982. Assessment of Multichemical Contamination. Proceedings of an International Workshop, Milan, April 28-30, 1981. National Academy Press, Washington, D.C.

National Research Council. 1983a. Acid Deposition in Eastern North America. National Academy Press, Washington, D.C.

National Research Council. 1983b. Effects of atmospheric transformations of polycyclic aromatic hydrocarbons. Pp. 3-1 to 3-48 in Polycyclic Aromatic Hydrocarbons: Evaluation of Sources and Effects. National Academy Press, Washington, D.C.

National Research Council. 1983c. Changing Climate. Report of the Carbon Dioxide Assessment Committee. National Academy Press, Washington, D.C.

National Research Council. 1984. Groundwater Contamination. National Academy Press, Washington, D.C.

National Research Council. In press, a. Monitoring and Assessment of Trends in Acid Deposition. National Academy Press, Washington, D.C.

National Research Council. In press, b. Methods for Assessing the Effects of Mixtures of Chemicals on Humans and Non-Human Biota. National Academy Press, Washington, D.C.

Nicholson, A. J., and V. A. Bailey. 1935. The balance of animal populations. Proc. Zool. Soc. Lond. 3:551-598.

Niering, W. A., and F. E. Egler. 1955. A shrub community of *Viburnum lentagon*, stable for twenty-five years. Ecology 36:356-360.

Orians, G. H. 1975. Diversity, stability and maturity in natural ecosystems. Pp. 139-150 in W. H. Van Dobben and R. H. Lowe-McConnell, eds. Unifying Concepts in Ecology. W. Junk, The Hague.

Paine, R. T. 1974. Intertidal community structure: Experimental studies on the relationship between a dominant competitor and its principal predator. Oecologia 15:93-120.

Paine, R. T., and R. L. Vadas. 1969. The effects of grazing by sea urchins, *Strongylocentrotus*, on benthic algal populations. Limnol. Oceanogr. 14:710-719.

Reif, A. E. 1984. Synergism in carcinogenesis. J. Natl. Cancer Inst. 73:25-39.

Salisbury, F. B., and C. W. Ross. 1978. Plant Physiology. 2nd ed. Wadsworth, Belmont, Calif.

Schindler, D. W. 1976. The impact statement boondoggle. Science 192:509.

Soulé, M. E., and B. A. Wilcox, eds. 1980. Conservation Biology. Sinauer Associates, Sunderland, Mass.

Sunadararaj, B. I., and S. Vasali. 1976. Photoperiod and temperature control in the regulation of reproduction in the female catfish. J. Fish. Res. Bd. Can. 33:959-971.

Suter, G. W., II. 1981. Commentary: Ecosystem theory and NEPA assessment. Bull. Ecol. Soc. Am. 62:186-192.

Utida, S. 1957. Population fluctuation: An experimental and theoretical approach. Cold Spring Harbor Symp. Quant. Biol. 22:139-151.

Whitham, T. G., A. G. Williams, and A. M. Robinson. 1984. The variation principle: Individual plants as temporal and spatial mosaics of resistance to rapidly evolving pests. Pp. 15-52 in P. W. Price, C. N. Slobodchikoff, and W. S. Gaud, eds. A New Ecology. Wiley-Interscience, New York.

Wiens, J. A. 1976. Population responses to patchy environments. Annu. Rev. Ecol. Syst. 7:81-120.

I

---

*Kinds of Ecological Knowledge and
Their Applications*

1

Individuals and Single Populations

Environmental concern commonly focuses on populations of organisms, whether the goal is protection of valued species, harvest of economically profitable species, or control of economically destructive species. We are interested in predicting and controlling changes in size and structure of populations that occur in response to environmental change, whether anthropogenic or not. The ecology of individuals and populations is of particular relevance to this interest. Population biology, the subject of this chapter, encompasses many kinds of research, from the study of the details of life history and behavior to the construction of mathematical models of the dynamics and genetics of multiple populations of a species over a large area.

A life history encompasses an individual's interactions with its physical and biological environment throughout its lifetime. Research into life histories has taken several paths, all of them valuable in the management of populations: detailed studies of the ecology of individual species (e.g., Chapters 12-17), comparative studies of groups of species (e.g., Chapters 13 and 15), and theoretical treatments of the evolution of life-history patterns (e.g., Stearns, 1976; Chapters 15 and 17). Studies of particular species yield the detailed information needed for management with respect to nutrient and habitat requirements, important interactions with other species, reproductive requirements, and significant behavioral idiosyncrasies. Moreover, research on organisms with complex life cycles has shown the importance of choosing the appropriate stages in the life cycle for management intervention. Comparative studies reveal general patterns that help focus individual studies and provide managers with guidance in the absence of detailed information. Theoretical research focuses attention on the elements of life history important in solving long-term management problems (e.g., Beddington, 1974; Lewontin, 1965; May, 1980). Thorough knowledge of species' life histories has a broad range of applicability to problems of population management, including captive propagation programs for endangered species (Frankel and Soulé, 1981), pest and disease control (Chapters 13-15), species protection (Chapter 17), harvesting (Chapter 12), predicting environmental impacts (Chapter 16), and restoring plant communities (Chapter 18).

As valuable as life-history information is for the prediction and control of population behavior, it provides only partial insight into the causes and consequences of changes in population numbers and composition. To determine how a population will respond to an increase in mortality due to harvesting or stress or how effective a given procedure might be for improving reproductive output, we need some understanding of population dynamics. Research in population dynamics has ranged from field studies designed to determine what factors affect population sizes in an area to theoretical studies of how such factors can act together to "regulate" population size over long periods. Some simple models use an "accounting" formulation to calculate future population size on the basis of current size and rates of growth, death, and birth. More complex models deal with such phenomena as dispersal, breeding structure, interchange between populations, environmental variability, and the effects of intraspecific interactions on population behavior.

Traditional models of population dynamics are based on the assumption that organisms do not change genetically during a period of management. However, genetic changes do occur when populations are exposed to repeated manipulations, and evolution can take place with startling rapidity—e.g., the evolution of insect resistance to pesticides and of bacterial resistance to antibiotics. Both insects and bacteria have the short life spans and high reproductive rates that speed evolution (May and Dobson, in press), but long-lived organisms can also evolve quickly if management results in large differences in mortality or reproduction among individuals. Size-selective harvesting of fish can lead to a reduction in the average age and size at maturity (Ricker, 1981), possibly as a result of changes in competition among age classes and in the frequencies of particular genes. When populations are small, harvesting practices can lead also to loss of genetic variability and can increase the deleterious effects of inbreeding. Preserving a species that is distributed into many small and isolated populations—as many endangered species are—involves an understanding of both population dynamics and population genetics (Frankel and Soulé, 1981).

Dynamic and evolutionary effects of manipulations are not always separable. Changing such population characteristics as total numbers, distributions of ages and sizes, and sex ratio not only changes the dynamics of a population, but also establishes a potential for evolutionary change in traits that exhibit genetic variation. Population management that is based only on dynamic considerations can in the long run produce results opposite to those intended. DDT works miracles on untreated populations of mosquitoes that transmit malaria, but within 5-50 generations the evolution of resistance might largely negate the effectiveness of the chemical; at the same time, human resistance to malaria can decrease in the absence of the disease (Chapter 15). Using large-mesh nets to ensure the harvest of male but not female migrating salmon can lead to an increase in the proportion of early-maturing small male salmon (jacks) (Gross, 1984, 1985). An increase in the proportion of jacks that escape the nets can lead to an increase in the proportion that breed and thus in the proportion that are hatched in the next generation.

IDENTIFYING KEY FACTORS

Organisms are influenced by many components, or factors, of their physical and biological environments. Those factors are not equally important; often, a few dominate the dynamics of populations and need careful study for successful management. Of particular interest are factors that exert especially powerful effects on population size (usually by influencing birth and death rates).

Key-factor analysis is a method for identifying and understanding the stages of an organism's life history in which critical controlling processes occur. The method has been used for designing pest-control strategies (Clark *et al.*, 1967). The basic data used in key-factor analysis are the number of individuals in each developmental stage or age group, survival from one stage to the next, and fecundity rates—i.e., the standard elements of life tables (Ricklefs, 1979). When the variance in these elements is partitioned among environmental causes, the key factors are the ones that cause high mortality (Harcourt and Leroux, 1967). Key-factor analysis is less applicable in heterogeneous than in uniform environments (Hassell, 1985).

In organisms with complex life histories (e.g., salmon), different stages use resources differently and sometimes live in different habitats. The behavior of the different stages might therefore be controlled by largely independent events. Too few individuals of such a species might survive one stage to saturate the habitat of the next. In addition, the more independent the habitats are, the less likely it is that their carrying capacities will vary concurrently. Therefore, populations of species with complex life histories often vary more widely in density than they would if all resources were gathered in one habitat (Istock, 1967), and study of a population in only a single stage is unlikely to reveal the causes of fluctuations.

BEHAVIOR

Biological processes of animals are tied to the behavior of individuals as they acquire nutrients, select habitats and mates, avoid predators, and interact socially. Some behavior can be predicted from knowledge about general environmental conditions and the types of species under consideration, but in many cases understanding behavior requires detailed knowledge about particular species and how they make behavioral "decisions." We discuss below decisions made by individual organisms that are important in dealing with environmental problems.

Habitat Selection

By natural selection, species evolve to prefer environments in which survival and reproduction are greatest. Responses to habitats are influenced not only by characteristics of the environment, but also by the presence or absence of individuals of the same species. Other individuals provide information about choices made by previous settlers, and they modify the environment for later settlers (Orians, 1980; Partridge, 1978).

Individuals might not settle in an otherwise suitable area if no other individuals are there. Young individuals might need to learn the locations of suitable sites from older ones, and slight changes in a habitat can lead to rejection, even if it is otherwise suitable for the species. For example, a management plan for the spotted owl (Chapter 17) avoids such rejection by recognizing that the owls require undisturbed old-growth forests for hunting and are reluctant to cross open areas. The collapse of the whitefish fishery in Southern Indian Lake was due not to a decline in the fish population, but to changes in selection of habitats as a result of raising the level of the lake (Chapter 21). The control of malaria was less successful than expected, because of a behavioral polymorphism in habitat selection (Chapter 15).

Attempts to find suitable natural predators to control agricultural and forest pests usually involve searches in areas with climates similar to those of the areas into which the control agent is to be introduced (Chapter 14).

Particular attention is paid to climate, rather than biological interactions, because the required food source—the pest to be controlled—is known to exist in the areas of introduction and because experience has shown that predators and parasites of a single host species typically change over the range of the host. It is rare for a single predator to be an important control agent over the entire range of its host. It is difficult, in practice, to find an agent that can control over even part of a pest's range, because many introduced predators fail to establish viable populations or fail to become common enough to achieve effective control (e.g., Chapter 14). Attempts at biological control therefore often involve importing a variety of potential control agents, in the hope that some will prove effective (Chapter 14).

Mating Systems

Mating systems vary widely. In the simplest, sex cells are shed into the surrounding medium, usually water, where fertilization takes place. In more complex animal systems, reproductively mature adults gather, choose mates from the available pool, and associate with one another for various periods after fertilization to care for each other or their offspring. In many species, mate selection is combined with habitat selection for breeding, especially in species in which one adult holds space that contains resources for the reproductive cycle. Much research into the evolution of mate choice is directed at determining the relative roles of mate quality and habitat quality and the criteria used for selection of each (O'Donald, 1967; Searcy, 1982).

A successful biological control program that has capitalized on the use of laboratory-reared sterile males is the control of screwworm populations in the southern United States (Perkins, 1982; Scruggs, 1975). The program depends both on the ability to produce enough sterile males to ensure that most males in the field are sterile and on the mating behavior of the flies. No long-term pair bonds are formed, but the first male to copulate with a female blocks her reproductive tract with a plug that prevents other males from copulating with her. If multiple matings were the rule, a much higher ratio of sterile to fertile males would be required to achieve the same degree of population control. A similar mating system is the basis of strategies to control Mediterranean fruit flies (*Dacus*).

Social Interactions

Social grooming is a component of social behavior that sometimes accompanies mate selection, but also occurs in other contexts. The observation that vampire bats groom each other extensively suggested a control strategy that facilitated transfer of a control agent applied to the fur of only a small number of bats (Chapter 13). The control agent was highly specific and had no effects on other species living in the same caves. The vampire bat case study is an excellent example of the value of detailed natural-history observation for population management.

POPULATION DYNAMICS

Population Regulation

Environmental changes can influence a population in two basic ways, each having important management implications. Fires, bad weather, and other events can reduce a population by removing a fraction of it that is independent of its density. If a population—or a part of it, such as an age group—is limited primarily by such density-independent factors, it can be reduced with little effect on the remaining individuals.

Many environmental changes, however, influence individual growth, birth, and death rates in a manner that depends on population density. Even though populations grow rapidly at low densities, growth often slows as density-dependent effects appear through individual interactions (e.g., competition for food or nest sites) or through the actions of predators, disease, and other factors that increase deaths or reduce births. When individuals are removed from a population, density-dependent compensation can occur—for example, an increase in the birth rate as population size decreases. The operation of density-dependent factors can stabilize population densities and thus cause displaced populations to tend to return to equilibrium, in which the rate of loss of individuals of a species equals the replacement rate. The greatest number of individuals of a species that the environment can support is called the carrying capacity (K); it too can vary. May (1973) has discussed mathematical studies that showed how density dependence can lead to cyclic or even chaotic population change.

Species with a very high reproductive potential (r) in the absence of competition are potentially able to recover quickly from population reduction, but they are more likely than species with a low r to show extreme population fluctuations. Many pest species, such as weeds, are poor competitors, but have good dispersal abilities and high r (Baker, 1974). Species with low r, such as whales and spotted owls (Chapter 17), often have stable populations consisting of long-lived individuals. Overharvesting of low-r species can easily lead to species extinction. Conversely, low-r species can more easily be controlled with measures used only periodically (e.g., vampire bats, as discussed in Chapter 13).

Commercial or recreational harvesting reduces populations to some value below the carrying capacity. Managers often try to maintain populations near the size that permits the greatest yield of individuals or biomass. Because yield is a result of both the reproductive rate and the number of individuals reproducing, yield is highest when the population is between low density (few individuals but high individual reproductive rate) and the high density of natural equilibrium (strong density-dependent reduction of average individual reproductive output or survival).

In theory, if environments are constant, every population has a "maximal sustainable yield" (MSY); but in practice, the MSY is very difficult to identify, let alone achieve. Accurate estimates of abundances are required, especially when an unstable equilibrium exists near the MSY (see below). Both yields and population sizes fluctuate more as the MSY is approached, the effect being most pronounced in large mammals, such as whales, in which density-dependent effects are strongest near K and r is low (May, 1980).

Reducing the risk of errors resulting from population management requires at least estimation of relative changes in population size and composition; good estimates of absolute population size are more desirable. But accurate estimates of the age distributions and sizes of populations are often difficult to obtain (Eberhardt, 1976; Seber, 1982). Although density and age of forest stands can be estimated accurately, errors in density and age estimates in bird, mammal, and fish populations are often large or unknown. Most marine fish populations are managed without knowledge of population sizes (e.g., Pacific halibut, as discussed in Chapter 12). Shepherd (1984) discusses the advantages and shortcomings of techniques for managing fish populations in the face of limited information; some difficulties in obtaining the information have been set forth clearly by Larkin (1978).

Populations of many species are often managed indirectly by managing the habitat. Habitat quality is commonly judged by population density, on the assumption that high density indicates high quality of habitat. This assumption is often valid, but data on survival and reproductive success are generally needed to understand differences in population densities. High densities can occur in suboptimal habitats composed mostly of subdominant individuals that have been forced out of other habitats (Van Horne, 1983).

Population Stability

A population with a zero growth rate is in equilibrium. The equilibrium is stable if a population displaced above or below the equilibrium point tends to return to it. However, a population can have more than one stable equilibrium point, and some equilibria are unstable (Berryman, 1981). Unstable equilibria often occur at low population densities of large, long-lived, slowly maturing species with low *r* (Southwood *et al.*, 1974), as well as in a variety of heavily harvested fish populations (Beverton, 1984). Increased difficulty in finding mates, reduced effectiveness of cooperative defense against predators, and other phenomena can result in a lower threshold of density below which the population collapses unless there is immigration. Density-dependent compensatory mechanisms might also break down at very low densities (Beverton, 1984). That possibility is of particular concern for rare and endangered species with few, isolated populations that are already small. Environmental fluctuations, disease, and other factors can easily reduce such a population to below its stability threshold. Some fish stocks have collapsed to extremely small numbers after overexploitation, and a few have not recovered even after the release of fishing pressure (e.g., Beverton, 1984; Peterman, 1978). In other species, the maximal yield could be very close to the harvest magnitude that would cause the population to collapse (Ricker, 1963). In any case, environmental variation appears to interact with fishing pressure to complicate the fluctuations in stock size (Chapter 8).

It is difficult to estimate minimal safe population sizes. M. L. Shaffer (unpublished manuscript) has modeled minimal population sizes and areas for grizzly bears in Yellowstone National Park by using successive computer simulations based on estimated population parameters with random variation. Using several scenarios, he estimated that 35-70 bears were needed to have a 95% probability of population survival for 100 years in that ecosystem. Studies of this type can be helpful in focusing attention on the need to choose the time scales at which to judge the acceptability of a given probability of extinction, but the underlying assumptions are not subject to rigorous testing.

Long-term management plans can accommodate the possible presence of low-density thresholds by providing a substantial safety margin, particularly for isolated populations with little immigration. It might be difficult to predict these thresholds with the available data, and guesses based on experience with similar species could be necessary (Chapter 17; Soulé and Wilcox, 1980).

Many defoliating insects appear to have multiple stable equilibria. Such species appear to be controlled at low densities in a density-dependent fashion by predators, parasites, or pathogens. These agents cannot "control" the pest populations above particular sizes, because other factors prevent the populations of the controlling agent from increasing enough (e.g., Campbell and Sloan, 1977; McNamee *et al.*, 1981). Traditional approaches to controlling these species have relied on the massive application of insecticides during outbreaks. However, control might be more effectively applied to subpopulations threatening to escape from the low-density control (Campbell and Sloan, 1978; MacLeod, 1977). Environmentally induced reductions in the vigor of trees can trigger outbreaks of some forest pests, and control techniques can be applied to areas where such conditions appear imminent (Berryman, 1981).

Such techniques use a system of "risk classification" of the habitats of the pests in determining when and where to apply controls. Another successful approach is the use of a biological control agent that has a shorter generation time than the pest and that can respond to outbreaks by increasing its own population. An example is the partial control of gypsy moths in the northeastern United States with bacteria and viruses (Leonard, 1974; Massachussetts Department of Environmental Management, 1981). McNamee et al. (1981) have provided a useful framework for classifying defoliating forest insects so that appropriate control strategies can be chosen.

Dispersion and Population Movements

Species vary in their patterns of dispersion in space (clumped, random, or even), density, dispersal, and migratory behavior. Most terrestrial species occupy habitat patches of various sizes and degrees of isolation. These variations have implications for management. Management of a migratory species must take into account the relationships between populations in breeding and nonbreeding habitats. For example, if the population of salmon at sea is near K, then increasing hatching success in the rivers might have little benefit (Peterman, 1978, 1984).

Fretwell (1972) has shown theoretically and by examples that populations in strongly seasonal environments are ultimately controlled in one "bottleneck" season. For example, winter resources might limit the number of individuals that survive to breed, so an increase in the preceding summer reproduction could cause little change in the number of individuals breeding in the following summer. This idea is particularly important in managing bird species that winter in tropical habitats undergoing extensive deforestation. Recognizing that the protection and enhancement of populations might not be sufficient, in itself, to prevent extinction, legislators of many countries have cooperated in formulating treaties to protect migratory bird species. Migratory patterns can also make species susceptible to local disturbances (e.g., Chapter 16). Salmon can be affected by a single dam in a migration of thousands of miles, and some migratory birds (such as geese, cranes, and swans) depend on a very small number of resting and feeding areas along their migration routes.

Animals that organize into dense aggregations can easily be harvested to very low numbers, because the profitable unit for harvest is the aggregation. Fish stocks that form aggregations (e.g., many clupeoids) apparently are subject to catastrophic collapse when fishing effort is continuously increased (May, 1980, 1984; Murphy, 1977). The situation with some very large whales is analogous, in that it can be profitable to hunt and harvest a single animal. May (1980) has discussed various problems encountered in managing the harvest of fish that form schools.

Isolated populations have less potential for recovering from temporary reductions when immigration rates are low, as happens when an occupied habitat becomes an island surrounded by unsuitable habitat. Maintaining travel corridors of appropriate habitat to connect populations, particularly of species at low density and with strict habitat requirements and poor dispersal ability, might be a key component of management (MacClintock *et al.*, 1977; Willis, 1974). The spotted owl (Chapter 17), a strict obligate of old-growth coniferous forest, is an example of such a species (Forsman *et al.*, 1984). However, if control is the objective, disruption of corridors can be effective in preventing dispersal.

The size of the habitat patch necessary to support even a single breeding female or pair depends on the size, diet, and behavior of the species (Galli *et al.*, 1976; Schonewald-Cox *et al.*, 1983). Carnivores generally require more area than herbivores, and area requirements increase with body size (McNab, 1963; Schoener, 1968). Required areas for managing populations of large social carnivores, such as wolves and bears, can be huge (M. L. Shaffer, unpublished manuscript; Soulé, 1980).

The extinction of bird species on Barro Colorado Island constitutes a particularly revealing example of spatial isolation. The island became isolated from the Central American mainland during the construction of the Panama Canal. One of the best predictors of local extinction was susceptibility to ground predators, such as snakes (Karr, 1982); the populations of these predators were increasing, probably because large predators were decreasing as a result of human activities.

Growth Rates, Age, and Size

The operation of forestry, fisheries, and agriculture depends on an understanding of how growth rates are influenced by interactions with the physical environment and with other organisms. Size is often used as an indicator of age in a continuously growing species, because it is easy to measure and is usually reliable. When age and size are poorly correlated (e.g., Caswell and Werner, 1978; Policansky, 1983), the use of size as an indicator of age can lead to inappropriate management decisions, as it did in a population of roe deer (*Capreolus capreolus*) in Scotland. The ages of culled deer in that population were estimated on the basis of body size and weight, as is common for deer (Ratcliffe, 1984), but the calves were growing so fast that the technique underestimated their number and overestimated the number of fertile females. The result was overharvesting.

If resources, such as food and light, are limiting to animals or plants, removing large individuals usually increases the growth rate of those remaining. These interactions are well known to wildlife and fishery managers, foresters, and agriculturalists. In the New Brunswick forest-management case (Chapter 19), the inverse relationship between planting density and growth rates of individual trees was an important part of the management plan. An understanding of the factors affecting growth rates among competing species allowed the successful regeneration of land in the derelict-lands case (Chapter 18).

Environmental manipulations affecting fecundity or survivorship have the greatest effect on population dynamics if the perturbations are introduced at early reproductive stages (Beddington, 1974; Emlen, 1970; Lewontin, 1965). Expected reproductive output is smaller in late reproductive stages than in early stages, so late stages have a smaller per capita effect on population production. Similarly, many individuals in prereproductive stages will not survive to breed, so their removal affects population reproduction less than the removal of early breeders. This is the basis for fishing and hunting regulations that specify minimal sizes or ages.

Age Structure

Variations in life expectancy or age structure in a population can influence its responses to management. Length of life and age structure are associated with many variables—such as spatial distribution, reproductive potential, and feeding habits—that are hard to estimate, so generalizations as to the implications of age structure for management can be unreliable. Nonetheless, a few points should be mentioned.

Beverton (1984) classified the expected response of fish species to fishing pressure on the basis of their ecological characteristics (including life history). Fish populations comprising long-lived individuals in many reproductive age classes are generally more stable and easier to manage than those having only few age classes comprising short-lived individuals. There is also a tendency for species that are densely aggregated when young to be more stable in the face of fishing pressure. Beverton classified halibut as "reliable, steady, and robust," and these life-history characteristics probably have been partly responsible for success in halibut management (Chapter 12).

It is important to think about population stability in terms of the life spans of individuals. A population of long-lived organisms might appear to be maintaining itself when in fact it is in danger of serious decline. If habitat degradation by pollution or other factors prevents young organisms from surviving, the long life span of adults can mask the loss of those age classes. The June sucker (*Chamistes liorus*) in Utah has maintained populations without recruitment for 15 years (U.S. Fish and Wildlife Service, 1984) and saguaro cactuses (*Carnegiea gigantea*) in the Sonoran desert remain common in areas that have seen no recruitment for even longer periods (Turner *et al.*, 1969). The other side of this coin is that such populations are able to survive long periods of adverse conditions, even if conditions favorable for reproduction occur only infrequently.

Sex Ratios and Sex Biases

Either intentionally or unintentionally, human activities often produce mortality differences between the sexes and thus biased sex ratios. The effects on population reproduction and dynamics depend on the form of the mating system. The degree to which differences in mortality can be controlled depends on the difficulty of distinguishing the sexes in the field.

In many animal species with polygynous mating systems, one male can fertilize the eggs of many females, so a reduction in the number of males has a relatively small effect on reproductive rates. Management of polygynous species in which the sexes are easily distinguished (e.g., deer, crabs, and pheasants) often involves permitting the harvest only of males. In the sockeye salmon (*Oncorhynchus nerka*), males are larger than females, so gillnets, which select for larger fish, take disproportionately many males. In spite of this skewing of sex ratio, Mathisen (1962) has shown that a 15:1 ratio of females to males led to egg hatching only 5% lower than that with a 1:1 ratio. In contrast, differences in mortality can have large effects in species whose reproduction is limited by whichever sex is the rarer, such as species in which both parents (e.g., geese) or only the males (e.g., rheas and seahorses) care for the young.

Sex ratios can be manipulated to increase productivity, including food production, as in the case of dairy cattle. Only the female structures of many commercially important crop plants are eaten (seeds and fruits). In plants with separate sexes, fruit yields can be increased by manipulating the sex ratio. In cosexual species—those with both sexual functions in the same individual—manipulations are directed toward altering the allocation of investment in female versus male function. Such schemes work when allocation patterns can be genetically selected for or when they are influenced by environmental conditions that can be manipulated. In the Cucurbitaceae (cucumbers, melons, squash, etc.), some genes shift the allocation of resources from male to female function (Kubicki, 1969a,b,c; Robinson *et al.*, 1976), and cultivars that favor female function have been selected (Velich and Satyko, 1974).

GENETIC AND EVOLUTIONARY CONCERNS

Artificial selection has been successful in producing desirable traits in animals and plants, and the genetic resources of domestic species can be maintained by intentional cross-breeding. Manipulations of the population dynamics of nondomesticated organisms, however, can produce the evolution of undesirable genetic consequences, often within a few generations. Conspicuous examples include the evolution of resistance to pesticides and antibiotics (e.g., Anderson and May, 1982; Peters, 1984, in press) and perhaps increased pathogen virulence (Ewald, 1983), decrease in size (Ricker, 1981), and increased inbreeding depression (see below).

The Evolution of Resistance to Pesticides

One of the best-documented cases of undesired evolutionary change is the development of resistance to DDT (Chapter 24). In retrospect, it should have been obvious that it would happen; in fact, the use of poisons to produce a class of resistant organisms for experimental purposes has been a major technique of microbiology for more than 50 years. The short generation times, high reproductive rates, and large populations of most pests favor rapid evolution (May and Dobson, 1985). The race between chemists and the evolutionary capacities of organisms parallels the race between the evolution of new antiherbivore devices by plants and the evolution of resistance to them by insects (Rhoades, 1983).

A number of approaches can be taken to reduce the rate of evolution of resistance. First, agents that interfere with fundamental and invariant processes, such as eating and oxygen uptake, can be used. Changes in fundamental processes are unlikely to evolve over a short period. However, such agents can affect nontarget species, because the processes in question are shared by many of them. The use of anticoagulants against vampire bats (Chapter 13) depended on the spread of the materials by intraspecific social grooming, a process that kept the materials within the target species.

Second, two or more chemicals can be used at the same time or in close alternation (e.g., Peters, 1984, in press; Smith, 1982). This reduces the number of survivors, because survival would require the evolution of two or more types of resistance simultaneously, which is unlikely in view of the small probability that several mutations will occur at once.

Third, agents can be used intermittently. This allows the frequency of evolved resistant types to decrease through competition with nonresistant types (e.g., Levin and Lenski, 1983). However, *Plasmodium* resistance to chloroquinone actually seems to be associated with a general biological advantage over nonresistant strains (Doberstyn, 1984), so this approach might not be universally applicable.

Genetic Consequences of Differential Harvesting by Sex and Size

Harvesting according to age or sex is a feature of many management schemes. Animals of one sex might be singled out for harvest (e.g., in deer and crabs) or the capture of animals might be sex-biased for other reasons, such as differences in size or activity (consider, for example, the evidence of size and sex selectivity of fishing gear). Nonrandom harvests could have produced such changes—perhaps genetic—as early maturity at smaller size in fish populations, e.g., salmon (Mathisen, 1962; Ricker, 1981), Atlantic cod (Borisov, 1978), an African cichlid (Silliman, 1975), and lake whitefish (Handford *et al.*, 1977); Moav *et al.* (1978) have suggested that this is a general phenomenon.

Many large male salmon are harvested with size-selective nets before they have a chance to spawn; this harvesting favors the survival of early-maturing, small males (jacks). Increased proportions of jacks, which have low commercial and sport value, are a matter of growing concern to the managers of exploited salmon populations (Ricker, 1981). Theoretical studies (Gross, 1984, 1985) have suggested that jacks are not abnormal, as previously thought, and that they and larger males exist in an equilibrium that depends on both differences in mortality and conditions in spawning areas. The theory suggests ways of reducing the proportion of jacks despite heavy harvesting of larger males. For example, spawning sites could be manipulated to reduce the cover in which the jacks seek protection from larger males, thus lowering their reproductive success and the proportion of jacks in the next generation.

The jack story shows how management can lead to undesirable evolution. For example, trophy hunting, which removes the finest specimens from the breeding population, leads to the evolution of deer genotypes with smaller antlers. Hunting in general is known to make animals furtive, and genetic changes in dispersal, foraging, and social behavior are possible. In short, the evolutionary effects of management must be considered if it can lead to differences in reproductive success of different genotypes (Law, 1979).

Genetic Consequences of Small Population Size

Genetic deterioration has often led to extinctions in captive (including laboratory) populations, and small natural populations might also be in such danger, because of inbreeding depression and loss of genetic variability (Frankel and Soulé, 1981; Schonewald-Cox et al., 1983; Soulé and Wilcox, 1980)—the former reduces fitness, and the latter reduces adaptive potential. Efforts to protect species threatened by the genetic effects of small population size involve estimating a minimal "effective" population size (minimal N_e) that is based on estimates of tolerable degrees of inbreeding, which in turn are based on experience with domestic and laboratory animals. (N_e is the number of potential breeding individuals in an "ideal" population—one with random mating, a sex ratio of 1:1, discrete generations, constant size, and a Poisson distribution of family size—that retains the same amount of selectively neutral genetic variability as the population under consideration. If the population departs from "ideal" conditions, as occurs commonly in nature, N_e is lower than actual N.)

Selecting tolerable rates of loss of genetic variability is much more difficult than choosing tolerable degrees of inbreeding, because the long-term consequences of reduced variability are not known and could depend on the species involved. Soulé (1980) crudely estimated a minimal N_e of 50 on the basis of inbreeding alone, according to an expectation of surviving extinction for 1.5 N_e, or 75, generations. Franklin (1980) suggested that, for long-term adaptive potential, N_e should be 10 times as high, i.e., about 500. The consequences of inbreeding might be less serious in species that normally engage in much inbreeding or that are already strongly homozygous (e.g., many self-fertilizing plants).

Sex-biased harvesting can also lower genetic variability, perhaps enough to have serious consequences. Ryman et al. (1981) have shown how various moose- and deer-hunting policies can lead to loss of genetic variability, with N_e perhaps as low as 5% of N.

2

Population Interactions

Every species in an ecological community is connected to many others. Each is both a predator (if plants can be regarded as predators on photons) and a prey, each can compete with other species that use the same resources, and each can engage in mutualistic interactions with other species. In a particular environmental problem, usually one of the roles that a species can play is of prime concern, but neglect of other interactions might cause management efforts to fail or produce unwanted side effects.

The processes in which species are involved are measured in different units. Rates of photosynthesis are measured in units of mass change; feeding rates are measured in energetic units. Competitive interactions might decrease reproductive rates or cause species to be absent from some areas. Not all these effects can be expressed in energetic terms. If disruption of mutualisms lowers seed set or dispersal, the energy content of the lost seeds is only a small component of the significance of those changes. The overall unit for expressing these effects is fitness (relative reproductive success), but total fitness rarely can be measured in the field, and even partial measures can be difficult to obtain. As a result, various units are used by environmental problem-solvers, the most appropriate one depending on the problem and the objectives of manipulation.

Human uses of ecological systems commonly involve altering the nature and extent of interactions among populations, whether or not this is the prime objective of the management programs. We commonly eliminate large carnivores, because they are dangerous and because they prey on wild or domesticated species that we wish to exploit. Herbivorous insects compete with us for the tissues of crop plants, and weeds compete with those plants for light, water, and nutrients. Our efforts to remove these species or reduce their abundances are based on the assumption that doing so will increase the yields of valued products from the species we wish to protect. The assumption is probably true in most cases, but there are remarkably few quantitative data on the effects of weeds on yields of crop plants (Mortimer, 1984). There are many more estimates of losses of crops to herbivores. Damage is often severe enough not only to justify control, but to cause economic hardship to agriculturalists (Barrons, 1981; May, 1977; Pimentel et al., 1980). It is clear, however, that control measures are often initiated when populations of pests are so small that economically significant damage is unlikely to occur (Pimentel et al., 1980). Restricting the use of control measures to times and places when they are cost-effective is important, because many of the materials used are toxic and exert other, usually detrimental, effects on ecological communities (Barrons, 1981; Bull, 1982; Carson, 1962; Dunlap, 1981) and because resistance to toxic chemicals evolves as a function of the frequency with which they are used (Chapter 1).

Green plants account for about 97% of all the carbon fixed in terrestrial and aquatic ecosystems, and their photosynthetic activity supports almost all other components of those systems. Plants carry out photosynthesis by only three different mechanisms, which differ in a variety of ways, including optimal temperature and amount of carbon that can be fixed per unit of water lost (Berry, 1975; Bjorkman and Berry, 1973). Nearly all woody plants at all latitudes and herbaceous plants at high and middle latitudes use the same pathway. Because of the relative uniformity of mechanisms of photosynthesis, plants are generally replaceable in their role as photosynthesizers. That is the basis of single-crop agriculture and the reason why forest productivity is largely independent of the number of species of trees growing on the site. Plants differ greatly, however, in physical structure, chemistry, depth of soil from which they extract water and minerals, and kinds and numbers of mutualistic interactions. Those differences make it possible for particular species of plants to play different roles.

All animals require high-energy organic compounds as food, so much of their diversity is related to what they eat and how they find it. Animals are involved in many kinds of competitive interactions, and they participate in many coevolved mutualisms with plants (e.g., pollination, fruit dispersal, and plant protection), other animals (e.g., mimicry and protection), and microorganisms (e.g., digestion and bioluminescence). Animals exert most of their effects through consumption of prey. Because energy-rich molecules are usually found at low to moderate densities in ecological communities, most animals are motile and move around to find food. In special environments, however, such as coral reefs and the rocky intertidal zone, each wave brings a new supply of food that cannot be depleted by the foraging activities of the attached animals, and animals compete for space instead of food and actually form the dominant structural elements of the ecological communities.

Like animals, microorganisms are diverse in their modes of obtaining energy. They use various substrates for their synthetic abilities, use various photosynthetic pigments, and derive energy by oxidizing many simple substrates, such as gaseous hydrogen, inorganic nitrogen, sulfur, and iron. Microorganisms live in the bodies of all species of larger organisms. They are able to synthesize compounds required by plants and animals (e.g., vitamins and nitrates), and they are the only organisms that can break down many biologically important molecules (e.g., cellulose, waxes, and lignins).

PREDATOR-PREY INTERACTIONS

It is useful to think about predator-prey interactions from the perspective of the relative sizes of predators and prey. "Typical" predators are larger than their prey, kill and consume all or most of each prey item they capture, and must find many different prey items daily or over their life spans; in the extreme, predators like baleen whales feeding on krill are so much larger than their prey that the prey are handled en masse. However, predators can be smaller than their prey and consume only parts of them. If they live externally on their prey, these predators are called herbivores or external parasites. If they live internally, they are usually called parasites or parasitoids (Price, 1980; Thompson, 1982).

These differences are important from a management perspective. Predators that eat most types of prey captured are unlikely to be sensitive to the loss of any particular prey species. Predators that are specialists, eating only a single species or a few closely related species of prey, are desirable for many biological control purposes (Chapter 14), because they are unlikely to feed on nontarget prey species (which would cause unexpected and unwanted side effects). Knowledge of size relations can be helpful in directing research efforts, because specialist predators are often smaller than their prey. Predator species being considered for biological control must, of course, be screened individually to determine their diets.

Narrow diets can evolve in the absence of size differences between predators and prey. For example, dietary specialists should be more prevalent in stable habitats, where the availability of specific resources fluctuates over rather narrow limits and particular prey types are therefore reliably available throughout the year (Charnov and Orians, 1973; Cowie, 1977; Pyke *et al.*, 1977; Schoener, 1971; Tinbergen, 1981; Werner and Hall, 1974). Monophagy is prevalent among predators with special adaptations for capturing and eating prey with unusual defenses, such as armor, spines, and toxic chemicals. Predators can be dietary specialists because their simple nervous systems leave them unable to use other than simple criteria for recognizing prey (Levins and MacArthur, 1969). Monophagy can be so narrow that predators eat tissue of only some types found in their prey; herbivores of large, woody plants are predators of this kind (Crawley, 1983; Strong *et al.*, 1984).

The consequences for an ecological system of adding or removing predators depend on whether the system is predator-controlled or "donor-controlled" (Pimm, 1979, 1980). In a donor-controlled system, the supply of prey is determined mainly by factors other than predation; therefore, the predators have little influence on their food supplies. Familiar examples of donor-controlled systems are communities of sessile, planktivorous animals of rocky intertidal shores, where a fresh supply of food is delivered with each successive wave; consumers of dead plants and animals; and consumers of fruits and seeds. In predator-controlled systems, the predators, by their feeding, reduce the supply of prey and their reproductive ability (i.e., the future supply). The distinction is important, because removal of predators in donor-controlled systems has little effect on population dynamics and interactions among their prey species, whereas removal of predators in predator-controlled systems often results in outbreaks of the prey and large changes in the relative abundances of the prey species (Estes *et al.*, 1982; Simenstad *et al.*, 1978). The ability to predict such changes is obviously important for the design of management schemes.

An important and rapidly growing application of predator-prey relationships is the use of the natural defenses of plants against their predators as a substitute for synthetic toxic chemicals. A major advantage of using natural defenses is that effects on nontarget organisms are minimized and the plant itself synthesizes and distributes the defenses. Breeding of pest-resistant crop plants has a long tradition, but improved knowledge of the chemical bases of resistance now enables managers to search for specific kinds of chemical and physical defenses in plants to deal with particular herbivores, rather than relying simply on randomized field trials to see what happens to work (Maxwell and Jennings, 1980).

Defensive chemicals are in two major classes—acute toxins and digestibility-reducing substances. Acute toxins are present in plant tissues in small amounts and exert their effects in the bodies of the herbivores by interfering with a basic metabolic process, such as nerve transmission, protein synthesis, or hormone balance. They are effective because animals have many tissue types not present in plants—tissue types that are targets for chemicals not toxic to the plants producing them. Digestibility-reducing substances act in the guts of the herbivores (technically outside their bodies, inasmuch as no cell membrane has to be crossed), by combining with proteins in the food to reduce their availability to the herbivore. Tannins and resins are common substances of this type. Acute toxins are most effective against generalized herbivores, because specialized herbivores rapidly evolve the ability to detoxify chemicals in their host plants. It is more difficult to evolve counteradaptations to digestibility-reducing substances, so they are often effective against specialized herbivores (Feeny, 1976; Rhoades, 1979; Rhoades and Cates, 1976). Herbivores can, however, evolve resistance to hydrolyzable tannins (Fox and Macauley, 1977). The use of such information in screening for strains of crop plants to use in different situations is still largely a task of the future, but the necessary ecological knowledge is being accumulated.

An important recent discovery is that plants can respond to herbivore attacks by turning on inducible defenses. These induced defenses can cause marked reductions in growth rates, survival, and pupal weights among insects feeding on foliage of previously attacked branches or trees (Haukioja and Hakala, 1975; Haukioja and Neimelä, 1979; Ryan and Green, 1974; Schultz and Baldwin, 1982); Rhoades (1979) has reviewed the evidence. The full significance of this discovery is not yet clear, but once the mechanisms underlying such responses are identified, they might be manipulated to induce defenses in advance of any attacks. Moreover, damaged plants might release volatile materials that induce neighboring plants to increase their defenses (Baldwin and Schultz, 1983; Rhoades, 1982); if so, that opens up still other possibilities to manipulate plant defenses against herbivores to reduce reliance on toxic substances.

COMPETITIVE INTERACTIONS

Competition occurs when a number of individuals use common resources for which the demand exceeds the supply or when resources are not scarce and organisms harm one another in the process of seeking those resources (Birch, 1957). Competition can occur among individuals of a single species (intraspecific competition) or among individuals of different species (interspecific competition). Interactions can be especially strong between two species or spread out over a larger number of species, each of which contributes a small part of the total (diffuse competition). Understanding competition can be helpful in solving environmental problems.

The most direct method of assessing the importance of competition is to remove individuals of one species and measure the responses of the remaining species. Often such a manipulation cannot be made, however, and even if it can, it reveals only the short-term responses to the reduction of competition. Longer-term responses, which might involve genetic changes in the component species, cannot be measured this way, except in the case of very short-lived species, such as bacteria, protists, algae, and some insects.

Competition is much more difficult to study than predation. Unlike predation, a more or less continuous process (all predators must eat regularly), competition can be intermittent, with periods of competition alternating with long periods in which competition is weak or absent. Nonetheless, patterns due in part to competition might persist through periods when competition is absent. Indeed, if species evolve traits that are adaptations to environments with particular sets of competitors, their behavior might reflect competition that occurred long ago—a phenomenon referred to as the "ghost of competition past" (Connell, 1980).

Most studies of competition have not involved measuring competition directly, but have attempted to predict and find the patterns that should be found in nature if competition were occurring. Necessary though this procedure might be in many cases, it increases the probability of mistakenly assuming either the presence or absence of competition because predictions about resulting patterns have been incorrect. The literature is biased, owing to underrepresentation of negative results, and, not surprisingly, there is much controversy among ecologists about the importance of competition in nature and about the extent to which patterns observed in ecological communities can be attributed to it (Cody, 1974; Diamond and Gilpin, 1982; Gilpin and Diamond, 1982; Strong *et al.*, 1979, 1984; Tilman, 1982). All observers do agree, however, that competition occurs and that it must be understood better if we are to understand ecological communities.

Except for competition for space in marine communities, where animals and plants compete vigorously (Paine, 1969, 1980), competition generally occurs among organisms at the same trophic level in a community. Some authors have suggested that competition is stronger at some trophic levels than at others. For example, Hairston, Smith, and Slobodkin (1960), observing that the world is green, suggested that competition was intense among plants, rare among herbivores, common among carnivores, and common among detritivores (coal and oil do not appear to be forming at appreciable rates today). They stressed that their conclusions were trophic-level generalizations and did not imply that all carnivores or no herbivores compete. Menge and Sutherland (1976), working in rocky intertidal environments, suggested that the importance of competition should rise with trophic level and that competition should be especially noticeable in trophically simple communities. They differed from Hairston *et al.* in predicting low competition among plants.

These and other predictions about competition have recently been examined by Connell (1983) and Schoener (1983). Connell's criteria for inclusion of studies in his sample were more stringent than Schoener's, so the two reached different conclusions. Schoener found strong support for the Hairston *et al.* hypothesis in both terrestrial and freshwater communities, but only weak support in marine communities. An important exception was that some groups of terrestrial carnivores, such as spiders and predatory insects, often do not compete strongly. The reason might lie in the fact that these small vulnerable carnivores are preyed on by many potential competitors for their prey (Schoener, 1983). Connell found little support for the Hairston *et al.* hypothesis.

Thus, the general importance of competition in nature is not settled; but there is general agreement about some patterns in competitive interactions. First, small animals are more vulnerable than large ones to vagaries in the physical environment and predators and, as a result, are less likely to compete. Second, most competitive interactions are highly asymmetrical in their effects (Connell, 1983; Lawton and Hassell, 1981; Schoener, 1983), i.e., the removal of one species has a strong effect on the other, but the reverse is not true. Indeed, many cases are so asymmetrical that the effects on one of the species are almost undetectable. Such a situation was predicted by theoretical considerations: a slight difference in competitive abilities between two species can lead to the elimination of one of them in simple environments (Gause, 1934; Park, 1948); therefore, in nature, even a slightly subordinate species might be much less common and found in many fewer habitat types than if the dominant species were not present. Indeed, the experimental removal of the larger of two competitors usually has a much greater effect than the removal of the smaller—this was true in 27 of 32 cases in Schoener's sample.

Knowledge that removal of the larger of two competing species is likely to result in substantial population increases among the smaller species and that the larger, despite its competitive dominance, is more likely to be eliminated by human-caused perturbations might suggest likely consequences to which responses should be prepared, especially in cases of biological control where two or more predators are introduced and they differ in size and hence in behavioral dominance. Human activities often introduce into ecological communities new species that, through their competitive and predatory activities, strongly affect indigenous species. Indeed, some of the most important biological problems have been caused by the introduction of species, whether inadvertent or deliberate. The general response to such problems has been the enactment of strict legislation governing importation of species. However, because importation is certain to continue, better understanding of its likely consequences will be of great value to the environmental problem-solver.

MUTUALISTIC INTERACTIONS

Mutualism occurs when two species benefit one another. As in competition, the reciprocal effects are rarely of equal strength. Mutualisms, which are found among all major groups of living organisms, range from obligate (as in the association of some algae and fungi to form lichens) to facultative. Most mutualisms are based on the transfer of energy and materials between the partners, and many mutualisms are believed to have evolved from predator-prey interactions (Thompson, 1982). Mutualistic interactions are of particular importance for the environmental problem-solver, because loss or severe reduction of a mutualist can have a major impact on a target species that would be unexpected if the nature of the mutualistic interaction were not known.

Most frugivores eat many different kinds of fruits over the course of the year, and many include animals in their diets (Wheelwright, 1983). However, many plants have brief fruiting seasons and are visited by only a few species of frugivores. Therefore, plants might be more seriously affected by the loss of one species of frugivore than would a frugivore by the loss of one of the species of plants whose fruits it eats. In contrast, most plants are visited by many species of potential pollinators, whereas many species of pollinators, especially bees, restrict their foraging to one or a few species of plants. (Agriculture, however, depends heavily on the honey bee, *Apis mellifera*, which has a catholic diet and ready access to the flowers of most cultivated plants. The loss of bees, as sometimes occurs when pesticides are used extensively in a region, can cause serious losses of fruits and seeds, even when plant growth remains normal and the harvesting activity of the bees is a rather minor component of the total energy flux in the plant community.)

Interactions among plants, their pollinators, and disseminators of their fruits and seeds are generally rather obvious, and managers are usually sensitive to the need to maintain these mutualistic relationships. Relationships between plants or animals and microorganisms, although less apparent, are equally important. Fixation of nitrogen in terrestrial ecosystems depends on the presence of a few species of bacteria and cyanobacteria associated with the roots of some species of plants, especially legumes (Alexander, 1971; Bond, 1967). Digestion of cellulose by most animals depends on the presence of specific microorganisms in their guts.

Generally, however, environmental problem-solvers need not be seriously concerned about the preservation of these mutualisms, because the microorganisms appear to be nearly universally distributed and the probability that a plant or animal will fail to establish its required mutualistic relationships is very low, even in the presence of severe environmental perturbations.

INDIRECT EFFECTS

We have considered direct effects of interactions, but species often influence other species with which they have no direct contact. For example, adult *Heliconius* butterflies in the American tropics depend for food primarily on the flowers of vines in the genera *Anguria* and *Gurania*, both in the cucumber family (Cucurbitaceae). The ability of the butterflies to lay eggs and hence their rate of attack on the larval food plant, passionflower (*Passiflora* spp.) vines, can depend on the abundance of cucurbits, even though the two groups of vines do not interact directly. Starfish (*Pisaster*), by preying on competitively dominant mussels in rocky intertidal environments, make possible the presence of many species of animals that they do not eat or otherwise directly affect (Paine, 1980). Such indirect effects are less well documented than are direct effects of interactions among species, because more careful experimentation would be needed; and such effects are likely to be missed, simply because investigators do not think to monitor the relevant species in the system likely to be affected by indirect interaction.

CONCLUSIONS

A component of many environmental problems is alteration in the interactions among species. Such a perturbation is a likely candidate for unexpected side effects of a project, most of which are undesirable from a human perspective. Assessing which species are likely to be lost from systems as a result of planned perturbations and which species are likely to be affected (whether beneficially or adversely) by those losses is an important part of preproject analysis and one in which ecological knowledge is central. The knowledge needed to avoid serious problems is often easy to obtain, but no pertinent data are likely to be gathered unless the effects of interactions are considered.

3

Community Ecology

INTRODUCTION

Every species population is part of an assemblage of species—plants, animals, and microorganisms—that share space and interact. We speak of this group of interacting organisms as an "ecological community." It is difficult to define that term precisely, for two reasons. First, whether communities are discrete entities with higher-order "emergent" properties (not readily derivable from an analysis of constituent species) or simply groups of species from the available pool is controversial (Krebs, 1985; Simberloff, in press). Second, whether sympatric species (occupying the same area) should be considered parts of a community if they do not interact with many of the other species present is debatable. The debate, in part, is over the strengths of interactions and how they contribute to community structure. Many environmental problems arise because the alteration of some of these interactions leads to new arrangements of species populations that, from a human perspective, are less desirable than the former ones.

Because the population dynamics of species depend on the kind and intensity of their interactions with other species—species that prey on them and compete with them and on which they prey—a knowledge of these interactions is often valuable in managing individual species. Chapter 2 discusses population interactions as though they are generally independent of each other. Yet we know that each species interacts with many others and that the interactions are often indirect. For example, grazers

47

depend for food on the productivity of grasses, whose growth depends on the activity of earthworms in the soil, which can be affected by the addition of toxic materials to the soil. Human manipulations of the environment can therefore affect individual species, not only directly, but also in indirect ways that can be understood only in the context of the functional structure of the whole community (Davidson *et al.*, 1984).

Ecological communities have numerous properties that transcend those of their constituent species and that require study themselves—trophic structure, rates of energy and nutrient flow, growth form and physical structure, number of species and their numerical distribution, stability characteristics, and ecotones (zones of transition between two habitat types), to name a few. The composition of a community changes in space and time as a result of physical and biological processes. This variability makes it difficult to detect changes caused by human intervention, even if long-term data are available.

Communities are usually named for the commonest or "most important" kinds of organisms found in them. Thus, we speak of "chaparral communities" and "blue mussel communities." The organisms chosen to identify communities are usually the ones that provide physical structure for them, such as plants in terrestrial environments, or that are the bases of food chains in systems lacking fixed structural elements, such as plankton in the ocean.

In addition to its value in managing individual species, a knowledge of community ecology is essential when the object is to manage communities themselves. For example, we often wish to maintain a diversity of organisms in an area, as mandated by law in the U.S. national forests (Chapter 15), or to maintain a particular set of species together, as in parks or agricultural areas. Restoring degraded habitats requires not only a knowledge of the physical conditions that favor the growth of individual species, but also an understanding of how the species interact under different conditions (Chapter 18). We might need to know how the introduction of a species can alter community makeup or how well a new species can substitute for another in maintaining community stability. Or we might wish to know how much disturbance a community can tolerate before undergoing important compositional change.

SPECIES COMPOSITION

The species composition of a community—the number and kinds of species present—is one of its most obvious features. A knowledgeable ecologist can deduce a great deal about environmental conditions in an area simply by looking at a site and inspecting a list of species present

(Dayton, in press). Systems under the influence of strong perturbations typically show reductions in the number of species that are numerically dominant. For example, continued eutrophication of lakes usually leads to an excess of some nutrients, a deterioration in water clarity due to algal proliferation, and a reduction in available oxygen due to increased rates of decay. Species that depend on clear water for foraging or on high oxygen content for respiration disappear from the lakes and are replaced by species that can use the increase in nutrients under relatively anoxic conditions. In Lake Washington (Chapter 20), continued addition of sewage led to the appearance of the blue-green alga, *Oscillatoria rubescens*, now known to be characteristic of severe eutrophication; when the sewage was reduced, *Oscillatoria* disappeared and later the zooplankton community changed.

The presence or absence of an indicator species (Chapter 7) is in itself, however, not always a reliable sign of conditions resulting from human-induced perturbations. Species "typical" of some communities are not invariably present. Local populations can die out for various reasons, including disease, high predation rates, and unusually harsh weather. A species might have failed to colonize an area because of its isolation. Consequently, reliable use of species as indicators of perturbations requires knowledge of their distribution under "normal" conditions and knowledge of past conditions at the site. In addition, knowledge of the composition of an entire community can substantially improve the usefulness of the indicator approach. Patrick *et al.* (1967), for example, related a variety of stream pollutants to changes in the relative abundance of groups of algal species that had different tolerances to the pollutants.

The number of species in a community (species richness) often changes in response to disturbance, and species richness has been used as an indicator of disturbance. Some types of stream pollution simplify the stream environment and reduce the number of available niches; others kill off many species outright (Patrick *et al.*, 1967). But moderate disturbance can produce less clear-cut effects, and some perturbations can alter the relative abundance of species without changing the number of species present (Dickman, 1968).

More sophisticated measures and indexes of community composition weight the number of species with the relative abundance or biomass of each (Krebs, 1985). A diversity index increases both as the number of species increases and as the numerical distribution of species becomes more even. A large environmental change often leads to local extinction of many sensitive species and to the predominance of a few "disturbance-tolerant" organisms or organisms capable of using the new conditions for increased growth, so diversity indexes have been used as measures of disturbance in a community. Such uses of diversity indexes have been controversial, because they have been applied with little regard for the functional changes that occur in disturbed ecosystems.

Many ecologists hoped that diversity measures would capture key changes in fundamental community processes and thus obviate more detailed analyses of species composition. These hopes have been largely unrealized, and there is increasing recognition that the most important information is often discarded in calculating these indexes (May, 1985). A diversity index should be used with caution, for several reasons:

• Many factors other than the disturbance of concern can cause a change in the index. This problem is particularly acute when communities in different areas are compared only once. Site differences in physical and biological factors—nutrient availability, presence or absence of key species, climatic differences, etc.—can cause differences in diversity.
• The species present in a community can change substantially without any significant change in diversity indexes.
• Some disturbances can increase diversity if they increase habitat heterogeneity, reduce the influence of competitively dominant species, or create opportunities for new species to invade (discussed below).

Most of the complexities of the processes that change diversity are not captured in diversity indexes, which are appropriately used only when we are confident that they reflect the behavior of the system being measured. However, because comparison of long species lists from several communities is cumbersome and trends can be difficult to communicate with such lists, changes in diversity indexes within a community can be used to capture the most salient features of that community. The complete lists of species are still needed, but they need not always be presented in full.

FACTORS AFFECTING SPECIES DIVERSITY

Several processes that contribute to change in the number of species in a community can be of concern when the goal of management is to preserve or increase diversity. A primary determinant of species diversity is environmental heterogeneity. On a large scale, species diversity increases as habitat types are added to the environment (Chapter 5). More niches are also added as structural complexity increases within a habitat. For example, vertical complexity in the form of an increase in the number of foliage layers is associated with an increase in bird species diversity, because birds often partition a habitat by occupying different horizontal strata (MacArthur, 1965). Plantings that change the number of foliage layers can change the number of bird species that a park can support (Gavereski, 1976).

Bird species diversity is correlated much less with plant species diversity than with the structural heterogeneity of vegetation (Karr and Roth, 1971; MacArthur and MacArthur, 1961; Orians, 1969; Recher, 1969). Birds forage widely every day and visit many plants in their search for food, responding differently to plant species primarily when those species are structurally very different, as are coniferous and broad-leaved trees (MacArthur and MacArthur, 1961) or spiny and nonspiny desert shrubs (Orians and Solbrig, 1977). In contrast, many herbivorous insects spend their entire foraging lives on one plant and choose their host plants on the basis of chemical characteristics, which often vary substantially with species (Caswell et al., 1973; Ehrlich and Raven, 1965; Fox, 1981; Rhoades, 1979; Westoby, 1978).

Ground-dwelling vertebrates often partition habitats horizontally, and species diversity increases with increase in horizontal heterogeneity (Pianka, 1966). Horizontal heterogeneity also contributes to bird species diversity: patchier habitats support more species (Roth, 1976). Structural heterogeneity is positively associated with insect diversity, with many species supported on a single plant, each specialized for foraging or hiding on a different substrate, such as upper leaf surface, twig, and trunk (Heinrich, 1979; Ricklefs and O'Rourke, 1975; Schultz, 1983a). In addition, chemical variations within a plant cause insects to move more often (Schultz, 1983b). Manipulation of physical and chemical structures of crop plants and mixtures of plants can thus be an effective way of combating pests (Crawley, 1983; Hare, 1983; Whitham et al., 1984).

Repeated perturbations in a community change its makeup as species undergo local extinction and reinvade. The number of species is usually relatively small in highly disturbed communities, because few populations are able to re-establish themselves before they are reduced by later disturbances. In contrast, a low rate of disturbance provides few opportunities for pioneering species and might allow competitively dominant species to usurp limiting resources, particularly in space-limited systems, such as rocky intertidal zones or some terrestrial plant communities. Therefore, the number of species in a community is often greater at intermediate rates of disturbance (Connell, 1978; Huston, 1979) than at either low or high rates.

Some climax plant communities seem to require periodic disturbance for long-term maintenance. For example, some California chaparral communities (Biswell, 1974) and some grasslands (Wells, 1965) appear to be maintained by periodic fires. When fire is controlled, these communities are replaced by others.

Predation

Predation and periodic disturbance of other types influence species diversity in similar ways. By removing competitively dominant species, predators can increase species diversity. For example, experimental removal of starfish from a rocky intertidal zone allows competitively dominant mussels to usurp space from other species; when present, starfish open up space for other species by selectively removing mussels (Paine, 1966, 1974). However, although mussels in the absence of starfish predation can take over space and reduce the diversity of macroinvertebrates in the rocky intertidal zone, the mussel beds provide vertical structure that actually increases total diversity when microinvertebrates are also considered; this shows how the effects of spatial heterogeneity and predation can interact (Suchanek, 1979).

Herbivores can exert similarly powerful effects on community structure in terrestrial grazing systems. The effects of grazing on the diversity of plants depends on whether herbivores selectively graze on the competitively dominant species, which increases diversity, or on poorer competitors, which decreases diversity (Harper, 1969). The selective grazing of herbivores can lead to the replacement of naturally dominant but palatable species with species that are spiny or toxic. The influence of predators on species diversity seems to be most powerful in space-limited systems.

Freshwater predators often select their prey by size, and that can result in large changes in the makeup of plankton communities (Zaret, 1980). In some cases, the introduction of planktivorous fish into fish-free lakes can reduce the numbers of larger, competitively dominant zooplankton and lead to increases in smaller species (Brooks and Dodson, 1965), although the changes can be complex and difficult to predict (DeMott and Kerfoot, 1982).

Competition

Competition can influence community structure by causing the elimination of some species from local regions or habitats and by reducing the abundances of species in the habitats in which they occur. There are reasons for expecting competition to be strongest among closely related species (Darwin, 1859; Lack, 1954), and many such cases of competition have been documented. However, competition has been especially looked for among closely related species, and the frequency of competition among more distantly related organisms could be much higher than currently believed. Competition among distantly related species is especially prevalent when space is the limiting resource. Plants of all taxonomic groups

53

compete strongly with one another and, in the rocky intertidal zone, animals of different phyla compete with one another and with algae (Connell, 1975; Paine, 1966; Underwood and Denley, 1984).

Territorial exclusion, a form of competition for space, can also occur between species. Interspecific territoriality is most common among closely related organisms, but does occur among more distantly related ones as well, especially among fish (Ebersole, 1977; Myerberg and Thresher, 1974). Some cases of interspecific territoriality among birds also involve distantly related species (Cody, 1969; Moore, 1978; Orians and Willson, 1964).

Competition occurs among distantly related grazing mammals, such as moose and hares (Belovsky, 1984). In desert ecosystems, ants and rodents compete strongly for seeds (Davidson et al., 1980, 1984; Kodric-Brown and Brown, 1979). Many more cases of competition among distantly related species are likely to be uncovered as ecologists devote more effort to the study of such competition.

Competition seems to be rare among herbivorous arthropods (Strong, 1984). This suggests that management practices that exert their effects precisely on the target insect species are unlikely to result in unintended side effects on many other species in the community. The use of toxic substances that adversely affect many species has led to greatly magnified influence on population dynamics of other species (Chapter 24). Careful targeting of management toward the focal species decreases the likelihood of side effects that undermine the goals of management.

Productivity

Field studies of more or less natural ecosystems have shown a positive relationship between the number of species in an ecosystem and its productivity (Connell and Orias, 1964). The most common interpretation is that in productive ecosystems more resources are above the minimal abundance required to support users than in unproductive ecosystems (Connell and Orias, 1964; MacArthur, 1972) and that animals in productive ecosystems can therefore specialize on resources that in less productive environments can be used only by generalists. Also, if productivity affects the number of species of plants, then the richness of species of animals that depend on the plants automatically increases as a consequence.

In contrast, it is commonly observed that eutrophication of lakes reduces, rather than increases, species richness (Rosenzweig, 1972). There is no generally accepted explanation of this response. One model assumes that increasing productivity makes predators more effective in eliminating some of their prey or in inducing wide oscillations in their abundances, which are likely to lead to extinction from a number of causes (Rosenzweig, 1972). Another suggests that competitive exclusion is more important in enriched environments; enriched environments favor "weedy" species that dominate typical members of less productive environments (Huston, 1979). Managers must be alert to the possibility that changes in ecosystem productivity, a common goal or by-product of human intervention, will lead to unexpected changes in abundances and distributions of many species in a system. Among the affected species are likely to be some of aesthetic or commercial value.

Spatial Factors

Habitat patch size and isolation, two factors that can have strong effects on diversity, are discussed in detail in Chapter 5. Species diversity increases as the area of an "island" of habitat increases, and species diversity in a given habitat patch generally decreases as the patch becomes more isolated either by distance or by the unsuitability of intervening habitat. How these factors affect design and management of ecological reserves and biological control programs is important, and there is much discussion over how to apply understanding of them to the long-term preservation of species and communities (Frankel and Soulé, 1981; Simberloff and Abele, 1976). For example, to conserve some communities of species, it is necessary to maintain a mosaic of various successional stages (Pickett and Thompson, 1980). The resulting habitat patchiness allows species adapted to each stage to find suitable areas. A mosaic of stages can also afford temporary refuge to prey or host species; they will eventually be eliminated in any particular patch by predators or parasites, but by then they will have colonized other patches (Dodd, 1959).

The spatial configuration of habitat patches also determines the extent and nature of ecotones. Ecotones support many species that would not be present in pure communities. Some forest management plans attempt to maximize diversity by creating configurations of habitat patches with much ecotone while retaining patch sizes and proximities that can support species that rely on single community types (Thomas, 1979). What constitutes the most appropriate configuration of patch size, shape, and spacing depends on the requirements of the species to be maintained.

Summary

The exact roles of the factors that influence biogeographic patterns of species are controversial (Brown, 1984), but the factors reviewed above are all known to affect diversity. An understanding of these factors can provide an environmental manager with a powerful set of tools for manipulating the environment to bring about or limit changes in diversity. Because management is usually targeted to particular species or groups of species and not toward diversity itself, however, additional knowledge of the natural history and population dynamics of the species of interest is required.

Competition, predation, and mutualistic interactions combine with components of the environment to influence species richness in various ways. Plant species richness is often high in the presence of low soil fertility and periodic disturbance, both of which interact to slow down the takeover of a site by competitively dominant species (Huston, 1979; Tilman, 1982; Chapter 18). A similar phenomenon occurs in rocky intertidal habitats, where animals are the dominant competitors for space (Menge and Sutherland, 1976; Paine, 1966).

COMMUNITY ORGANIZATION

Species richness is only one property of a community that influences its structure and dynamics. Abiotic factors, such as moisture and temperature (Holdridge, 1967), and the biological processes of competition (Strong et al., 1984), predation (Paine, 1980), evolution (Orians, 1975), and trophic structure (May, 1983; Paine, 1980; Pimm, 1982, 1984) act to influence the structure of a community by determining its makeup and the constraints under which its constituent species live. Because the linkages between species can be at once complex, indirect, and strong (Paine, 1984), investigating the effects of perturbations by studying single species can be misleading (Kimball and Levin, 1985).

Nonetheless, a small number of factors often dominate the organization of a given community and determine its response to particular stresses. For example, soil nutrients in many tropical forests are primarily tied up in the vegetation and superficial soil layers. When all the vegetation of these forests is removed for cultivation and the soils are unprotected from nutrient leaching, the soil can lose its capacity for regeneration for a long period (Gomez-Pompa et al., 1972). Particular species often dominate the visual appearance and structure of communities, providing physical structure for the existence of many other species. For example, coral reef communities depend critically on the reef-building activities of living corals. In terrestrial communities, vascular plants provide the dominant substrate on which most biological interactions are carried out.

A single species can be critical to the maintenance of a community in its ''normal'' state, such as starfish in some rocky intertidal communities. Those ''keystone predators'' exert an influence on community makeup out of proportion to their numbers or biomass. The idea of keystone species was first applied to predators that have sessile prey, especially when the preferred prey is a dominant competitor for space in the absence of predation. For example, the elimination of the sea otter along the Pacific coast of North America contributed to a large increase in the number of sea urchins, a major food item of otters; and the proliferation of urchins resulted in the decline of kelp beds through excessive urchin grazing (Duggins, 1980; Estes *et al.*, 1982). Lobsters can function similarly as keystone species by preying on urchins off the Atlantic coast (Mann and Breen, 1972). The African elephant exerts a large effect on the landscape by destroying shrubs and trees; that results in the proliferation of grasses, an increase in the frequency of fires, and the conversion of woodland to grassland (Krebs, 1985).

STABILITY AND RESILIENCE OF ECOLOGICAL COMMUNITIES

Ecologists hold diverse opinions about the relationships between the numbers of species in an ecosystem and the complexity of their interactions and about the system's responses to perturbations. Some have asserted that simple ecosystems are much less stable than complicated ecosystems (Elton, 1958; Hutchinson, 1959; MacArthur, 1955; Watt, 1964), and others have asserted the opposite (Gilpin, 1975; Goodman, 1975; Horn, 1974; May, 1973; Pimm, 1979). This diversity of opinion reflects inadequacies in information and the use of different definitions of stability and different types of communities and perturbations. "Stability" has been used to refer to lack of fluctuations (constancy), resistance to being changed by external perturbations (inertia), speed of recovery from perturbations (resilience), and other ideas (Goodman, 1975; Holling, 1973; Orians, 1975). Frank (1968) has pointed out that a community of long-lived species can appear to have some aspects of stability merely because the component species live a long time.

A general relationship between stability in any general sense and species richness is unlikely. Many natural ecosystems are species-poor, but nonetheless stable by some definition mentioned above (e.g., Arctic tundra). And some species-rich systems are sensitive to disturbance, because of the intricacies of the connections among their component species (e.g., tropical rain forests). Moreover, human-induced perturbations not only change species richness, but also create new patterns of interactions (e.g., Cairns, 1980). Until the species have adjusted through evolution to those new patterns, the systems might behave in ways that reflect not simply their altered richness, but the evolutionary novelty of the interactions (May, 1973).

For management purposes, it is important that the meaning of "stability" most appropriate for the problem at hand be clearly specified. In some cases, such as preservation of valued species, it could be most important to prevent the system from being changed very much by the planned actions (Chapter 16). In other cases, such as control of erosion, it could be more important to quicken the return to a former condition of the community, because the problem depends primarily on the duration of a disturbance. Ecological knowledge probably will never be able to provide answers that are general and yet precise enough to replace the need for understanding specific systems and perturbations. Such knowledge can be expected, however, to help in focusing research more narrowly on the most important interactions.

INVADABILITY

An early survey of invasion by plants and animals was carried out by Elton (1958). He concluded that invaders were more likely to establish populations in cultivated and otherwise disturbed environments than in pristine environments, and he noted that islands were more susceptible to invasion than mainland areas. This general perspective has been supported by recent research, although the reasons for the relationships are not much clearer than they were 30 years ago. Determining whether a species might invade new areas requires knowledge about its life history, relationships with other species, and responses to various agents that perturb ecosystems.

Herbaceous plants have been among the most successful invaders of new environments. The flora of California now contains nearly 1,000 exotic plants, and much of the intermountain west is dominated by European and Asian annuals (Mack, in press; Mack and Thompson, 1982). Communities of freshwater fish also appear to be unusually susceptible to invasion by exotic species (Courtenay and Stauffer, 1984). Birds do not invade new areas as easily. Only three natural invasions of North America by birds have occurred during the last century: those of the cattle egret and two gulls. All three exploit food resources that have greatly expanded in recent decades (Orians, in press). Deliberately introduced species, such as starlings and house sparrows, primarily exploit human-modified environments. Many of the escaped captive birds that have established feral populations in North America also exploit new food resources, particularly those provided by extensive plantings of ornamental trees and shrubs in southern cities.

Invasions by herbivorous insects are complex, but most species feed on plants that are closely related to the species on which they feed in their native range (Furniss and Carolin, 1980). Many insects colonize introduced plants in all parts of the world, but their natural food plants are generally unknown (Strong *et al.*, 1984).

SUBSTITUTABILITY

When a species is removed from an ecological community, its roles are sometimes taken up entirely or in part by other species. The degree to which this occurs is referred to as substitutability. Because all species are involved in many different interactions, substitutability probably varies with the particular role being considered. For example, the blight-caused loss of the American chestnut in Appalachian forests resulted in only a temporary reduction in rates of photosynthesis in those forests, because other trees replaced chestnuts in the canopy. However, species that are specialists on the tissues of chestnut trees (folivores, frugivores) must have suffered major losses that will continue as long as chestnuts are rare.

The likely effects of species losses on community dynamics depend on the details of current interactions, so an important part of project planning is a survey of competitive, predator-prey, and mutualistic interactions of an obligate and specialized nature. Such information can help in predicting which species losses are most likely to affect other species in the system. The significance of the potential effects can be evaluated, and steps to reduce the likelihood of their occurrence can be included in the project plan.

ECOLOGICAL SUCCESSION

As long as physical conditions do not change greatly, more or less distinct communities tend to replace others after disturbance in a predictable way. Although ecological succession was originally thought of as a community process, examination of particular successions has shown that abrupt, wholesale extinction of the constituent species of one community with concurrent colonization by the species of another is rare (Drury and Nisbet, 1973). The fates of some pairs or groups of species are inextricably intertwined, as are the fates of some mutualists, but these linkages are in a minority. Typically, the times of appearance and disappearance of most species in a succession are generally independent of those of others, and some species that seem late are present early, but in an inconspicuous form.

The nature of the interactions among species that determine their turn-over during a succession and the relative stability of the climax stage are poorly understood for many successions, partly because no community is exactly like any other. Recent research has suggested that processes and patterns of succession differ among communities and depend on which species are present at the start and are available to colonize later and on their life histories (Horn, 1976). Some early species modify the environ-ment to facilitate growth and recruitment of other species, as colonizers of sand dunes stabilize the soil and so allow others to become established (Olson, 1958). Many pioneering plant species are so intolerant of shade that the shade they create inhibits growth of their own seedlings. Some species inhibit others chemically (Rice, 1974). Some late successional species persist because they are more tolerant of potential sources of mortality, such as fire or grazing (Harper, 1969; Sousa, 1984).

Although successions are highly variable in detail, most have some characteristics in common. Odum (1969) listed many patterns of change in energy flow, biomass, and physical structure that are predictable. Some of these patterns, such as the ratio of gross production to respiration (P_G/R) for the community as a whole, can indicate the stage of a succession and how long a given stage is likely to persist without intervention.

Early successional stages typically have relatively high P_G/R, whereas later stages have ratios approaching 1:1. Part of the reason is that early species are usually herbaceous, with most of each plant's resources devoted to growth and reproduction; many later species, which persist longer, support more woody tissue and devote more resources to competition than to reproduction. Thus, early stages do not lose in respiration most of the matter produced by photosynthesis, as do later stages, and usable (net) production is relatively high. The high net production of early succession is harvestable for human use, and this is taken advantage of in agriculture and forestry.

Human societies usually try to maintain early successional stages pre-cisely because they are more productive, but maintaining them in the face of the natural tendency for change requires large expenditures of energy, effort, and materials. Prolonging normally short-lived early successional stages by calculated disturbances (such as plowing and weeding) or the use of chemicals (such as herbicides and pesticides) entails environmental and health problems (e.g., see Chapters 14, 23, and 24). Simply harvesting in the same site for a long period can result in slow degradation of the soil by erosion and leaching of nutrients. As the soil loses its capacity for production, the economic and environmental costs of maintenance grow with the use of fertilizers.

The challenge for ecologists is to help to identify ways of minimizing the large expenditures and environmental costs of maintaining early successional communities. Integrated pest-management programs aim at reducing the role of pesticides by integrating the use of pesticides with other modes of control (e.g., crop rotation and biological control) on the basis of detailed studies of pest life history and ecology. Soil erosion can be reduced by such techniques as reducing tillage and selecting optional contours (Greenland and Lal, 1977). Herbicide applications to powerline rights of way can be reduced by planting shrubs that impede succession (Niering and Egler, 1955).

CONCLUSIONS

All human-induced environmental disturbances alter interactions among species in some way that leads to direct and indirect affects on the composition of ecological communities and their dynamics. Generally, the direct effects on species of concern are more readily identified and anticipated than are the indirect effects, especially the effects that influence community properties that are the summation of activities of many species. The major question is sometimes how long a community will remain in an altered state. At other times, the main question is how seriously a community is changed. Major changes might be intolerable, even if the community eventually returns to its predisturbance state. The problems described in this chapter are among the most difficult to deal with and are accordingly those for which careful planning and monitoring of a project are especially important, if unexpected and undesired ecological changes are to be avoided or reduced.

4

Materials and Energy

Organisms depend on the input of energy (in the form of sunlight or high-energy molecules), water, and mineral nutrients for metabolism and growth. Human societies depend on harvesting organisms or parts of them, so many of our perturbations of ecosystems are intended to maintain or increase production of organisms. Methods of increasing production of organisms useful to people include domestication, increasing the proportion of production that comes from species of economic value, removing predators, reducing competition, and increasing supplies of the resources that support production (e.g., by fertilization, irrigation, and modification of microclimates). Some valued ecosystem traits, such as processing of wastes and generation of aesthetically attractive landscapes, depend on the total production of species in the system, but often only a portion of total ecosystem production is of immediate value. Environmental problem-solving is often aimed at increasing allocation of production to particular forms, such as wood production in forests or seed production in agriculture.

The overall determinants of biological production in terrestrial ecosystems are generally well known. The rate of photosynthesis depends on solar radiation, the availability of water in the soil, abundance of mineral nutrients, and temperature. All these are commonly manipulated to increase productivity of agroecosystems.

Productivity is tied to fluxes of energy and matter. Sometimes it is measured in units of mass per unit time, as when one assesses the growth of vegetation or livestock. Sometimes it is measured in units of energy per unit time, especially when the objective is to estimate the efficiency with which one form of biomass (say, a plant) is converted to another (an animal). Viewed as energy transformation, productivity conforms to the laws of thermodynamics; viewed as material flux, it exhibits conservation of mass. Hence, the measurement of nutrients and nutrient fluxes is a logical companion to productivity studies.

Production, as opposed to productivity, is the material produced and hence is measured in units of biomass. It can be differentiated into gross production and net production. Not all gross production (P_G) is available for harvest, because materials and energy are expended metabolically by living organisms through respiration (R) to maintain themselves. Net production (P_N) is the portion of gross production that exceeds maintenance ($P_G = P_N + R$). In most ecosystems, *NP* corresponds to the amount of material that can be harvested or processed by detritivores.

Measuring the production of entire ecosystems is difficult, and only incomplete data are available on even the most intensively studied ecosystems. Nonetheless, enough is known to enable us to see general patterns in the allocation of the products of photosynthesis to compartments in the systems. For example, interesting patterns are revealed by examining ratios of energy stored in long-lived tissues to energy stored in short-lived tissues of plants. If the ratio is high, much energy is tied up in tissues that remain intact for most of the life of the plant, such as trunks, branches, and large roots of trees and shrubs and rhizomes of some perennial herbs and grasses. The patterns revealed by such a comparison are as follows (Jordan, 1971):

- The ratio of production of wood (long-lived) to production of litter (short-lived) in forests generally increases as solar energy available during the growing season decreases—that suggests the increasing importance of energy storage in low-energy environments.
- The ratio of wood production to litter production decreases as precipitation decreases—that suggests that increases in size are less valuable than the production of photosynthetically active tissue in dry environments.

An important implication of these patterns is that a higher proportion of energy tied up in long-lived tissues means that less energy is available for use by other organisms in the community, because long-lived tissues are not heavily used as food.

Consideration of ratios directs attention to trade-offs between different patterns of allocation. A plant, for example, can allocate a quantity of energy to wood, leaves, flowers, pollen, nectar, fruits, or defensive chemicals. An increase in the allocation to one comes at the expense of allocation to at least one of the others. Domestication involves major changes in allocation patterns to increase the energy devoted to products of value to people (Snaydon, 1984). The trade-off often expresses itself as a reduction of resistance to harsh climatic conditions, predators, parasites, and pathogens. Protection from these agents must be provided by people—e.g., through weeding, application of pesticides, and killing of predators.

PERTURBATIONS AND PRODUCTIVITY

Natural perturbations—such as floods, droughts, windstorms, and fire—can cause extensive and profound changes in an environment, sometimes even the elimination of living organisms. Perturbations caused by people can be just as extensive geographically, and their cumulative effects can be profound, e.g., conversion of forest to croplands or cropland to residential areas.

Productivity often increases as a result of perturbation, but decreases as perturbation becomes more severe (Odum *et al.*, 1979). Sulfur dioxide and nitrous oxide at low concentrations in the air over a cornfield can increase production, because of the fertilizing effect of the sulfur and nitrogen. But, as their concentrations increase, productivity of the cornfield decreases, eventually to far below the preperturbation level.

Common physical perturbations are those imposed deliberately in land management, e.g., logging or selective harvesting of forests (Chapters 19 and 23), channeling or damming to alter water flow (Chapter 21), and mining (Chapter 18). The disturbance might be simply the harvesting of the species of interest (Chapter 12). Development projects can also inadvertently affect species (Chapter 16).

Effects of perturbations caused by chemicals can be more subtle than those caused by physical alterations. Pollutants that have been taken up by photosynthesizers can be passed on to grazers or detritivores, stored in those organisms, and concentrated (as in the case of some radionuclides and fat-soluble substances). When the grazers are eaten by predators or scavengers, the pollutant burden is passed along. Other predators or scavengers might eat those and thus continue the process of concentration. Eventually, pollutant concentration can be high enough to cause significant biological effects (Chapter 22).

CHEMICAL PATHWAYS AND BIOLOGICAL CONCENTRATION

Biological concentration of organic chemicals, especially hydrophobic, lipid-soluble ones, can be extreme. For example, polychlorinated biphenyls (PCBs) can accumulate to more than 100,000 times the concentration in water. Eating 0.5 kg of Lake Erie fish can cause as much PCB intake as drinking 1.5×10^6 L of Lake Erie water. Such pollutants cycle both through the inorganic parts of the environment and through the biota, and their presence can affect primary and secondary productivity directly or indirectly. The DDT case study (Chapter 24) shows how a pollutant can affect populations by altering reproductive performance.

The pollutants that follow a decomposition pathway might appear, in some form, in the sediments of a water body or in the soils. Some metals released into the environment are transformed by bacterial action and become either more or less toxic to higher organisms. For example, relatively insoluble inorganic mercury, which is not highly toxic, can be methylated and become biologically mobile and highly toxic.

Effects of chemical perturbations become evident when death or illness results from excessive concentrations of toxic materials. Chemical perturbations can also affect ecosystem processes and patterns. For example, increased input of phosphorus often results in eutrophication of lakes, with substantial increases in primary productivity and changes in species composition (Chapter 20).

NUTRIENT FLUXES

The major elements of organisms—e.g., carbon, nitrogen, oxygen, sulfur, and phosphorus—are involved in massive global cycles that involve both living and nonliving components of the biosphere. Except for phosphorus, these elements can be present as gases, and the magnitudes of their movement are being substantially modified by human activity. For example, the anthropogenic fluxes of sulfur today are approximately equal to those of the natural sulfur cycle (Andreae and Raemdonck, 1983; Franey *et al.*, 1983). Changes in global biogeochemical cycles affect environmental problem-solving at specific sites and are beyond the control of managers. But they are the most important of the cumulative effects of individual projects, because the perturbations are largely the sum of the inputs of specific projects. One of the most challenging problems in maintaining high environmental quality is to find ways of reducing undesirable effects of individual projects on biogeochemical cycles. This requires dealing with the cumulative impacts of many projects.

Patterns and rates of nutrient fluxes have traditionally been studied along with biological productivity, especially in agricultural systems where relations between nutrients and productivity were first demonstrated. Construction of "nutrient budgets" for ecosystems by tabulating the incomes and outputs of various elements is an application of the first law of thermodynamics, i.e., net energy changes in a system are determined by the inputs and outputs, whatever the internal pathways or mechanisms. Nutrient conditions in both Lake Washington (Chapter 20) and Southern

Indian Lake (Chapter 21) were predicted essentially in this way. Attaching biological relevance to nutrient values, however, requires an understanding of physiological processes. The importance of phosphorus as a fundamental and controllable limiting nutrient was central to arguments about Lake Washington, for instance. Among the elements that constitute major proportions of cellular mass (carbon, nitrogen, oxygen, and phosphorus), phosphorus is unique in not having a gaseous atmospheric phase. For this reason, plant production in lakes is often limited by rates of supply of phosphorus and can be manipulated over long periods. Because the potential for plant production rises and falls with the availability of phosphorus in the water, growth decreases if phosphorus is removed from the water and its input is lessened. (The rate of decrease depends on the size and shape of the lake and the flushing rate of water.) Although early work on phosphorus in lake waters discounted its importance as a regulating nutrient (Juday and Birge, 1931; Juday et al., 1928), experience with eutrophic systems showed its pivotal role by the middle of the twentieth century (Hasler, 1947). Domestic sewage is especially rich in phosphorus from animal wastes and the polyphosphate complexes that are commonly used as builders or surfactants in laundry detergents.

Relationships between production and light are sometimes more important than those between production and nutrients. For example, Southern Indian Lake remained thoroughly mixed after impoundment, and suspended solids from an eroding shoreline increased abiotic turbidity and diminished light penetration (Chapter 21). Lake Washington was moving to a somewhat similar situation of low water transparency during 1963 and 1964, the peak years of enrichment (Chapter 20).

In Southern Indian Lake, the element that received most attention after impoundment was not the one that had been targeted by the impact assessments. Impact studies had considered phosphorus because it is often a limiting nutrient in freshwater lakes, but mercury emerged as more important. The environmental problem became chemically much less like that of Lake Washington than like the problem with DDT (Chapter 24). In Southern Indian Lake, the deleterious material was not manufactured and added artificially to the system, but was released as a consequence of the manipulation.

INTERACTIONS AMONG PRODUCTIVITY, BIOMASS, AND NUTRIENTS

Development of our understanding of productivity, biomass, and their relations to nutrients was spurred in part by practical concerns. During the 1960s, a debate arose about the causes of lake eutrophication and the most potent management options (Edmondson, 1974; Likens, 1972; National Research Council, 1969). That phosphorus and phosphorus loading were basic for controlling productivity in eutrophic basins was shown by convincing logic and careful analyses of cases in which loading was altered by design (e.g., Edmondson, 1969, 1972; Schindler, 1974).

Current models are based on findings that are embodied in quantitative relations among hydrological conditions, basin structure, and nutrient and chlorophyll concentrations in lakes (Chapra and Reckhow, 1983; Dillon and Rigler, 1974; Vollenweider, 1969, 1976). These models help to relate physical and chemical circumstances to productivity and trophic state. They are relevant both to applied issues and to studies of the constraints on energy transfer among trophic levels and of the magnitude of the production base. A key feature of the models, however, is not so much what they explain, but what they do not. They cannot predict algal biomass, chlorophyll, or related entities even to within an order of magnitude when a set of heterogeneous basins are compared. Substantial residual variability in lake productivity cannot be explained by nutrient loading (Carpenter and Kitchell, 1984; Harris, 1980). For single lake basins or lake districts, the fits are usually far better, particularly when variation in externally derived nutrient loading is large (Edmondson and Lehman, 1981; Schindler et al., 1978), but even in such basins, changes in chlorophyll and water transparency can be extreme, despite relatively constant nutrient loads (Edmondson and Litt, 1982).

INDEXES OF ECOSYSTEM FUNCTIONING

Productivity is used as an index of perturbation in a number of ways. In aquatic ecosystems, following the ideas of Eppley and Peterson (1979) and Harrison (1980), primary production can be considered as a sum of production based on "new" nutrients and production based on "regenerated" nutrients. The "new" portion of production is the primary carbon fixation that is supported by externally supplied nutrients, e.g., from upwelling, chemical weathering, and stream or overland runoff. The "regenerated" portion of production is the availability of nutrients in situ from excretion and decomposition of cells. Production that is based on regeneration cannot be harvested without compromising the production base itself. Measuring "new" and "regenerated" nutrients is not easy, but the data needed are not much different from those needed to construct quantitative nutrient loading models (Vollenweider, 1969, 1975, 1976). The techniques for doing so have become very sophisticated (Reckhow, 1979).

The cycling index (Finn, 1976, 1978) is a useful measure of the effects of perturbations on nutrient fluxes in ecosystems, especially terrestrial ones. The cycling index is the ratio of the amount of nutrients recycled in an ecosystem per unit time to the amount of nutrients moving through the system. For most undisturbed terrestrial ecosystems, the index is between 0.6 and 0.8. The 20-40% of the nutrients lost must be made up either from the atmosphere or from weathering of bedrock, if the ecosystem is to remain in a steady state. In agricultural or other disturbed systems, the index is often much lower, and productivity is maintained by fertilization.

CONCLUSIONS

Our efforts to extract materials from ecosystems, to use them as waste processors, and to manage them for productivity or maintenance of species richness often conflict. Managing the conflict requires understanding of the determinants of biological productivity and of how those determinants are influenced by human-induced perturbations. The most difficult changes to predict are those involving dynamics of important nutrients, and most surprises result from incorrect prediction of the effects of projects on the behavior of nutrient elements. The same problem is central in understanding the influence of human activities on global air circulation patterns and climate—issues addressed only peripherally here. Changes in behavior of nutrients seldom attract attention, because they are often not directly perceivable by the unaided senses, although they do become obvious, for example, when they impair visibility in scenic areas.

Other conflicts arise over decisions to divert energy flow patterns in ecosystems toward species of indirect value to people and away from species with no current value. Indeed, much of the difficulty in preserving species richness is related not to overexploitation of species, but to conversion of the ecosystems to uses that are incompatible with preservation of the community of species originally present.

5

Scales in Space and Time

When we alter the size, shape, and spatial distribution of patches of particular ecological communities, we alter the population dynamics of the species that live in them. The ability of a population to withstand environmental fluctuations, for example, depends not only on the life history of the species, but also on population size and the availability of immigrants from other populations. Analogously, the temporal characteristics of environmental manipulations can influence the kind and strength of their effects. Perturbations of longer duration or greater frequency might exceed the capacity of a community or species to absorb or recover from them, whereas short or single disturbances might not. Focusing on the consequences of single perturbations can lead to a failure to perceive the patterns and cumulative effects of those perturbations over time and space. If a population or community is repeatedly disturbed for long enough, changes qualitatively different from and more serious than the effects of single perturbations often occur. An appropriate choice of scale for thinking about, analyzing, and manipulating these processes is crucial.

PATCHINESS AND COMMUNITY COMPOSITION

Species-Area Relationship

Large areas tend to have more species than small areas of similar habitat type (Cain, 1938; Connor and McCoy, 1979; Gleason, 1922; MacArthur and Wilson, 1967; Preston, 1960), partly because larger areas typically have more habitat variation within the general habitat type. Each variant contains species adapted specially to it. However, even when habitats in a region are uniform, there is a relationship between area and number of species. A second reason for this relationship, which is especially pronounced in small areas, is that each species has a minimal viable population size for a given probability of extinction (Shaffer, 1981). As the area of a habitat decreases, local populations get smaller and extinction becomes more likely. In addition, the species-area relationship is partly a result of sampling (Connor and McCoy, 1979), i.e., large areas receive more immigrants than small ones and therefore obtain larger samples from the species pool. These three reasons for the species-area relationship are not mutually exclusive and can be difficult to distinguish (Connor and McCoy, 1979).

Extinction of Small Populations

At least five forces increase the probability of extinction of small populations:

• *Demographic stochasticity.* Random fluctuations of demographic events (birth, death, and determination of sex) endanger small populations. For example, the probability that all individuals in a generation will be male is much greater in a small than in a large population.

• *Genetic stochasticity*, consisting of inbreeding depression and production of homozygotes for lethal or severely deleterious recessives. Inbreeding depression is the general decrease in traits that contribute to fitness, such as fertility. It has been documented in many plant and animal species and appears to be associated with the increased homozygosity (in which the two copies of a particular gene are identical) that results when near relatives mate. Because mating with relatives is more frequent in small populations, inbreeding depression is greater. In addition, greater homozygosity for the population as a whole might be associated with reduced genetic variability and lead to reduced ability of the population to adapt to environmental change (Soulé, 1980). Mating with near relatives also increases the likelihood of producing individuals homozygous for recessive traits that are lethal or severely deleterious.

• *Environmental stochasticity.* Random variation in the physical or biotic environment of a species affects demographic values, whether the population is large or small. Such variation, even if not severe, can threaten the very existence of a small population, however, because the probability that all individuals will be killed is much greater for a small than for a large population.

• *Disasters and catastrophes.* The once-in-a-century flood or fire, for instance, can destroy a local population.

• *Social behavior.* Some animal species have stylized forms of social behavior (e.g., predation, defense, thermoregulation, and mating displays) that break down if there are too few individuals. A breakdown can lead to breeding failure and endanger the population.

Minimal viable population size varies widely among species, for a number of reasons. In general, the minimal number of individuals necessary to support a population for a long period increases as average population density decreases. The average area necessary to support an individual animal is greater for predators than herbivores, and in general it increases with body size within groups of similar species (McNab, 1963; Schoener, 1968). Species differ in their ability to tolerate the increase in inbreeding that occurs in small populations. In general, plant populations appear to be able to survive longer on smaller sites than animal populations. It is easier to conserve plants than animals by artificial means (e.g., cold storage and seed banks).

Patch Geometry and Edge Effects

The shapes of habitat patches cause effects similar to those due to patch size. As patches deviate from circular to linear, the proportion of their area close to an edge increases, as it does when they decrease in size. Species adapted to conditions found at the interfaces between patches of different types can exploit an increasing fraction of the areas of small patches. They can compete with species adapted to the interiors of patches, parasitize them (Brittingham and Temple, 1983), or function as predators against which interior species are not well adapted (Wilcove, 1985). Human modifications of environments often create patches that are much longer than they are wide (Godron and Forman, 1983), thereby exacerbating the effects of patchiness itself.

The shape and orientation of patches can have important ecological consequences. Cutting of forests into strips causes less erosion if the strips follow the contours of the terrain, rather than being oriented at right angles to them (Hornbeck et al., 1975). At high latitudes, direct sunlight might penetrate to the ground only at dawn and dusk in narrow clearcuts oriented in an east-west direction, whereas in north-south patches direct sunlight is present at ground level at midday—the time of most intense solar radiation.

DISTRIBUTION OF PATCHES IN SPACE AND TIME

Spatial or temporal patchiness is sometimes obvious, but more often difficult to detect. For example, the distribution of herb species in a field might appear random when actually it is determined by the microspatial heterogeneity of soil nutrient conditions (Tilman, 1982). Patchy distribution of organisms can result from variability in the physical environment (such as soil types), physical disturbance, and patchiness in biological interactions (see also Chapter 3). Studies of intertidal and subtidal communities have shown the importance of local heterogeneity generated by physical and biological disturbances in both temperate areas (Dayton, 1971; Menge and Sutherland, 1976; Paine, 1966; Paine and Levin, 1981) and tropical areas (Connell, 1978; Porter *et al.*, 1981). The spatial and temporal variability of lake and deep-sea benthos is well known (Berg, 1938; Brinkhurst, 1974; Grassle *et al.*, 1975; Jonasson, 1972).

Because marine systems are dominated by species that do not derive nutrients from the substrate (animals and nonvascular plants), substrate-related variability is due primarily to the physical environment and stability of the substrate and the type of anchorage it provides (e.g., Dayton, 1984, 1985). In terrestrial systems, however, soils differ primarily in their ability to supply nutrients and water, so variability in distribution of plants, burrowing animals, and species that depend on plants is related to these properties. Soil scientists could be included in terrestrial research teams more often than they usually are.

Spatial Considerations

Population dynamics in patchy environments are determined primarily by the rates of individual movements between patches and rates of local population extinction. The properties that enable populations to maintain themselves in patchy environments include high dispersal rates, tendencies to cross unsuitable habitats, high growth rates, early reproduction, and high reproductive rates (Baker and Stebbins, 1965; MacArthur and Wilson, 1967). These traits increase the probability that isolated patches will be found, that reproduction will occur before the patch becomes unsuitable, and that new colonists will be generated. Species with the traits increase in abundance in environments that are heavily modified by people.

Patches are not static, but change with time, primarily as a result of the growth of and interactions among colonizing organisms. The succession of organisms over time is driven by several processes that are not mutually exclusive, but that differ in their relative importance under different conditions (Horn, 1974, 1976; Shugart and West, 1980). Species are adapted to conditions at different stages before populations become extinct as a result of within-patch changes. Early successional stages typically are much shorter than later successional ones. Therefore, even if rates of disturbance are low, populations of early successional species are more likely to become extinct. However, current rates of human-caused disturbance are so high in most areas of the world that species requiring late stages of succession, such as old-growth forests, are in the most precarious positions (Chapter 17). In addition, many organisms have complex life histories in which different stages require distinct habitats. Not only must individuals find all the necessary habitat types at the correct time, but populations are affected by fluctuations in the availability of habitats for each stage; these fluctuations can be independent of each other (Istock, 1967). The supply of appropriate habitat must be adequate at the correct time during the year; its abundance at other times can be irrelevant.

Even within an apparently uniform patch, interactions among subunits can be complex. This complexity was not a factor in the case of Lake Washington (Chapter 20), because the high inflow of water during winter and early spring, when the lake is isothermal, causes free circulation throughout the lake. Southern Indian Lake (Chapter 21), however, contains several subbasins, and the major outlet has been close to the inlet since the diversion of the Churchill River. The lake is thus not a well-mixed body, and its properties differ between regions, whether or not they are affected by the flow. The high quality of the whitefish fishery before impoundment was maintained by confining fishing to areas where the stocks were virtually free from cestode infestation. After impoundment and diversion of the normal river flow, the stocks became redistributed, and the fish were concentrated in areas with high infestation rates.

Local populations occasionally become extinct because of predation, disease, or physical disturbance. The rate of recolonization is inversely related to the distance from other occupied sites that are sources of immigrants. Increasing isolation of patches increases the probability that locally extinct populations will not be replaced, and creation of communities that would normally develop without help might need to be managed because of a lack of immigrants (Chapter 18). In general, species characteristic of later successional stages are poorer dispersers than "weedy" species of earlier stages. In plants, late stages show a striking increase in the average size of seeds (Harper *et al.*, 1970; Salisbury, 1942); in animals, the later stages show less tendency to disperse and greater reluctance to cross stretches of unsuitable habitat.

Temporal Considerations

Community dynamics are strongly affected by interactions among perturbation characteristics (intensity, duration, and frequency), succession, and the rate at which a community recovers. Very different outcomes are possible if relative rates of disturbance and recovery are altered. For example, tropical slash-burn agriculture is compatible with long-term soil fertility if plots are small, are farmed for only a few years, and are allowed to remain fallow for several decades (Gomez-Pompa *et al.*, 1972). However, if plots are made larger and are recut after shorter fallow periods, soil fertility cannot recover within the period of a single cycle and rapidly declines (Myers, 1984).

If pesticides are used infrequently, time might be available between applications for susceptible genotypes of pests to replace resistant types that are at a disadvantage in the absence of the pesticides. If pesticides are used more often, this process does not go to completion, and many resistant genotypes are still present when the pesticide is used again. As a result, the pesticide becomes progressively less effective, applications are increased in frequency and magnitude, and the evolution of resistance among the pests is accelerated (Chapters 1 and 24).

Repeated or continuous perturbations can lead to qualitative changes in community structure, because the ability of the system to remove or recover from the disturbance is exceeded. Striking changes took place in the plankton communities of Lake Washington when the addition of sewage at multiple points exceeded the flushing rate of the lake (Chapter 20).

In addition to the deleterious genetic effects that often occur in small populations (Franklin, 1980; Selander, 1983), there are long-term evolutionary consequences of patch size and distribution. Evolution in response to spatial or temporal change in the environment is retarded by small population size, because genetic variability is reduced in small populations (see Endler, 1977, for a discussion of the effect of migration on evolution). For example, dividing agricultural plots into different sections and applying a different pesticide to each should retard the evolution of resistance.

CONCLUSIONS

The great importance of size and spatial relationships in the working of ecological processes points to the importance of dealing explicitly with scales in space and time in all efforts to solve environmental problems. The major changes in processes and products that accompany changes in spatial and temporal scales can escape attention, if efforts are not directed specifically at them in all phases of environmental problem-solving. Well-intentioned efforts can be undermined if they are planned for too short a term or for areas that are too small (Soulé and Wilcox, 1980). However, key ecological processes might be obscured if inappropriately large temporal and spatial scales are used. Averaging over large areas can mask the importance of local patchiness for the survival of particular species. Individual trees could be especially susceptible to attack by herbivores, and the maintenance of large populations of trees could depend critically on patches. Similarly, patches of high concentrations of nutrients resulting from defecation of zooplankton might be essential to survival of algae, and patches of high concentrations of plankton might be important to the survival of marine fish larvae (Sissenwine, 1984).

6

Analog, Generic, and Pilot Studies and Treatment of a Project as an Experiment

INTRODUCTION

The previous five chapters emphasized the application of ecological knowledge of many general types, but they did not deal directly with knowledge specifically applicable to particular problems. However, ecological systems often respond to perturbations in a site-specific fashion, and large or novel perturbations can produce effects not predictable from basic ecological principles, so specific knowledge is often necessary for the understanding and prediction of many environmental effects. This chapter briefly discusses the value and limitations of using specific information and of treating projects and other environmental actions as experiments when potentially serious effects cannot be adequately predicted from information acquired in advance. Observation of the actual effects of an action can sometimes be the only source of information for management decisions.

ANALOG STUDIES

An analog study concerns an action similar to one of immediate interest. The most useful analog study would be conducted in the same area, on the same species or ecosystem, on a similar scale, and under similar conditions and would be thoroughly analyzed. Even if the analogs are not perfect, study of them can be useful in scoping a problem, identifying major potential effects, and designing appropriate studies.

When a project is similar to many others, analog studies typically accumulate, especially if they are legally required. For example, in the legally mandated Maryland Power Plant Siting Program for managing cumulative impacts, study of the environmental effects of each power plant should increase the ability to predict the effects of future power plants (e.g., Maryland Department of Natural Resources, 1984). But the use of data on one power plant to predict effects of another also requires knowledge of the differences between the receiving ecological systems and the differences between power plants. Variations in physical and ecological conditions between areas can lead to serious errors in extrapolation. For example, in the Southern Indian Lake project (Chapter 21), no specific analog studies dealt with changing of the level of a shallow lake in a permafrost zone. Because the best analogs—deep lakes in Siberia—differed in important respects from Southern Indian Lake, investigators failed to predict the magnitude of bank erosion that occurred and the consequent changes in water turbidity and mercury release.

In the Lake Washington case (Chapter 20), valuable analogs were available from experience with nutrient enrichment in temperate lakes in Europe. In the derelict lands case (Chapter 18), no studies of methods for diversifying plant communities on previously mined land were available, but attempts to diversify roadside vegetation informed and motivated the research studies. Analog studies were of value in the caribou case (Chapter 16) in identifying the major issues, developing field techniques, formulating a study strategy, and locating important resource people who were familiar with the problem.

Some cases not only relied on analog studies for their design, but were designed so that they became valuable analogs themselves. The Lake Washington experience now serves as a model for many lake eutrophication studies and has contributed importantly to our understanding of nutrient loading in lakes. The comprehensive review and analysis of the Southern Indian Lake project increased its instructional value. Analog studies are most valuable if their results are analyzed and summarized in an easily accessible form.

GENERIC STUDIES

A generic study is designed to increase knowledge of the physical and biological phenomena common to a group of environmental problems. It might be designed to reveal the processes that underlie ecological responses to manipulations in a particular environment—e.g., a study of the effects of marine oil pollution along coastal beaches or of oil pollution in aquatic environments in general. It might deal with a restricted subset of environmental perturbations, such as the uses of DDT, or try to provide a model of the effects of a large class of perturbations, such as pollution by organochlorides in general. Because ecological effects vary with the environmental manipulation or insult, the location, the biological communities present, the scale, and other factors, a major question concerning the usefulness of generic (and analog) studies is the extent to which their results can be applied to a particular site and problem. The most useful generic studies identify, characterize, and explain the major ecological effects of a particular environmental insult or manipulation and provide a framework for understanding the effects of variations in location and other factors.

The distinction between analog and generic studies is not always clear. So much was learned from the Lake Washington experience, for example, that it might serve as a generic model. A major practical difference between analog and generic studies is that generic studies are usually well known to biologists, whereas analog studies are often buried in government or industry files. Also, generic studies are usually designed to address a specific question, which means that the postproject monitoring is often good enough to provide comparative data.

When government and industry cooperate in addressing problems of mutual concern, studies can be coordinated and research resources used efficiently. For example, before issuing a permit for exploratory drilling in state-owned offshore lease areas, California requires extensive studies designed by a committee of representatives of academe, government, and industry. Coordination of studies among a number of applicants for leases in the Santa Barbara Channel allows each to conduct studies on specific problems common to all drilling programs, such as the effects of discharge fluids on hard-bottom organisms; results from all the studies can be used to develop permit stipulations for other applicants. The advantage of such coordinated endeavors is that duplication is minimized, so more research is possible.

Environmental studies, for practical reasons, cover limited areas, relatively short periods, and conditions dictated by the time and place chosen. Moreover, many generic studies involve simulation of the environmental change of interest, often on a smaller scale and with a narrower range of variables and variation than encountered in the actual situation. For example, the aquatic effects of thermal effluents from a nuclear power plant might be studied by injection of hot water into a river (if the appropriate permits required by the Clean Water Act were obtained first). But such an injection would be made over a much shorter time and a much smaller area than the usual discharge of effluent, and the injected water might lack some contaminants present in the normal effluent.

Some other limitations of generic studies are discussed in the review of clearcutting studies (Chapter 23). The Hubbard Brook clearcutting experiments of the 1960s (Bormann *et al.*, 1968) were for a long time accepted as general models of the effects of clearcutting on stream eutrophication, soil nutrient loss, and the regeneration potential of land. However, herbicides were applied after the clearcut in the Hubbard Brook experiments, whereas rapid revegetation is encouraged in many forestry operations. Errors in extrapolation from a generic model to a specific situation can be reduced if generic studies focus on a particular application and location.

Generic reviews can yield ecological knowledge of great potential value. They might focus on various topics of potential use to environmental managers: a group of related problems; natural history and ecology of particular species, areas, or communities; general approaches to problems; and broadly useful methods or techniques. A list of useful generic reviews and related literature can be found in the annotated bibliography on ecological impact assessment prepared by Duinker and Beanlands (1983).

PILOT-SCALE EXPERIMENTS

Pilot-scale experiments are designed to investigate the effects of specific kinds of environmental perturbations on time and space scales much smaller than those of a planned project or action. They are appropriate when there is little confidence in the hypotheses underlying the prediction of effects and when realistic simulation of the perturbation is possible.

The major uncertainty in connection with pilot-scale experiments is whether their results can be generalized to similar large-scale phenomena (Hilborn and Walters, 1981). Although the overall effects of large developments, such as hydroelectric dams, cannot be realistically simulated on a small scale, some individual effects of large developments can be studied experimentally with the traditional scientific procedure of dividing a problem into its components for analysis.

Experiments were used to advantage in the derelict lands case (Chapter 18), to investigate the performance of groups of plants under different soil conditions; there is no obvious reason why the results of these studies cannot be generalized to larger-scale applications. Pilot-scale experiments were conducted to study the potential effects of the Alaska pipeline on caribou movements (Cameron *et al.*, 1979). A variety of small-scale experiments were conducted on the effects of DDT (Chapter 24). Laboratory and field experiments were used to investigate toxicity in several species and to determine residence time of DDT under different environmental conditions. Field and laboratory experiments confirmed the occurrence of bioaccumulation. The Atomic Energy Commission program for studying the effects of nuclear radiation (Chapter 22) is a good example of the combination of a variety of laboratory and field experiments to improve understanding of a complex problem.

TREATING A PROJECT OR ACTION AS AN EXPERIMENT

If reliable prediction of the ecological effects of a project or action is not feasible, even with extensive field investigations, the limits of predictive ability should be openly recognized, so that planners, decisionmakers, and managers can take into account the uncertainty of the ecological outcomes of particular actions (Paine, 1981). Viewing a project or action as an experiment can aid in designing a program to monitor effects. Such monitoring can have two major advantages: the detection of unexpected effects can be used as a basis for altering procedures (Holling, 1978), and the monitoring information can be used in the planning and design of similar projects or actions.

The idea of a project as an experiment can be applied to a variety of environmental problems, particularly in impact assessment and resource management. As Beanlands and Duinker (1983) pointed out, however, most impact assessments concentrate on the prediction of impact almost to the exclusion of followup studies to determine actual impact. Even in postproject studies, little effort is spent in determining whether predictions were incorrect and why (Larkin, 1984). An attempt by a study group to audit the results of a large number of impact assessments in the United Kingdom revealed several major impediments to learning from such experiences (Anonymous, undated): predictions were absent or vague, monitoring schemes were inadequate, documentation was poor, and relevant parties were unwilling to cooperate. Chapter 10 offers some general suggestions for designing monitoring studies to make them more useful for planning projects or actions. Among the projects described in case studies, the best examples of the idea of a project as an experiment are the Garki malaria project (Chapter 15), the project to raise the level of Southern Indian Lake (Chapter 21), the eutrophication and cleanup of Lake Washington (Chapter 20), and the efforts to protect caribou during development of a hydroelectric dam in Newfoundland (Chapter 16).

Walters and Hilborn (1976) and Walters (1984) have suggested that natural resources can be managed experimentally, not only by following up the consequences of management actions, but by basing management decisions on the need for increased knowledge. In that paradigm, managers must weigh the trade-offs between the value of potential immediate yield and the value of information. The value of information is greatest when understanding of the system is poor and the consequences of incorrect decisions most severe.

7

Indicator Species and Biological Monitoring

Living organisms can be used to monitor movements, accumulations, and modifications of materials in their environments and to monitor the biological effects of those materials. They can also be used to indicate the effects of habitat alterations and fragmentation and the effectiveness of management schemes designed to preserve or change individual species or community-level patterns.

Toxic substances are used by people because they confer economic or other benefits that are believed to outweigh the dangers they pose. The regulations devised for their use are guided by knowledge of their biological effects. Chemical and physical monitoring can tell the quantities of materials entering or already in the environment and sometimes the fraction that is anthropogenic. But only biological monitoring can tell us what those materials are doing to organisms. Living organisms not only are essential for determining the biological effects of pollutants, but have several advantages over physical and chemical monitoring:

• Living organisms can function as *continuous monitors*. They accumulate records of past conditions in their tissues. These records can sometimes be read and monitored more economically than can records obtained by establishing stations that continuously monitor environmental conditions directly. But living organisms can also distort or obscure the records by altering chemicals in their bodies; therefore, detailed information is needed to determine which organisms are best suited to provide continuous records of various contaminants.

81

• Living organisms can increase the *sensitivity of monitoring*. Many species can concentrate materials in their tissues and so amplify weak environmental signals. Lipid-soluble pesticides are concentrated in body fat and further concentrated when one organism eats another. Foraging bees bring pollen and nectar to their hives from many nearby sources. Sampling of pollen from bodies of bees has provided a much more complete picture of the distribution of several pollutants in the Puget Sound region than was obtained by more expensive chemical monitoring (Bromenshenk *et al.*, 1985).

• Living organisms can provide information about *complex mixtures of materials*. Laboratory studies, for reasons of expense and ability to control concentrations, typically expose organisms to potential toxicants one at a time. In nature, however, organisms often encounter toxicants in complex mixtures. By observing changes in such functions as locomotion, growth, and reproduction in living organisms, we can determine the effects of those mixtures, even when their exact nature is unknown. (Fortunately, appropriate control measures can often be instituted without knowledge of all the details of a mixture.)

• Living organisms can provide information about how pollutants affect performance under *natural conditions*. Test organisms in a laboratory are typically kept in ideal conditions and are seldom subjected to inclement weather, disease, predators, crowding, or food shortages. In nature, however, these insults can occur simultaneously with the presence of the toxic materials of concern. Under such conditions, concentrations of toxicants that would have no effects in the laboratory might be detrimental. Because our concern is to protect organisms under natural conditions, biological monitoring must be an integral part of monitoring schemes, although it must be kept in mind that the variability of natural systems often makes it hard to be certain that an apparent effect is a genuine one.

• The use of communities of living organisms for monitoring can provide information about how toxic materials influence *patterns of interactions among organisms*, community patterns, and processes of concern (see Chapter 3).

CHOICE OF ORGANISMS TO USE FOR BIOLOGICAL MONITORING

Groups of organisms can be chosen for particular monitoring purposes on the basis of the required speed of response (Cairns *et al.*, 1973) or on the basis of sensitivity to temperature changes (Cairns, 1977), for instance.

In all cases, the monitoring systems should be chosen with a clear understanding of the goals to be achieved, the status of current knowledge, and the ability and commitment to act on the results obtained.

Life-history characteristics of organisms strongly influence their utility for various types of biological monitoring. For example, short-lived organisms respond quickly to environmental changes, whereas long-lived ones might integrate stresses over years, decades, or even centuries. Species with high metabolic rates and, hence, usually high growth rates are often more sensitive to contaminants than are species with low metabolic rates. Sessile organisms are exposed to all the contaminants that enter their immediate environments, whereas mobile organisms can escape many of them by leaving the area. Many organisms stop reproducing under stressful conditions, so changes in fecundity rates can be important signals of environmental change; but special care is required here, because stress stimulates reproduction in some organisms, particularly plants.

Even within a category, some organisms are better indicators of environmental change than others. Vascular plants are effective detectors of air pollution, because particular species are especially sensitive to particular pollutants. For example, the toxic components of photochemical smog in California were unknown until they were revealed by plant assays. Lesions on beans, spinach, and grape leaves indicated the presence of ozone (Middleton, 1956); lesions on annual bluegrass leaves indicated the presence of peroxyacetylnitrate (Bobrov, 1955). The ways in which different plants respond to major air pollutants are documented by Weinstein and McCune (1970).

Simpler terrestrial plants, such as lichens and mosses, are often more sensitive to airborne pollutants than are vascular plants, because they absorb water and nutrients directly from air and rainwater. As a result, they concentrate pollutants and exhibit toxic effects more quickly than vascular plants, even though they are not generally more sensitive (Hawksworth, 1971; Lawrey and Hale, 1979). A measure of the magnitude of pollution is the lichen species diversity index, which combines the number of species present and their relative abundances (the abundances are based on extent of coverage, growth form, or degree of luxuriance).

In aquatic environments, algae and cyanobacteria are useful indicators of various changes. Increases in nutrients (eutrophication) can be assessed according to increases in biomass of algae (*Spirogyra, Oedogonium, Stigeoclonium,* and *Cladophora*) and several genera of cyanobacteria (such as *Oscillatoria* in Lake Washington, as described in Chapter 20). Diatoms and some other algae concentrate heavy metals and radioactive materials by a factor of several thousand.

Mollusks have been used as monitors in aquatic environments with a monitoring network known as Mussel Watch (Goldberg *et al.*, 1978). Species that have been used include some in the genera *Mytilus, Crassostrea*, and *Ostrea* in outer estuaries and coastal marshes; in *Geukensia* in tidal marshes; and in *Anodonta* and related genera in freshwater lakes, rivers, and streams (Goldberg *et al.*, 1978; Patrick and Kiry, 1976). Much is known about how pollutants are deposited in mollusk shells and preserved for long periods. Metals, pesticides, and hydrocarbons are also deposited in soft tissue; they are often rapidly flushed from some soft tissue and can be accumulated there, too. Much more work needs to be done before we will know whether concentrations of most materials in soft tissue can be used effectively to indicate current environmental changes or past environmental contamination.

The utility of organisms as biological monitors of environmental changes is greatest when the functional relationships between perturbation and response are understood. Nonetheless, organisms can be useful as indicators even if causal relationships are obscure. If a valued ecosystem component disappears or is reduced, attention is directed to a change that might be important. Caution is always necessary, because natural environments are highly variable. Species that are "typical" of some types of environment can be absent from specific locations for reasons having nothing to do with human disturbance, and jumping to a premature conclusion might retard progress toward the goal of understanding the causes of environmental changes and how they affect organisms.

Some general guidelines in selecting species most useful for particular monitoring purposes are nonetheless possible. For example, top carnivores (the organisms that develop the highest concentrations of persistent, fat-soluble pesticides) are often of special value in detecting environmental changes. They have large home ranges and they typically forage over wide areas, where they can be exposed to materials with a patchy distribution that are difficult to detect with fixed monitoring stations. Sea birds are valuable for monitoring, because species that nest in a single location might forage at widely differing distances from the breeding colony; therefore, knowing which species are encountering which chemicals (as revealed by analysis of regurgitated food) helps in locating sources of and areas contaminated by toxic materials (Boersma, in press; Gilman *et al.*, 1979). Sea birds are useful also because the relative ease of measuring their distributions and abundances makes them good monitors of changes in populations of fish and aquatic invertebrates, whose populations are much more difficult to assess. The herring gull (*Larus argentatus*) was chosen to monitor pollutants in the Great Lakes of Canada and the United States, because it accumulates pollutants only from the lakes, rather than from land, and because it is a year-round resident (pollutant concentrations in its tissues and eggs are not influenced by pollution in distant places). Several environmentally important compounds were first identified in herring gull tissues, and the gull has helped to monitor the fate of persistent organochlorines after control measures were established (Mineau *et al.*, 1984).

Butterflies and moths are well enough known for changes in their distributions and abundances to be detected, and they can evolve rapidly enough for genetic changes induced by environmental changes to appear within a few decades. The familiar cases of industrial melanism among lepidoptera in western Europe and northeastern North America are among the best-documented cases of evolution by natural selection in which environmental pollution was the primary factor in changing the survival rates of some genotypes (Kettlewell, 1973; May and Dobson, in press).

Many animals are used by veterinarians, toxicologists, and physicians to monitor human toxicants in the environment (Buck, 1979; Harshbarger and Black, in press; Schwabe, 1984). The expertise of those specialists could usefully be combined with that of ecologists to improve the choice and use of animals as environmental monitors.

MONITORING AND ENVIRONMENTAL SPECIMEN BANKING

Living organisms can preserve records of environmental materials in their tissues, but these records might change as the organisms age. Even more important, the death and decay of an organism eliminates such records. It is therefore useful for a program of biological monitoring to include specimen banking. The first indication of the utility of specimens as records of former environments was the documented thinning of eggshells as a result of the widespread use of DDT (Cooke, 1973; Peakall, 1975; Chapter 24). This benefit was, of course, fortuitous—the eggs were collected for other purposes. The potential value of preservation and methods for preserving and deciding what to preserve are treated in detail by Lewis, Stein, and Lewis (1984), who point out that appropriate specimens can provide records of trends of pollution over long periods. Moreover, specimens remain available as analytical techniques improve, so we can use new methods retrospectively. Similarly, as new chemicals become of concern, banked specimens can be examined for them, even though no attention was paid to them when the specimens were collected. Banked specimens can also serve as records to determine the effectiveness of pollution control programs. Nonbiological materials, such as lake and marine sediments, also have some of these advantages and should therefore be included in a global banking scheme; but they cannot provide the full range of information obtainable from specimens of living organisms collected at specified times and places.

The results of monitoring are difficult to evaluate unless the organisms collected can be identified. Faulty identifications could lead to erroneous conclusions about environmental changes and their causes. For this reason, in spite of the cost, it would be useful for monitoring programs to be associated with regional taxonomic centers that could care for the collections and identify the materials. The lack of adequate taxonomic collections and trained persons able to care for them is a serious scientific problem in the United States, as elsewhere, and it adversely affects our ability to follow and interpret human-caused environmental changes.

MONITORING OF BIOLOGICAL RESOURCES

We have discussed the role of biological monitoring in the detection of pollutants and their effects. Biological monitoring is also needed to provide inventories of biological resources. Human modifications of habitats are profoundly influencing distributions and abundances of species. Species that thrive in disturbed habitats—such as croplands, pastures, early successional habitats, and urban environments—are increasing at the expense of species that require old-growth forests, riparian environments, wetlands, estuaries, and flowing waters. Indeed, many experts believe that current rates of deforestation in the tropics could cause the extinction of as many as a million species within the next half-century, many of them before they have even been named and described (Ehrlich and Ehrlich, 1981; Lovejoy, 1979; Myers, 1979, 1980). An important role of biological monitoring is to determine which species are increasing and which are decreasing in abundance, where losses of species are most serious, and hence where conservation efforts should be directed. Much of this information is compiled in the ''Redbooks'' on rare and endangered vertebrates and invertebrates published by the International Union for the Conservation of Nature.

Biological monitoring is especially valuable in helping to identify the effects of habitat fragmentation, a common form of alteration of terrestrial environments that results from human activity. Monitoring can help to determine the minimal sizes of patches required by species and how the rate of occupancy of suitable sites declines as distance between habitats increases. For example, the selective loss of neotropical migrant birds with decreasing size of forest patches in eastern North America has been documented by monitoring.

Birds are good organisms for biological monitoring, because there are relatively few species, they are nearly all easily recognized, and there are large numbers of amateurs who know them well and observe them regularly. Christmas bird counts and breeding-bird censuses have been carried out for many decades and are standardized. They provide excellent long-term records of patterns of distribution and abundances. The northward spread of a number of species of nonmigratory birds as a result of extensive winter feeding of birds in eastern North America has been well documented by Christmas bird counts. The striking spread of the house finch (*Carpodacus mexicanus*) in the northwestern United States during the last 20 years can be seen clearly in the breeding-bird censuses, including changes in abundances of species with which the house finch might compete (Mindinger and Hope, 1982). These changes probably indicate as yet unknown environmental changes.

MONITORING AND THE IDEA OF A PROJECT AS AN EXPERIMENT

A major theme of this report is that many projects intended to produce something of value to human society take place on spatial and temporal scales much greater than can be duplicated in experiments designed purely for scientific purposes. Treating these projects as scientific experiments is a key component of effective environmental decision-making (Chapter 10). Monitoring can be a vital part of the use of such projects as scientific experiments. The scales of projects are often large, so monitoring often needs to be extensive. In addition, the results of monitoring are more valuable if they are reported and analyzed in forms readily available to scientists and managers. As in other experiments, specimens collected as part of such monitoring should be retained in repositories, where they will be available to future researchers.

8

Dealing With Uncertainty

This chapter concerns the management of ecological systems in the face of uncertainty. Some uncertainty is unavoidable in ecologists' predictions about ecological systems, but decisions that might lead to unexpected environmental changes have to be made. To deal with uncertainty in ecological prediction, we must first identify its sources and consequences—this information is often the most useful that an ecologist can give to an environmental manager.

SOURCES OF UNCERTAINTY

Many ecological systems—among them those most affected by human activities—are poorly understood. We often lack the detailed information necessary for making accurate predictions, particularly under the unusual conditions that we impose on ecological systems. Several broad categories of uncertainty make the precise prediction of ecological changes difficult.

Complexity

Because the relationships among species are complex, changes in one can lead to unexpected changes in another. Indirect effects—those propagated through a number of species links—are particularly difficult to predict (Brown, in press). For example, when DDT was introduced into terrestrial environments, no one imagined that it would eventually be found in marine fish (Chapter 24). Equally hard to predict are the effects of multiple perturbations, such as the introduction of different kinds of pollutants from multiple sources into a river (Chapter 9).

The response of a population or community can become nonlinear at some degree of perturbation, perhaps with a threshold beyond which the response changes qualitatively. Fish populations can undergo rapid collapse when overharvested (Murphy, 1977; Ricker, 1963), and lake communities can undergo distinct changes in composition when nutrient concentrations change (Chapter 20). Continued reduction and fragmentation of habitat can also lead to the extinction of local populations and whole species (Chapter 17). In some cases, nonlinearity of population dynamics can lead to chaotic behavior even in environments that are homogeneous in space and time (May, 1981; Chapter 2).

Natural Variability

Populations and communities vary in space and time because of both intrinsic processes and changes in their physical and biological environments (see also Chapter 5). In most species, reproduction is seasonal, and mortality is neither constant throughout the year nor equally distributed with respect to sex and age. An estimate of the makeup of a population or community at any particular time is like a snapshot—often one with poor resolution—of one state among many possible states. The responses of populations and communities are all too often influenced by their site. The presence or absence of a particular species, for example, can change a community's response to environmental change. Ecological systems commonly respond in a site-specific or situation-specific fashion, so our predictive capability is likely to be poor when appropriate analog studies for nearby or similar sites are unavailable.

Temporal variability in ecological systems can have several implications. Some systems are highly variable because of periodic natural disturbances or intrinsic cyclic behavior. For example, the marine intertidal communities on the western coast of North America are subject to severe storms that can temporarily destroy them. Many organisms in such systems are adapted to those disturbances and have characteristics that facilitate recovery (Sousa, 1984). Changes in the systems must be interpreted through an understanding of the natural history of the organisms present. Similarly, changes in plankton communities that undergo seasonal turnover must be interpreted in the light of population changes in natural conditions.

Changes in climate can also complicate assessment of human-caused environmental changes. For example, it is often unclear whether changes in fish populations are due to fishing pressure or to changes in ocean conditions (e.g., Chapter 12; Parrish and MacCall, 1978; Steele, 1984;

Wooster, 1983). Environmental changes can interact with biological phenomena to produce large and complex changes in population numbers and perhaps alternative stable states (Steele and Henderson, 1984; Wooster, 1983). Changes in climate can be severe but brief, such as El Niños (e.g., Barber and Chavez, 1983), or they can take place over hundreds of years with large effects on the animals and plants distributed over wide areas (Bryson and Murray, 1977).

Natural variability makes it difficult to establish baseline conditions. The possibility that changes in populations or communities are due to natural changes in the abiotic environment should always be kept in mind in making predictions about ecological systems.

Random Variation

Because many forces acting on populations are more or less random, populations behave in probabilistic rather than deterministic ways, and their responses to perturbations can never be predicted precisely. Population numbers can fluctuate randomly and on a periodic or cyclic basis—seasonally or over longer or shorter periods. If the scale of observation does not match the scale of natural fluctuations, even cyclic (e.g., seasonal) fluctuations can be perceived as random noise that makes determination of the state of a system difficult. Because variation is sometimes random and sometimes systematic, we need to know about the variability of important population or community measures—mean values are not enough—if we are to manage them adequately and predict the effects of changes.

Errors of Estimation

In addition to errors stemming from random variation, measurement error is unavoidable, and it can be large. For example, independent observers counting migrating animals can disagree substantially if the animals pass them rapidly and in great numbers. Estimates of population size must often be made indirectly, as when fish catch statistics are used to estimate stock size (e.g., Chapter 12). Whether estimation is direct or indirect, inevitable errors of estimation add to the uncertainty of management projections.

Errors of estimation commonly combine with natural variability to make the detection of ecological effects challenging. Sampling programs necessary to distinguish an effect from background noise become more elaborate, more time-consuming, and more expensive as the accuracy of estimation required and the extent of natural variability increase. In view of the practical difficulty of distinguishing among effects, it is surprising that we traditionally worry about concluding that an effect has occurred when it has not (type 1 error), but rarely worry about concluding that an effect has not occurred when it has (type 2 error) (Zar, 1976). If multiple type 2 errors are made for a given class of environmental change, serious cumulative ecological effects can occur (Chapter 9).

Lack of Knowledge

We often lack experience with a particular type of environmental perturbation. When neither appropriate studies nor reliable theories or models are available, accurate prediction is not possible, and it might be necessary to conduct extensive ecological studies as an aid to management decisions. However, if even the best possible ecological studies are inadequate to support prediction of acceptable accuracy, a planned project or action must be viewed as an experiment in itself (Chapters 6 and 21).

MANAGING IN THE FACE OF UNCERTAINTY

The point of discussing the many obstacles to making accurate predictions is not to argue the futility of trying, but to show that the process of prediction must be viewed as complex and probabilistic. An appropriate approach to managing ecological systems recognizes the random component of population dynamics by dealing with measures of system variability, the possibility of nonlinearity, and the consequences of errors (e.g., Beddington 1984a,b; Walters, 1984). For example, attempting to achieve maximal sustainable yields in many fisheries can result in more variable yields, greater population fluctuations, and a greater likelihood of population collapse (May, 1980). Regulating a fishery at higher stock size can stabilize yield and decrease the likelihood of catastrophic failure of the fishery. When populations are very variable and exceeding a critical threshold would have severe consequences, the best strategy could be to hedge one's bets—to reduce risk by trying not to exceed the threshold (Beddington, 1984a; Jewell and Holt, 1981).

When we have little understanding of either the major potential effects of a perturbation or the dynamics of the receiving system, our predictive capability is likely to be extremely poor. Developing a project or initiating a management plan can then be viewed as conducting a large-scale experiment. Information from well-designed monitoring studies can allow management to adapt to the occurrence of unexpected effects (Holling, 1978) and can be valuable in the planning of similar actions (Chapter 12).

In some cases, particularly those involving the management of renewable biological resources, information can be gathered by managing experimentally—purposefully manipulating the system to learn about the effects (Bar-Shalom, 1976; Beddington, 1984a; Ludwig and Hilborn, 1983; Walters, 1984). Models can aid in conceptualizing a problem and in developing hypotheses that can be used to design research and monitoring studies (Holling, 1978). The use of multiple working hypotheses permits one to compare the potential effects of several possible perturbations.

A number of methods are available for dealing with randomness in ecological systems. Error analysis is a formal method of dealing with random effects when models are formulated explicitly (Meyer, 1975). Random error is inserted into a model to simulate variation, and the model is explored by Monte Carlo simulation or simple first-order analytical techniques (Walters, in press). This method of modeling has been used to determine the minimal viable population for endangered species (Shaffer, 1981). Survival is a probabilistic affair, and one who would determine the minimal tolerable population size must decide for how many generations a species is to be protected and the acceptable probability of extinction. Such modeling efforts are usually limited to analyzing the effects of random error, but it is also possible to model systematic errors, in an attempt to identify potential causes of extinction.

Sensitivity analysis is a method for assigning different values to the parameters of a model to explore the consequences of errors in choosing and measuring them (Meyer, 1975). The analysis helps to identify the variables to which the system is most sensitive. It permits exploration of alternative management plans and can be used to decide which parameters need to be estimated most accurately.

The obstacles to making accurate quantitative predictions of the behavior of populations and communities are formidable. The natural variability and complexity of ecological systems will always limit our ability to make precise predictions. In the long run, a better understanding of natural variability and of the factors that make some systems more responsive to change than others will improve our predictive ability. But as long as we continue to alter our environment in new ways, uncertainty will always be associated with the effects of our manipulations. Environmental manipulations will always be experimental to some extent, and our most promising course is to structure each one so that we can learn as much as possible from it.

9

The Special Problem of
Cumulative Effects

THE NATURE OF THE PROBLEM

The continuing degradation of ecological systems due to repeated perturbations is a difficult environmental and scientific problem. Because
current science and policy focus primarily on the environmental effects
of single projects or actions, little progress has been made in dealing with
cumulative effects. Recent awareness of such serious cumulative effects
as acid rain, the rapid loss of tropical forests, and the threatened extinction
of many species has brought increasing political and scientific attention
to cumulative environmental effects.

Many cumulative effects result from the movement of materials through
the environment. An analysis of the propensity for natural processes to
concentrate or disperse materials can help to determine where cumulative
effects will be most severe. Several processes lead to the concentration
of materials. The most important is movement of the medium itself, such
as atmospheric mixing and stratification, water currents, and downward
movement of water in soil. Other processes are relatively independent of
the medium itself, as when chemical interactions cause flocculation and
settling of suspended particles entering estuaries and when particles settle
because of gravity.

Living organisms are responsible for considerable concentration of
materials, and their patterns of movement and dispersal can result in
increased exposure to environmental contaminants. Some organisms
migrate great distances, and many species aggregate for feeding or

breeding. Passage of material through food chains can lead to great concentration; top carnivores are exposed to much higher concentrations of contaminants than are plants at the bottom of the food chain (Chapter 24).

Processes that lead to dispersal of materials are sometimes similar to those concentrating them. Water currents, wind, the migration of organisms, and the dispersal of their propagules can all serve either to disperse or to concentrate materials. Dispersal can either increase or decrease the ecological effects of materials, depending on its rate in relation to the biological activity of the materials. Dispersal of highly toxic substances that remain active for a long time might simply increase the area of effects. An inadequate understanding of how such processes work in specific cases can lead to undesirable results. Recent discoveries of areas of toxic waste buildup in Puget Sound, Washington, indicated that the processes of dilution originally counted on to disperse toxic materials and reduce them to harmless concentrations were not functioning as expected. The balance of processes that disperse and concentrate materials in different environments can result in the manifestation of effects over widely differing scales.

The effects of environmental perturbations accumulate when the frequency of individual perturbations is so high that one comes before the system has recovered from the previous one or when the ecological effects of perturbations in adjacent areas overlap. Thus, cumulative effects result from spatial and temporal crowding of environmental perturbations. The combined effects of repeated perturbations can be more severe than and sometimes qualitatively different from the sum of the effects of individual events.

Comparing the problem of cumulative effects with the "tyranny of small decisions" described by the economist Alfred Kahn, Odum (1982) pointed out that, when numerous small decisions on related environmental issues are made more or less independently, the combined consequences of the decisions are not addressed; therefore, no provision is made for analyzing the patterns of the perturbations or their effects over large areas or long periods.

A step toward identifying the major issues involved in the assessment of cumulative impact was made in a recent international workshop held in Toronto, Ontario, under the joint sponsorship of the present committee and the Canadian Environmental Assessment Research Council. Some of this chapter draws on the proceedings of the Toronto workshop (Beanlands *et al.*, in press).

KINDS OF CUMULATIVE EFFECTS

Several types of perturbation can produce cumulative effects. Many involve the addition of materials to the environment, for example, the addition of sewage effluent to Lake Washington from multiple sources over many years (Chapter 20). In that case, a major cumulative effect was the rapid deterioration of water quality with attendant changes in species composition. The materials added can also be toxic (Chapter 24).

A second major type of cumulative effect involves the repeated removal of materials or organisms from the environment. In forestry, for example, both trees and minerals are removed (Geppert et al., 1984). The effect of harvest rate on population dynamics is seldom a linear function of population density, and the response of a population to an incremental increase in harvesting rate can vary substantially with conditions. Increasing the intensity of harvesting can lead to population collapse if a critical threshold is passed, especially if the increase in harvesting is combined with natural environmental changes (Beverton, 1983; Murphy, 1977; Parrish and MacCall, 1978; Peterman, 1978).

A third kind of cumulative effect can occur when management decisions result in environmental changes over large areas and long periods. In the New Brunswick forest (Chapter 19), forestry was based on single stands, instead of a large area of forest. The cumulative effect was a decrease in average yields with an increase in annual variability. Harvesting practices that seem appropriate from the standpoint of population dynamics often result in cumulative genetic effects, such as loss of genetic variability or inadvertent selection for undesired traits (Chapter 1). A classic example of the cumulative genetic effects of repeated interventions is the evolution of pesticide resistance (Chapter 1).

A fourth, more complicated situation arises when stresses of different types combine to produce a single effect or suite of effects. For example, the construction of too many wells, developments, and drainage canals lowered the water table in the Florida Everglades substantially. Similarly, the construction of roads in national forests can trigger logging, recreational activities, and poaching, which might combine with acid rain and other influences to threaten sensitive populations of forest species.

Complex cumulative effects also occur when many individual areas in a region are repeatedly altered. The result can be dramatic changes in the mix, arrangement, and internal characteristics of the habitats of species. Large habitats can become fragmented into patches separated by areas that are inhospitable to many organisms. As habitat patches become smaller and more isolated, species that depend on them become less able to find them and to maintain populations in them (Chapter 17). The conservation of species in the face of severe habitat fragmentation could be our greatest terrestrial environmental problem.

Cumulative effects need not be adverse, but might be considered by many as improvements in the environment. Sometimes, repeated attempts to "improve" the environment have the greatest cumulative effects of all. The conversion of deserts to arable land by massive irrigation projects destroys desert habitats, but creates economic benefits and new habitats that favor species adapted to exploit irrigated farmland. Even the attempt to keep some environments in a constant state through repeated interventions can have unexpected consequences. Fire is often controlled to protect habitats, but some communities require periodic fires to maintain their typical species composition (Kozlowski and Ahlgren, 1974). Moreover, reducing the frequency of fires can increase the severity of fires that do occur by fostering the accumulation of flammable material.

DEFINITION OF CUMULATIVE ENVIRONMENTAL EFFECTS

• *Time-crowded perturbations.* Cumulative effects can occur because perturbations are so close in time that the effects of one are not dissipated before the next one occurs. An example is repeated harvesting of agricultural crops or forests that removes some nutrients faster than they are regenerated between harvests (Geppert *et al.*, 1984; Krebs, 1985, Ch. 28). Similarly, the evolution of resistance to pesticides occurs because the susceptible genotypes are repeatedly reduced in number each time the pesticide is applied (Georghiou *et al.*, 1983; May and Dobson, in press).

• *Space-crowded perturbations.* Cumulative effects can occur when perturbations are so close in space that their effects overlap. An example is power plants close enough that the heat plumes of their cooling water overlap (e.g., Slawson and Marcy, 1976).

• *Synergisms.* Different types of perturbations occurring in the same area can interact to produce qualitatively and quantitatively different responses by the receiving ecological communities. For example, several pollutants might interact to produce toxic mixtures (e.g., National Research Council, 1982, 1983b; for examples of mixtures toxic to humans, see Reif, 1984).

• *Indirect effects.* Cumulative effects can be produced after or away from the initial perturbation or by a complex pathway. For example, when the level of Southern Indian Lake in Manitoba was raised, the inundation of the lake shorelines resulted in the release of mercury into the lake (Bodaly *et al.*, 1984) and increased the turbidity of the water (Hecky, 1984). Neither of these consequences was predicted by knowledgeable limnologists (Hecky *et al.*, 1984).

• *Nibbling*. Incremental and decremental effects are often involved in each of the above categories, but not always. However, this "nibbling" of environments is so important that it should be given its own category. The numerous examples include time and space crowding (the addition of power plants to a river one at a time or the introduction of several pollutant sources into a lake), as well as removal of habitat piece by piece, as in the degradation of Chesapeake Bay (Flemer *et al.*, 1983).

• Other types of impact have sometimes been equated with cumulative effects, such as threshold developments that stimulate additional activity in a region (e.g., new energy developments in northern Canada) or projects whose environmental effects are delayed (time lags) or are felt over large distances (space lags). These types of effects can be cumulative if their impacts overlap in time or space or are synergistic with those of other developments.

DIFFICULTIES IN PREDICTING AND CONTROLLING CUMULATIVE EFFECTS

A fundamental problem in predicting and controlling cumulative effects is the frequently large mismatch between the scales or jurisdictional boundaries of management authority and the scales of the ecological phenomena involved or their effects. Affected environments—such as river basins, airsheds, and estuaries—usually cross local and state, and sometimes even national, boundaries. Conversely, an environmental insult that originates in several jurisdictions might exert its effects largely or wholly in one area. The history of the Delaware River Basin Commission project is a striking example of the complexity of managing such a situation scientifically and equitably (Ackerman *et al.*, 1974).

As difficult as management of cumulative effects can be for regional authorities, management by local authorities is even more so. Local authorities are ill equipped to deal with regional trends in environmental deterioration. They often lack the legislated responsibility, motivation, expertise, and resources to perform the necessary analyses on an appropriate scale. When effects appear far from their sources, the local jurisdictions with the sources might not even become involved.

A major impediment to the control of cumulative effects is the perception that the effects are minor. How often are projects stopped because they appear to be contributing (however slightly) to a trend in environmental deterioration? What happens if scientists recognize a trend, but cannot detect or distinguish the effects of individual actions? The problem of multiple "insignificant" effects is at the core of our failure to deal with the continued nibbling of coastal and other habitats. It is extremely difficult to predict the point of attrition that constitutes a critical threshold. Below what population size will a species become unable to recover (Chapter 14)? Beyond what degree of fragmentation will an ecosystem begin to lose species rapidly or change markedly in its functioning? At what concentration will an environmental contaminant cause qualitative changes in community organization (Chapter 20)?

Predicting the effects of perturbations is particularly difficult when many sources of different types act in concert. Accurate information on the nature of each source is difficult to obtain, and their interactions can be complex. The current controversy over the sources of acid rain and its effects in particular areas is a case in point (National Research Council, 1983a, in press; U.S. Environmental Protection Agency, 1980). The "proof" required in this socially and politically charged case is prohibitively expensive to acquire. Control of eutrophication in lakes has been the greatest success in solving a multiple-source problem, possibly because of the limited areas involved. When effects are global, as with the long-term buildup of atmospheric CO_2, assigning causality and regulating the sources are extremely complex, both scientifically and politically (National Research Council, 1983c).

It is clear that continued loss of communities and ecosystems will, in the long term, constitute a "cost" to society, but there is little agreement on how to express this cost. We lack an accepted framework for determining how the cost of habitat deterioration, with resulting fragmentation, is apportioned among nonresponsible parties, because the observable effects of each incremental loss are often small and local. Assigning a value to an increment of loss can be very difficult, but at some point these small and primarily local effects might be expressed in different forms on a regional scale. Knowing the nature and values of ecological thresholds would give us a benchmark for associating costs with changes, but would not solve the problem of valuing changes that do not threaten to exceed a threshold. Research that will help us to identify the threshold of rapid, observable deterioration is badly needed.

SCALE AND THE RATES OF CRITICAL PROCESSES

Choosing appropriate scales on which to analyze cumulative effects requires an understanding of the relative balance of important concentrating and dispersing phenomena and of the rates of other processes that influence the removal of or recovery from individual stresses in the environment being studied (see also Chapter 5). Because the rate of atmospheric mixing is so high and because the atmosphere is relatively uncompartmentalized, many materials that have low rates of removal from the atmosphere (by decomposition, chemical transformation, settling, etc.) are spread on a global scale. In the atmosphere, dispersing processes typically predominate over concentrating processes, with notable exceptions, as when inversions and scavenging of particles by precipitation lead to local concentration. Atmospheric pollution has been modeled largely on a regional and global scale appropriate to the scale of effects.

Mixing in the oceans is much more compartmentalized and less complete. Concentrating and dispersing processes are more evenly balanced, and the dominating process varies with local conditions. In many freshwater systems, such as lakes, concentrating processes predominate. Individual lakes often behave independently of one another, so the appropriate scale for analyzing cumulative effects is likely to be an individual lake and its associated watershed.

Terrestrial environments are particularly complex spatially and temporally. Spatial heterogeneity is marked, and the existence of many persistent and relatively impermeable structural components results in a high degree of compartmentalization. Because the basic substrate is solid, movement of materials is restricted, and many impacts remain local. Processes that affect the concentration or dispersal of materials are complex in terrestrial systems. Chemical reactions in the soil result in leaching of materials through the substrate and into groundwater or, if precipitation is less, in their concentration in different layers of the soil.

Spatial scales in terrestrial systems can be particularly important when communities or habitats are increasingly fragmented by nibbling and alteration. The sizes of habitat patches and the distance between them can profoundly influence the survival of species that depend on them. The probability of local extinction increases as population size decreases, and the chance that a given habitat patch will be recolonized depends on the dispersibility of the species, the proximity of sources of potential colonizers, and the ability of a population to be established by a small number of colonizing individuals (Chapters 1 and 5). Seminatural plant communities on derelict lands in Great Britain often fail to re-establish themselves naturally, because the sources of seeds of desired plants in undisturbed habitats are too widely scattered (Chapter 18). Adjacent areas provide mostly seeds of introduced weedy species. To maintain natural communities, managers must artificially increase the supply of propagules.

The common approach of treating cumulative effects in distinct environmental compartments is artificial and often inadequate. Atmospheric phenomena, such as wind and precipitation, are intimately involved in the transport of materials into and out of terrestrial systems. Contaminants in terrestrial systems eventually find their way into aquatic systems, and contaminants in aquatic systems often find their way into terrestrial organisms through food chains. The DDT case study (Chapter 24) is a classic example of the complex interchanges that occur between environments. DDT sprayed on crops drifts on the wind and is transported by runoff into aquatic systems, where it is biologically concentrated through food chains. Eggshell thinning problems have followed in terrestrial raptors that feed on contaminated fish or in other birds that eat contaminated aquatic organisms.

Whether an ecological system is adversely affected when subjected to repeated perturbations depends not only on the rate and intensity of perturbations, but also on the rate at which their effects are absorbed by or removed from the system. If toxic chemicals are rapidly degraded biologically to a harmless form, a receiving system will be capable of absorbing a relatively high rate of toxicant input. Ecological systems that are characterized by very low rates of growth and reproduction, such as the arctic tundra, are extremely vulnerable to repeated disturbance.

MANAGING CUMULATIVE EFFECTS: BEYOND A CASE-BY-CASE APPROACH

Improvement in our efforts to predict and to limit or reduce cumulative effects will require a restructuring of both scientific and institutional approaches and a considerable improvement in the interchange of information between scientists and managers. If one accepted the dictum that "everything in ecology is connected to everything else," the already formidable challenge of dealing with cumulative effects would become overwhelming. Fortunately for both scientists and managers, connections are not equally strong, and a first step in trying to predict cumulative effects is to determine which interactions are most important (to reduce the problem to as simple a set of relationships as possible) and which sources of perturbation are likely to affect the ecosystem components of concern (Chapter 10).

As part of the Sustainable Development of the Biosphere project being carried out at the International Institute for Applied Systems Analysis in Austria, scientists have been developing a promising matrix-based framework for scoping potential cumulative-effects problems in the atmosphere (Clark, in press). The approach is designed to identify the many kinds of perturbations that affect particular valued atmospheric components—and the multiple effects of particular perturbations—through an analysis of the mechanisms of interaction that underlie direct and indirect pathways of cause and effect. This approach can be applied to environments other than the atmosphere.

By indicating the relative strengths of the effects of all potential perturbations of a system, the framework can help an environmental manager to choose control variables that offer the most promise and to avoid attempts to control a system by manipulating variables with little potential influence. Because it also identifies the multiple effects of a given perturbation, the framework might also help to focus attention on actions that have the greatest general impact on the environment.

When major effects and their pathways have been identified, the primary tasks of scientists are to track the state variables of ecological systems, to identify trends in environmental quality, and to predict critical ecological thresholds. Considerable attention has been given to tracking environmental trends (Chapter 7), and some refreshingly novel low-cost approaches to biological monitoring have been developed (Bromenshenk *et al.*, 1985). But methods for predicting critical thresholds are sorely lacking. Our understanding of the factors involved in determining population thresholds has improved (Chapters 2, 5, and 17), but much more basic research is needed.

Dealing more adequately with cumulative effects will require changes in institutional approaches of both scientists and managers of cumulative effects. Assessment involves many scientific disciplines and requires both interdisciplinary and disciplinary scientists who are comfortable with and adept at working with experts in a variety of fields. Unfortunately, the reward systems for both faculty and graduate students at most universities discourage such activities.

The Toronto workshop mentioned above brought together representatives of applied and basic science, environmental management, business, and resource agencies. The most critical immediate need in management practices identified in the workshop was the need to improve the match in scales between the ecological phenomena involved in cumulative effects and the management jurisdiction. A good start in this direction would be to improve communication between scientists and managers, and the nomogram used by Erdle and Baskerville (Chapter 19) to indicate the long-term benefits and costs of a variety of forestry options is a useful tool for this purpose.

Comprehensive generic reviews of the scientific issues germane to particular cumulative-effects problems, including a discussion of scale, would be valuable to a manager. Local decision-makers, who often lack access to the expertise and resources necessary to deal with cumulative effects, also need syntheses of regional ecological knowledge. For example, Cooper and Zedler (1980) have proposed a system for land-use planning that allows development projects to be analyzed on the basis of regional ecological knowledge of ecosystem sensitivity to stress.

The Toronto workshop made it clear that the project-specific nature of environmental impact assessment is inadequate for the management of cumulative effects. The management of cumulative effects must in some way match the scale of the ecological phenomena involved.

When important cumulative effects have been identified in aquatic environments, regional management authority has often been proposed or implemented. When valued ecosystem components have been readily identified and have been recognized and generally agreed on by the public, this approach has sometimes been successful. For example, the formation of a metropolitan regulatory authority to control sewage pollution in Lake Washington was very effective, because the public readily comprehended the issues and agreed on the value of a clean lake (Chapter 20). The effort to establish regional control of multiple sources of pollution in the Delaware River Basin was much more complex (Ackerman *et al.*, 1974): many issues of conflict were involved, there was less public agreement on what was most valued, the scientific questions were more difficult, the regional authority was far less successful, and attempts to expand the scope of the metropolitan authority (like that formed to protect Lake Washington) to deal with other and less clear issues, such as transportation, failed to win public support.

Regional authority might need some ground swell of public support if it is to succeed. In a number of recent cases, ad hoc nongovernment organizations, such as "save the bay" groups, have helped greatly in bringing particular cumulative-effects issues to public attention and have created and supported the public debate that eventually provided the impetus and pressure for government agencies to act. Perhaps more important, such groups can serve as intermediaries between scientists and regulators, thereby fostering exchange of information and ideas on controversial topics.

RECOMMENDATIONS

Some of the recommendations produced at the Toronto workshop for improving the scientific and institutional approaches to managing cumulative effects follow; they are adapted from the proceedings of that workshop (Beanlands *et al.*, in press). Some of them are addressed to scientific research, and some to improving institutional handling of cumulative effects.

• Cumulative environmental effects should be placed in readily seen time and space scales, as described for atmospheric effects (Clark, in press). This should help to identify different susceptibilities of different environments and ecosystems to cumulative effects caused by many kinds of perturbations.

• A better match between scales of management and scales of environmental effects is needed. To this end, cases in which cumulative environmental problems have been dealt with—both successfully and unsuccessfully—should be reviewed to understand how the setting of management and ecological boundaries influenced their success or lack of it.

• Research should be conducted to determine what rates of addition of materials to environments and harvesting of resources from them are consistent with protection of valued components of various environmental systems, as well as the systems themselves. Studies of response rates and recovery times should be included in this research.

• Research should be conducted to determine the types of indicators, thresholds, and environments most likely to be useful for assessing and managing different kinds of cumulative effects.

• Monitoring should be built into the design of projects, and cumulative effects taken into account where appropriate. This requires an understanding of the appropriate boundaries. In addition, monitoring should be frequent enough and carried out for long enough to detect cumulative effects.

• Agreements between decision-makers, managers, and scientists with respect to the appropriate time and space boundaries for dealing with cumulative effects should be documented. This will force clear thinking about important issues and will provide a record so that procedures can be improved.

10

A Scientific Framework For
Environmental Problem-Solving

To manage the effects of environmental manipulations, we must be able to predict them. However, knowledge is seldom sufficient to allow accurate prediction, so studies are necessary to provide the information needed to make decisions. Such studies must be carefully planned, because they are expensive in time, money, and effort. This chapter presents a general framework for identifying, scoping, and planning studies of environmental problems. The framework is, in essence, an admonition to think before acting and to use established scientific principles. Table 1 makes it clear that deficiencies in environmental impact assessment are due not only to scientific difficulties—the ones with which this chapter is primarily concerned—but also to political, administrative, and economic difficulties.

Despite their bewildering variety, environmental problems share some basic features, including actions that result in environmental changes, public and scientific concern about those changes, a need for methods for predicting environmental responses to human actions, and limited resources for the acquisition and analysis of relevant ecological information. We draw heavily on a number of recent efforts to make environmental assessment and management scientifically more credible (Andrews *et al.* 1977; Anonymous, 1980; Council on Environmental Quality, 1978; Fritz *et al.*, 1980; Holling, 1978; Larkin, 1984; Rosenberg *et al.*, 1981; Sanders *et al.*, 1980; Sharma *et al.*, 1976; States *et al.*, 1978; Walters, in press; Ward, 1978) and in particular on a recent Canadian review (Beanlands and Duinker, 1983).

TABLE 1   Some Recent Criticisms of Ecological Impact Assessment, Based Primarily on Beanlands and Duinker (1983), Carpenter (1976), Rosenberg *et al.* (1981), and Skutch and Flowerdeu (1976)

Guidelines too elaborate and requirements too diverse
Time and money constraints not recognized
Unreasonable expectations of decision-makers
Tendency to start gathering baseline data immediately, at the expense of careful planning
Failure to formulate objectives clearly and to develop a study strategy
Unwarranted belief that ecological principles used in managed systems are as appropriate to
    unmanaged systems
Failure to recognize the value of early input from those who might later be involved in re-
    view, leading to an adversarial process
Failure to define project boundaries
Failure to consider cumulative effects
Failure to state the bases of value judgments
Lack of scientific standards for impact assessment
Lack of respect in academe for impact assessment
Vague and unverifiable predictions
Lack of a rigorous, quantitative approach, especially in monitoring
Lack of continuity in studies conducted during planning, developmental, and operational
    phases of a project
Failure to follow actions with adequate monitoring studies
Use of impact assessment for disclosure, rather than for learning
Failure to recognize the scientific value of experimentation and monitoring
Failure to consider the recovery potential of species and ecosystems
Poorly written reports in which major points are buried in enormous amounts of information
Inordinate expenditure of effort on descriptive studies with little potential for predictive value
Inaccessibility of reports and results of studies, making them difficult to evaluate and learn
    from

DEFINING ENVIRONMENTAL GOALS AND SCIENTIFIC QUESTIONS

In spite of the difficulties and controversies associated with identifying environmental goals, a clear statement of goals early on can help to focus research and can increase the chance of protecting components of the environment likely to be identified as valuable to society. The first step in defining such goals is to identify the components of the environment perceived as valuable, such as salmon in rivers of the northwestern and northeastern United States, a "natural-looking" community of plants on reclaimed land (Chapter 18), clean water (Chapter 20), clean air, forest productivity (Chapter 19), and fishery productivity (Chapter 12).

The second step is to determine the desired degree of protection, exploitation, or control. This decision usually involves choosing a state in which to maintain the ecological system in question and a length of time for which to maintain it. For example, we might wish to reduce the population of a pest or the amount of damage it causes, increase the yield of a harvested species, or maintain the species composition of a valued habitat. The period of management might be months, decades, or centuries.

The third step is to learn the environmental and financial costs of managing the system. Maintaining an ecological system in other than the ''natural'' condition usually requires some expenditure and might produce unwanted side effects. Harvesting a population for maximal yield can increase the variability of the yield and the likelihood of overharvesting (Chapter 1). Increasing production of an agricultural crop or forest often involves the use of hazardous pesticides (Chapter 24), which can have several deleterious cumulative side effects (Chapters 1, 3, and 4).

Identifying environmental goals is complex and requires input from the public and from scientists. The public is concerned primarily with the choice of environmental goals. Scientists can help to identify non-obvious goals and can indicate the environmental and economic costs involved. Scientists are also needed to translate environmental goals into scientific objectives, which show what information is needed to answer the major questions and hence help in the planning of studies. As in any research plan, scientific objectives are based not only on the need for particular information, but also on how easily that information can be obtained.

The issues on which environmental goals are based are specified in part by law and in part by public concern (see Table 2). The National Environmental Policy Act requires early public and professional input in identifying those issues (Council on Environmental Quality, 1978). The goal of protecting the Southern Indian Lake whitefish fishery was economically motivated (Chapter 21). Provincial wildlife biologists recognized caribou migration as a major public concern in the Newfoundland hydroelectric development case (Chapter 16). In the case of Lake Washington (Chapter 20), interested scientists and the public cooperated to define goals and develop an appropriate response; the DDT case (Chapter 24) shows how such interactions can lead to new understanding and to legislation.

Attempts to achieve a goal sometimes have unexpected results. The New Brunswick forest case study shows how attempting to maximize forest timber production on the basis of individual stands might not only fail to maximize yield over the whole forest, but fail to provide consistency in yields over a long period. In fisheries, managing for maximal sustainable yield often produces large variations in both yields and stock abundance (May 1980), making overexploitation and population collapse more likely

TABLE 2   Some Common Criteria for Identifying Important Issues and Valued Ecosystem Components in Impact Assessment

Legal requirements
   Air and water quality standards
   Public health
   Rare, threatened, and endangered species
   Protected areas or habitats
Aesthetic values
   Landscape appeal
   Attractive communities
   Appealing species (e,g., large ungulates, colorful birds, cacti)
   Species at higher trophic levels (e.g., eagles and tigers)
   Clear air and water
Economic concerns
   Species or habitats of recreational or commercial interest
   Ecosystem components
Environmental values and concerns
   Ecosystem rarity or uniqueness
   Sensitivity of species or ecosystems to stress
   Ecosystem "naturalness"
   Genetic resources
   Ecosystem services
   Recovery potential of ecosystems
   "Keystone" species

than more conservative management would (e.g., Murphy, 1977). Harvesting or managing populations over long periods can also produce undesired cumulative genetic changes (Chapter 1).

SCOPING THE PROBLEM

Scoping involves bringing together all interested parties—public, business, government, and scientific—so that they can interact and express their views before major actions or studies are initiated (Council on Environmental Quality, 1978). Early scoping can help to identify the important issues and potential environmental effects associated with planned actions. It can help to define scientific objectives and guide the design of ecological studies. Scoping can also be useful during a program or project to ensure that the most important issues are being addressed, that studies are producing useful results, and that important new issues are noted (Fritz et al., 1980; Sanders et al., 1980).

Once the valued ecosystem components, significant issues, and major potential effects have been identified, ecologists can establish scientific objectives. When prediction of environmental effects is a major purpose of studies, four questions are usefully posed (Beanlands and Duinker, 1983):

- Are valued ecosystem components expected to be affected either directly or indirectly by the project or action?
- Can the valued ecosystem components be studied directly? (In the absence of adequate guidance from experience or the literature, pilot investigations might be needed to indicate the feasibility of particular studies.)
- Is it possible to study valued ecosystem components indirectly? For example, because large carnivores are often difficult to study directly, the effects of an action on their prey base or their habitat could be studied instead. Studies of indirect effects are most appropriate when they are reliably associated with effects on the valued ecosystem components.
- Would the use of indicators of impact be helpful? (See Chapter 7.)

Formulating a conceptual model of the relationships between the proposed action and the receiving environment can help to identify pertinent questions and potential environmental effects. The purpose of such models is to identify the physical and biological pathways by which an action can produce ecological effects. By focusing on relationships important to the manifestation of effects, they help to develop specific, testable hypotheses to explain why particular effects should or should not occur. Conceptual models can also help to identify logical errors, to highlight factors that require special study, to synthesize ideas and knowledge, and to communicate information (Beanlands and Duinker, 1983); guidance for developing conceptual models can be found in Holling (1978), Ward (1978), Fritz et al. (1980), and Beanlands and Duinker (1983). Multidisciplinary workshops can be used to articulate a problem and plan studies (Holling, 1978) and have been used to advantage in this way (ESSA, 1982).

Given adequate time and resources, sophisticated modeling should be considered (Munn, 1979). Basic guidance in development and use of such models can be found in Frenkiel and Goodall (1978), Holling (1978), and Ward (1978). Quantitative modeling can help by forcing assumptions to be made explicit, by making their consequences clear, and by revealing the sensitivity of outcomes to details of various assumptions. Simulation models are necessarily based on numerous unverified assumptions and cannot predict quantitative changes very accurately (Hilborn, 1979; Walters, 1975). But they can be useful in identifying potential qualitative effects and exploring the consequences of alternative management plans. Sensitivity analysis allows an exploration of the consequences of altering the assumptions of a model. Simulation models are often used in connection with freshwater systems (Chapter 21), in which the driving physical and chemical processes are fairly well understood; fishery and wildlife management (Chapter 12); epidemiology (Chapter 15); and forest management (Chapter 19).

ESTABLISHING STUDY BOUNDARIES

One of the first and most important tasks in the design of research on environmental effects is to establish a set of boundaries—temporal, spatial, and ecological. When might effects appear, and how long might they last? How long must studies last to allow reasonable predictions and reliable diagnosis of effects? Over what area will effects occur? Are there any natural barriers to the transmission of effects? Are any physical or biological processes likely to spread effects to other areas? What ecosystem components will be affected? At what levels of biological organization will effects appear? What species or ecosystem processes need to be studied and over what area? Boundaries in open systems, such as the ocean or atmosphere, are the most difficult to define. Variations in ecosystem components of interest can strongly influence the time required for biological effects to appear (Holling, 1973).

Making judgments about boundaries is difficult, and many surprises have occurred. For example, large water impoundments can influence local climate or induce earthquakes (Baxter and Glaude, 1980). When DDT was first used as a pesticide, no one expected it to appear in animals in the ocean (Chapter 24). The DDT story and similar cases (e.g., that of acid rain) have shown that environmental effects can spread by subtle pathways. Assumptions implicit in management decisions might result in setting boundaries that omit critical processes. Attempts to increase stocks of anadromous fish by increasing reproduction in rivers might fail if survival in the ocean is already limited by food supply (Peterman, 1984). The cumulative effects of multiple actions have taught us that specific projects and actions must be viewed in the context of related actions (Odum, 1982). Recent efforts to protect and conserve species have shown how management of a population requires consideration of its relationship with other populations (Frankel and Soulé, 1981; Schonewald-Cox, 1983; Soulé and Wilcox, 1980). And only recently has it been recognized that systematic management procedures—e.g., sex-biased or size-biased harvesting—can lead to undesirable genetic changes over remarkably short periods (Chapter 1).

The establishment of boundaries is constrained by administrative, project-related, technical, and ecological factors (Beanlands and Duinker, 1983). Administrative constraints include jurisdictional limits, insufficient time or funding, and political factors. Spatial boundaries are often obvious, unless long-range transport phenomena are involved. Temporal boundaries might be less well defined, because of political and other uncertainties; as we go from short-term, local effects at the population level to long-term, regional effects at the ecosystem level, we are less able to predict them (Christensen *et al.*, 1976). Other technical constraints are imposed by environmental variability, project location, and logistical problems.

Setting appropriate temporal and spatial boundaries is important in the management of species populations, whether for protection, control, or harvest. When populations become small, patterns and rates of interchange of individuals and genes between populations become critical. The sizes of populations needed for the long-term maintenance of the spotted owl in the Pacific Northwest depend on whether the Columbia River is a dispersal barrier (Chapter 17). When timber is managed on a forest-wide basis, rather than by stands, yields are higher and more consistent (Chapter 19). Physical and chemical processes can be critical in defining boundaries in aquatic systems, particularly when the spread and accumulation of pollutants are involved. The control of eutrophication in Lake Washington depended on knowledge of flow rates through the lake and the low turnover of phosphate in lake sediments (Chapter 20).

DEVELOPING AND TESTING HYPOTHESES

Statements about relationships between proposed actions and ecosystem components or processes are, in effect, hypotheses that can be tested. Studies designed to test them can increase our ability to predict environmental effects. In addition, the explicit statement of hypotheses helps us to identify important assumptions and formulate specific objectives for ecological studies. However, despite the acknowledged value of testing hypotheses in solving environmental problems, many studies are not designed and conducted to do so. Many studies in wildlife management, for example, involve elaborate collection of field data with only after-the-fact attempts at explanation (Romesburg, 1981). What happened as a result of a project is rarely studied (Beanlands and Duinker, 1983; Larkin, 1984).

In practice, most general hypotheses are evaluated by testing specific predictions that arose from them. In environmental impact assessment, the predictions themselves are a major product of preproject research. It is often helpful to develop several hypotheses about possible effects and their causes, so that studies can be designed to distinguish among them.

A hypothesis can be tested by studies before a project and by treating the project itself as a test. Methods for testing ecological hypotheses in preproject studies include the use of microcosms (Crow and Taub, 1979; Heath, 1979), field and laboratory experiments (Giddings, 1980; Suter, 1982; Ward, 1978), and computer simulations (Frenkiel and Goodall, 1978). Despite the difficulty of assigning causality in field experiments (Sharp *et al.*, 1979) and of extrapolating from small studies to large problems (Hilborn and Walters, 1981), pilot-scale perturbation studies could be the most productive research tool for impact assessment, although underused (Beanlands and Duinker, 1983; Ward, 1978).

If projects are to be treated as large-scale experiments, baseline data must be collected before the project begins (Beanlands and Duinker, 1983; Hilborn and Walters, 1981; Larkin, 1984). The baseline can best be viewed as a description of the mean values and natural variability in the system (Hirsch, 1980). Judgments of how much information is needed are often difficult, because of periodic cycles, random events, and spatial heterogeneity and because many variables can change systematically during the baseline study period (e.g., owing to succession).

Statistical guidance is available for designing baseline and monitoring programs once the variables of interest have been identified (Cowell, 1978; Eberhardt, 1976, 1978; Green, 1979; Kumar, 1980; Lucas, 1976; Sharp *et al.*, 1979; Ward, 1978; Zar, 1976). Two common problems that make it difficult to design projects as experiments properly are the lack of adequate controls (Cowell, 1978) and the lack of true replicates (Eberhardt, 1976). Eberhardt (1976) suggested a "pseudodesign" with baseline data on a control area and the project site. They can be compared with data collected when the project is complete, with replicates in time substituting for spatial replicates.

Baseline and monitoring studies are most effective if they are statistically designed to detect changes of the magnitude expected (Zar, 1976). This expectation in turn determines the extent of sampling required (Hartzbank and McCusker, 1979). In highly variable systems, adequate sampling might be too expensive, and resources might be better used in carrying out less direct studies. Baseline information can sometimes be derived after impacts have already occurred (e.g., Cowell and Syratt, 1979).

The Lake Washington case (Chapter 20) is an excellent example of testing hypotheses concerning the effect of lake fertilization changes on the makeup of plankton communities. A great deal was learned from this case, because monitoring continued throughout the development and treatment of the problem. Similarly, scientists at Southern Indian Lake (Chapter 21) were able to test hypotheses derived from the results of other lake studies and current limnological models. Carefully designed studies before and after project development showed that some of the predictions were wrong because the models used were not based on knowledge of lakes in a permafrost zone.

Field and laboratory experiments can be used to test hypotheses. The Garki malaria project (Chapter 15) was itself a large-scale experiment to investigate a model for controlling malaria through a combination of drugs and mosquito control. Careful monitoring studies before, during, and after applications allowed the model to be tested, and an important phenomenon—exophily—was discovered.

SPECIFYING PREDICTIONS AND DETERMINING THE SIGNIFICANCE OF EFFECTS

A major purpose of developing general ecological hypotheses is to generate a set of specific predictions of ecological change that can be used in decision-making. The predictions should be as clearly and precisely stated as possible. The period over which a change is expected to occur, the bases of the prediction, and the degree of uncertainty should be specified.

Determining the significance of predicted or observed ecological changes is often very difficult, because ecological systems are not fully understood. A clear distinction, if it can be made, between the magnitude of a change and its biological importance is useful. The rates of change and recovery are often important components of ecological effects (e.g., Cairns and Dickson, 1980).

The overall significance of an effect is tied closely to the definition of environmental goals. The best course for scientists is to predict or describe changes precisely. Whether or not a change is "significant" is a judgment that transcends science and is best made by all interested parties.

Several of the cases studied were organized around tests of specific predictions. In the derelict lands restoration case (Chapter 18), predictions were derived from empirical results of other restoration efforts and basic plant ecological theory. The bases of these predictions were clearly stated, and tests of them produced results of value to other restoration projects. Predictions in the Atomic Energy Commission radiation studies were derived from knowledge of food-chain dynamics and laboratory studies, and hypotheses were continually revised as predictions were tested experimentally (Chapter 22).

In the Lake Washington case, scientists predicted not only specific changes in water quality, but also the periods over which deterioration and recovery would occur (Chapter 20). In the Southern Indian Lake studies (Chapter 21), predictions were based on analogs and limnological principles, but were mostly qualitative. In both cases, careful monitoring was incorporated into comprehensive postproject analyses to improve understanding.

MONITORING

Biological monitoring is used in ecological studies in two basic ways. Studies conducted during or after an action or project are designed to learn what ecological changes resulted. Anticipatory monitoring is designed to track the effects of activities that might be cumulative or pose hazards to human health (Baker, 1976). Properly done, monitoring provides continuous indexes of environmental quality that can signal environmental degradation or improvement (Chapter 7).

In the event of unexpected environmental changes, monitoring can facilitate adaptive changes in management and in the design of ecological studies (Hilborn *et al.*, 1980; Holling, 1978; Walters and Hilborn, 1976). From a broader perspective, followup monitoring and retrospective analysis are ways to learn from experience and improve the prediction of ecological effects. Monitoring is most effective when it is designed to test ecological hypotheses and when preproject studies have provided baseline information (see Beanlands and Duinker, 1983). Postproject studies of the accuracy of predictions are useful, but are not as useful as followup monitoring that coordinates preproject and postproject sampling and that tests relevant hypotheses.

Periodic analysis of results can help to detect unexpected changes and evaluate sampling programs, allowing them to be changed in a timely way. Thus, an iterative approach to monitoring—with results fed into study design—is often effective, particularly when methods have not been well tested and when effects are uncertain. Any changes in sampling, however, must be made carefully, to ensure that new data are statistically comparable with those already collected.

Baseline monitoring of characteristics with substantial variation has a low probability of helping to detect changes due to a project. Measurements of baseline variability can help to identify the characteristics that it will be useful to measure in followup studies (Green, 1979) and can be used, with estimates of the duration of effects, to determine how long followup should continue.

Even if all the above criteria are met, followup studies of ecological effects can help in planning only if they are made available in an easily digestible form, ideally as published summaries and as complete postproject analyses (Hilborn and Walters, 1981).

In several case studies, followup monitoring was shown to be part of an overall study design to test hypotheses (Chapters 15, 22, and 23). The hydroelectric development in Southern Indian Lake (Chapter 21) was treated as a large-scale experiment, and monitoring provided increased understanding of artificial lakes. Because the results were analyzed and published, they can be applied to other cases.

In the caribou case (Chapter 16), monitoring of caribou movements and herd productivity began before development and continued during construction. Daily information on caribou movement was incorporated into constraints on construction activities. Annual monitoring of catch and fishing effort is used by the International Pacific Halibut Commission to set fishing quotas. Monitoring of conditions in Lake Washington allowed scientists to detect changes, predict trends in eutrophication, and predict and document recovery of the lake after action was taken; because the work was published, its lessons are readily available to managers of similar projects. Monitoring for DDT in the environment first identified the spread of another important group of toxic chemicals, PCBs (Chapter 24).

SUMMARY: DEVELOPING A STUDY STRATEGY

A study strategy is a plan for conducting ecological studies to help to predict and manage ecological effects. It is motivated by the environmental goals identified in scoping and is organized around the scientific objectives defined on the basis of the goals. Scoping identifies what information is required, and the study strategy specifies how to acquire it. A problem must be carefully thought through before studies aimed at solving it begin (Beanlands and Duinker, 1983).

Potential study objectives should be evaluated, so that efforts can be devoted to studies with some chance of producing useful results. What information is needed? Why is it needed? Is it possible to acquire adequate information? How will the information be used to satisfy the ecological goals? How will it be used in decision-making? Highly accurate characterization of a variable is of little use if decisions are made on the basis of other considerations. Decision analysis helps to ensure that modeling and research remain focused on the objectives.

A basic first step in designing studies is a review of what is already known about the problem. Larkin (1984) believes that literature review can provide more than 50% of the information needed in most initial impact assessments, and as much as 75% when coupled with brief reconnaissance surveys.

To summarize, a broad ecological study strategy:

- Is based on thought-out environmental goals.
- Is organized around clearly defined scientific objectives designed to satisfy the environmental goals.
- Includes a description of the boundaries established for the problem, with demonstration of their appropriateness.
- Is designed to evaluate hypotheses about how the ecological system functions and will be affected by perturbations.
- Specifies predictions that will be tested, with the basis of the predictions and a statement of confidence in their accuracy.
- Defines the basis for choosing environmental goals and evaluating their significance.
- Explains clearly how each part of the study fits into the overall design.
- Provides for baseline and followup monitoring to determine the effects of the project or perturbation.
- Allows the results of the study to be used to evaluate the plan and to modify it if necessary.

11

References

Ackerman, B. A., S. Rose-Ackerman, J. W. Sawyer, and D. W. Henderson. 1974. The Uncertain Search for Environmental Quality. Free Press (Macmillan), London.

Alexander, M. 1971. Microbial Ecology. John Wiley & Sons, New York.

Anderson, R. M., and R. M. May, eds. 1982. Population Biology of Infectious Diseases. Dahlem Konferenzen. Springer, Berlin.

Andreae, M. O., and H. Raemdonck. 1983. Dimethyl sulfide in surface oceans and marine atmosphere: A global view. Science 221:744-747.

Andrews, R. N. L., P. Cromwell, G. A. Enk, E. G. Farworth, J. R. Hibbs, and V. L. Sharp. 1977. Substantive Guidance for Environmental Impact Assessment: An Exploratory Study. The Institute of Ecology, Washington, D.C.

Anonymous. 1980. Biological Evaluation of Environmental Impacts. FWS/OBS-80/26. U.S. Council on Environmental Quality and Fish and Wildlife Service, U.S. Department of the Interior, Washington, D.C.

Anonymous. Undated. Post-development Audits to Test the Effectiveness of Environmental Impact Prediction Methods and Techniques. Final Report. Project Appraisal for Development Control, Environmental Impact Assessment and Planning Unit, Department of Geography, University of Aberdeen, Aberdeen, Scotland.

Baker, H. G. 1974. The evolution of weeds. Annu. Rev. Ecol. Syst. 5:1-24.

Baker, H. G., and G. L. Stebbins. 1965. The Genetics of Colonizing Species. Academic Press, New York.

Baker, J. M. 1976. Biological monitoring—Principles, methods and difficulties. Pp. 41-53 in J. M. Baker, ed. Marine Ecology and Oil Pollution. John Wiley & Sons, New York.

Baldwin, I. T., and J. C. Schultz. 1983. Rapid changes in tree chemistry induced by damage: Evidence for communication between plants. Science 221:227-280.

Barber, R. T., and F. P. Chavez. 1983. Biological consequences of El Niño. Science 222:1203-1210.

Barrons, K. C. 1981. Are Pesticides Really Necessary? Regnery Gateway, Chicago.

Bar-Shalom, Y. 1976. Caution, probing, and the value of information in the control of uncertain systems. Ann. Econ. Soc. Meas. 5:323-337.

Baxter, R. M., and P. Glaude. 1980. Environmental Effects of Dams and Impoundments in Canada: Experience and Prospects. Can. Bull. Fish. Aquat. Sci. 205.

Beanlands, G. E., and P. N. Duinker. 1983. An Ecological Framework for Environmental Impact Assessment in Canada. Institute for Resource and Environmental Studies, Dalhousie University, Halifax, Nova Scotia, and Federal Environmental Assessment Review Office, Ottawa, Ont.

Beanlands, G. E., W. J. Erckmann, G. H. Orians, J. O'Riordan, D. Policansky, M. H. Sadar, and B. Sadler, eds. In press. Cumulative Environmental Effects: A Binational Perspective. Canadian Environmental Assessment Research Council, Ottawa, Ont., and National Research Council, Washington, D.C.

Beddington, J. R. 1974. Age structure, sex ratio, and population density in the harvesting of natural populations. J. Appl. Ecol. 11:915-924.

Beddington, J. R., rapporteur. 1984a. Management under uncertainty. Group report. Pp. 227-244 in R. M. May, ed. Exploitation of Marine Communities. Dahlem Konferenzen. Springer, Berlin.

Beddington, J. R. 1984b. The response of multispecies systems to perturbations. Pp. 209-225 in R. M. May, ed. Exploitation of Marine Communities. Dahlem Konferenzen. Springer, Berlin.

Belovsky, G. E. 1984. Moose and snowshoe hare competition and a mechanistic explanation from foraging theory. Oecologia 61:150-159.

Berg, K. 1938. Studies on the bottom animals of Esrom Lake. K. Danske Vidensk. Selsk. Skr. Nat. Mat. Afd. 9(8):1-255.

Berry, J. A. 1975. Adaptation of photosynthetic processes to stress. Science 188:644-650.

Berryman, A. A. 1981. Population Systems: A General Introduction. Plenum, New York.

Beverton, R. J. H. 1983. Science and decision-making in fisheries regulation. Pp. 919-938 in G. D. Sharpe and J. Csirke, eds. Proceedings of the Expert Consultation to Examine Changes in Abundance and Species Composition of Neritic Fish Resources. FAO Fishery Report 291. Food and Agriculture Organization, Rome.

Beverton, R. J. H., rapporteur. 1984. Dynamics of single species. Group report. Pp. 13-58 in R. M. May, ed. Exploitation of Marine Communities. Dahlem Konferenzen. Springer, Berlin.

Birch, L. C. 1957. The meanings of competition. Am. Nat. 91:5-18.

Biswell, H. H. 1974. Effects of fire on chaparral. Pp. 321-364 in T. T. Kozlowski and C. E. Ahlgren, eds. Fire and Ecosystems. Academic Press, New York.

Bjorkman, O., and J. Berry. 1973. High-efficiency photosynthesis. Sci. Am. 229(4):80-93.

Bobrov, R. A. 1955. Use of plants as biological indicators of smog in the air of Los Angeles County. Science 121:510-511.

Bodaly, R. A., R. E. Hecky, and R. J. P. Fudge. 1984. Increases in fish mercury levels in lakes flooded by the Churchill River diversion, northern Manitoba. Can. J. Fish. Aquat. Sci. 41:682-691.

Boersma, P. D. In press. Seabirds reflect petroleum pollution. Science.

Bond, G. 1967. Fixation of nitrogen by higher plants other than legumes. Annu. Rev. Plant Physiol. 18:107-126.

Borisov, V. M. 1978. The selective effect of fishing on the population structure of species with a long life cycle. J. Ichthyol. 18:896-904.

Bormann, F. H., G. E. Likens, D. W. Fisher, and R. S. Pierce. 1968. Nutrient loss accelerated by clear-cutting of a forest ecosystem. Science 159:882-884.

Brinkhurst, R. O. 1974. The Benthos of Lakes. St. Martin's Press, New York.

Brittingham, M. C., and S. A. Temple. 1983. Have cowbirds caused forest songbirds to decline? BioScience 33:31-35.

Bromenshenk, J. J., S. R. Carlson, J. C. Simpson, and J. M. Thomas. 1985. Pollution monitoring of Puget Sound with honey bees. Science 227:632-634.

Brooks, J. L., and S. I. Dodson. 1965. Predation, body size, and composition of the plankton. Science 150:28-35.

Brown, J. H. 1984. On the relationship between abundance and distribution of species. Am. Nat. 124:255-279.

Brown, J. H. In press. Discussion: Cumulative impacts in terrestrial communities. In G. E. Beanlands, W. J. Erckmann, G. H. Orians, J. O'Riordan, D. Policansky, M. H. Sadar, and B. Sadler, eds. Cumulative Environmental Effects: A Binational Perspective. Canadian Environmental Assessment Research Council, Ottawa, Ont., and National Research Council, Washington, D.C.

Bryson, R. A., and T. J. Murray. 1977. Climates of Hunger. University of Wisconsin Press, Madison.

Buck, W. B. 1979. Animals as monitors of environmental quality. Vet. Human Toxicol. 21:277-284.

Bull, D. 1982. A Growing Problem: Pesticides and the Third World Poor. Oxfam, London.

Cain, S. A. 1938. The species-area curve. Am. Midl. Nat. 20:573-581.

Cairns, J., Jr. 1977. Effects of temperature changes and chlorination upon the community structure of aquatic organisms. Pp. 129-144 in M. Marois, ed. Towards a Plan of Actions for Mankind. Vol. 3. Biological Balance and Thermal Modifications. Pergamon Press, Oxford, Eng.

Cairns, J., Jr. 1980. The Recovery Process in Damaged Ecosystems. Ann Arbor Sciences, Ann Arbor, Mich.

Cairns, J., Jr., and K. L. Dickson. 1980. Risk analysis for aquatic ecosystems. Pp. 73-83 in Biological Evaluation of Environmental Impacts. FWS/OBS-80/26. U.S. Council on Environmental Quality and Fish and Wildlife Service, U.S. Department of the Interior, Washington, D.C.

Cairns, J., Jr., G. R. Larra, R. E. Sparks, and W. T. Waller. 1973. Developing biological information systems for water quality management. Water Res. Bull. 9:81-99.

Cameron, R. D., K. R. Whitten, W. T. Smith, and D. D. Roby. 1979. Caribou distribution and group composition associated with construction of the trans-Alaska pipeline. Can. Field Nat. 93:155-162.

Campbell, R. W., and R. J. Sloan. 1977. Natural regulation of innocuous gypsy moth populations. Environ. Entomol. 6:315-322.

Campbell, R. W., and R. J. Sloan. 1978. Natural maintenance and decline of gypsy moth outbreaks. Environ. Entomol. 7:389-395.

Carpenter, R. A. 1976. The scientific basis of NEPA—Is it adequate? Environ. Law Reporter 6:50014-50019.

Carpenter, S. R., and J. F. Kitchell. 1984. Plankton community structure and limnetic primary production. Am. Nat. 124:159-172.

Carson, R. 1962. Silent Spring. Houghton Mifflin, Boston.

Caswell, H., and P. A. Werner. 1978. Transient behavior and life history of teasel (*Dipsacus sylvestris* Huds.). Ecology 59:53-66.

Caswell, H., F. Reed, S. N. Stephenson, and P. A. Werner. 1973. Photosynthetic pathways and selective herbivory, a hypothesis. Am. Nat. 107:465-480.

Chapra, S. C., and K. H. Reckhow. 1983. Engineering Approaches for Lake Management. Vol. 2. Ann Arbor Sciences, Ann Arbor, Mich.

Charnov, E. L., and G. H. Orians. 1973. Optimal Foraging: Some Theoretical Explorations. University of Utah Press, Salt Lake City.

Christensen, S. W., W. Van Winkle, and J. S. Mattice. 1976. Defining and determining the significance of impacts: Concepts and methods. Pp. 191-219 in R. K. Sharma, J. D. Buffington, and J. T. McFadden, eds. Proceedings of a Workshop on the Biological Significance of Environmental Impacts. NR-CONF-002. U.S. Nuclear Regulatory Commission, Washington, D.C.

Clark, L. R., P. W. Geier, R. D. Hughes, and R. F. Morris. 1967. The Ecology of Insect Populations in Theory and Practice. Methuen, London.

Clark, W. C. In press. The cumulative impacts of human activities on the atmosphere. In G. E. Beanlands, W. J. Erckmann, G. H. Orians, J. O'Riordan, D. Policansky, M. H. Sadar, and B. Sadler, eds. Cumulative Environmental Effects: A Binational Perspective. Canadian Environmental Assessment Research Council, Ottawa, Ont., and National Research Council, Washington, D.C.

Cody, M. L. 1969. Convergent characteristics in sympatric species: A possible relation to interspecific competition and aggression. Condor 71:223-239.

Cody, M. L. 1974. Competition and the Structure of Bird Communities. Princeton University Press, Princeton, N.J.

Connell, J. H. 1975. Some mechanisms producing structure in natural communities: A model and evidence from field experiments. Pp. 460-490 in M. L. Cody and J. M. Diamond, eds. Ecology and Evolution of Communities. Harvard University Press, Cambridge, Mass.

Connell, J. H. 1978. Diversity in tropical rain forests and coral reefs. Science 199:1302-1310.

Connell, J. H. 1980. Diversity and the coevolution of competitors, or the ghost of competition past. Oikos 35:131-138.

Connell, J. H. 1983. On the prevalence and relative importance of interspecific competition: Evidence from field experiments. Am. Nat. 122:661-696.

Connell, J. H., and E. Orias. 1964. The ecological regulation of species diversity. Am. Nat. 98:399-414.

Connister, R. Z. 1964. The mechanism of prey digestion in the spotted crab, *Cancer ocellatus*. J. Am. Syst. Assoc. 42:36-54.

Connor, E. F., and E. McCoy. 1979. The statistics and biology of the species-area relationship. Am. Nat. 113:791-833.

Cooke, A. S. 1973. Shell-thinning in avian eggs by environmental pollutants. Environ. Pollut. 4:85-157.

Cooper, C. F., and P. H. Zedler. 1980. Ecological assessment for regional development. J. Environ. Manage. 10:285-296.

Council on Environmental Quality. 1978. Regulations for Implementing the Procedural Provisions of the National Environmental Policy Act. Reprint 43 FR 55978-56007, November 29, 1978, 40CFR Parts 1500-1508. Washington, D.C.

Courtenay, W. R., Jr., and J. R. Stauffer, Jr., eds. 1984. Distribution, Biology, and Management of Exotic Fishes. Johns Hopkins University Press, Baltimore, Md.

Cowell, E. B. 1978. Ecological monitoring as a management tool in industry. Ocean Manage. 4:273-285.

Cowell, E. B., and W. J. Syratt. 1979. A technique for assessing ecological damage to the intertidal zone of rocky shores for which no previous baseline data is available. Pp. 29-39 in Proceedings of the Ecological Damage Assessment Conference. Society of Petroleum Industry Biologists, Los Angeles.

Cowie, R. J. 1977. Optimal foraging in great tits (*Parus major*). Nature 268:137-139.

Crawley, M. J. 1983. Herbivory: The Dynamics of Plant-Animal Interactions. Blackwell Scientific, Oxford, Eng.

Crow, M. E., and F. B. Taub. 1979. Designing a microcosm bioassay to detect ecosystem level effects. Int. J. Environ. Stud. 13:141-147.

Darwin, C. 1859. On the Origin of Species by Means of Natural Selection, or the Preservation of Favoured Races in the Struggle for Existence. John Murray, London.

Davidson, D. W., J. H. Brown, and R. S. Inouye. 1980. Competition and the structure of granivore communities. BioScience 30:233-238.

Davidson, D. W., R. S. Inouye, and J. H. Brown. 1984. Granivory in a desert ecosystem: Experimental evidence for indirect facilitation of ants by rodents. Ecology 65:1780-1786.

Dayton, P. K. 1971. Competition, disturbance, and community organization: The provision and subsequent utilization of space in a rocky intertidal community. Ecol. Monogr. 41:351-389.

Dayton, P. K. 1984. Processes structuring some marine communities: Are they general? Pp. 181-197 in D. R. Strong, Jr., D. Simberloff, L. Abele, and A. Thistle, eds. Ecological Communities: Conceptual Issues and the Evidence. Princeton University Press, Princeton, N.J.

Dayton, P. K. 1985. Ecology of kelp communities. Annu. Rev. Ecol. Syst. 16:215-245.

Dayton, P. K. In press. Cumulative impacts in the marine realm. In G. E. Beanlands, W. J. Erckmann, G. H. Orians, J. O'Riordan, D. Policansky, M. H. Sadar, and B. Sadler, eds. Cumulative Environmental Effects: A Binational Perspective. Canadian Environmental Assessment Research Council, Ottawa, Ont., and National Research Council, Washington, D.C.

DeMott, W. R., and W. C. Kerfoot. 1982. Competition among cladocerans: Coexistence of *Bosmina* and *Daphnia*. Ecology 63:1949-1966.

Diamond, J. M., and M. E. Gilpin. 1982. Examination of the "null" model of Connor and Simberloff for species co-occurrences on islands. Oecologia 52:64-74.

Dickman, M. 1968. The relation of freshwater plankton productivity to species composition during induced successions. Ph.D. thesis, University of British Columbia, Vancouver.

Dillon, P. J., and F. H. Rigler. 1974. The phosphorus-chlorophyll relationship in lakes. Limnol. Oceanogr. 19:767-773.

Doberstyn, E. B. 1984. Resistance of *Plasmodium falciparum*. Experientia 40:1311-1317.

Dodd, A. P. 1959. The biological control of prickly pear in Australia. Pp. 565-577 in A. Keast, R. L. Crocker, and C. S. Christian, eds. Biogeography and Ecology in Australia. Monographiae Biologicae. Vol. 8. W. Junk, The Hague.

Drury, W. B., and I. C. T. Nisbet. 1973. Succession. J. Arnold Arbor. 54:331-368.

Duggins, D. O. 1980. Kelp beds and sea otters: An experimental approach. Ecology 61:447-453.

Duinker, P. N., and G. E. Beanlands. 1983. Ecology and Environmental Assessment: An Annotated Bibliography. Institute for Resource and Environmental Studies, Dalhousie University, Halifax, Nova Scotia, and Federal Environmental Assessment Review Office, Ottawa, Ont.

Dunlap, T. R. 1981. DDT: Scientists, Citizens, and Public Policy. Princeton University Press, Princeton, N.J.

Eberhardt, L. L. 1976. Quantitative ecology and impact assessment. J. Environ. Manage. 4:27-70.

Eberhardt, L. L. 1978. Appraising variability in population studies. J. Wildl. Manage. 42:207-238.

Ebersole, J. P. 1977. The adaptive significance of territoriality in the reef fish *Eupomacentrus leucostictus*. Ecology 58:914-920.

Edmondson, W. T. 1969. Eutrophication in North America. Pp. 124-149 in Eutrophication: Causes, Consequences, Correctives. National Academy of Sciences, Washington, D. C.

Edmondson, W. T. 1972. Nutrients and phytoplankton in Lake Washington. Am. Soc. Limnol. Oceanogr. Spec. Symp. 1:172-193.

Edmondson, W. T. 1974. Environmental Phosphorus Handbook. Limnol. Oceanogr. 19:369-374. (book review)

Edmondson, W. T., and J. T. Lehman. 1981. The effect of changes in the nutrient income on the condition of Lake Washington. Limnol. Oceanogr. 26:1-29.

Edmondson, W. T., and A. H. Litt. 1982. *Daphnia* in Lake Washington. Limnol. Oceanogr. 27:272-293.

Ehrlich, P. R., and A. Ehrlich. 1981. Extinction: The Causes and Consequences of the Disappearance of Species. Random House, New York.

Ehrlich, P. R., and P. H. Raven. 1965. Butterflies and plants: A study in coevolution. Evolution 18:586-608.

Elton, C. S. 1958. The Ecology of Invasions by Animals and Plants. Methuen, London.

Emlen, J. M. 1970. Age specificity and ecological theory. Ecology 51:588-601.

Endler, J. A. 1977. Geographic Variation, Speciation, and Clines. Princeton University Press, Princeton, N.J.

Eppley, R. W., and B. Peterson. 1979. Particulate organic matter flux and planktonic new production in the deep ocean. Nature 282:677-680.

ESSA (Environmental and Social Systems Analysts Ltd.). 1982. Review and Evaluation of Adaptive Environmental Assessment and Management. Environment Canada, Vancouver, B.C.

Estes, J. A., R. J. Jameson, and E. B. Rhode. 1982. Activity and prey selection in the sea otter: Influence of population status on community structure. Am. Nat. 120:242-258.

Ewald, P. W. 1983. Host-parasite relations, vectors, and the evolution of disease severity. Annu. Rev. Ecol. Syst. 14:465-485.

Feeny, P. 1976. Plant apparency and chemical defense. Rec. Adv. Phytochem. 10:1-40.

Finn, J. T. 1976. Measures of ecosystem structure and function derived from analysis of flows. J. Theoret. Biol. 56:363-380.

Finn, J. T. 1978. Cycling index: A general definition for cycling in compartment models. Pp. 138-164 in D. C. Adriano and I. L. Brisbin, eds. Environmental Chemistry and Cycling Processes. CONF-760429. Technical Information Center, U.S. Department of Energy, Washington, D.C.

Flemer, D. A., G. B. Mackiernan, W. Nehlsen, R. B. Biggs, D. Blaylock, N. H. Burger, L. C. Davidson, D. Haberman, K. S. Price, and J. L Taft. 1983. Chesapeake Bay: A Profile of Environmental Change. Chesapeake Bay Program, U.S. Environmental Protection Agency, Annapolis, Md.

Forsman, E. D., E. C. Meslow, and H. M. Wight. 1984. Distribution and biology of the spotted owl in Oregon. Wildl. Monogr. 87:1-64.

Fox, L. R. 1981. Defense and dynamics in plant-herbivore systems. Am. Zool. 21:853-864.

Fox, L. R., and B. J. Macauley. 1977. Insect grazing on *Eucalyptus* in response to variation in leaf tanning and nitrogen. Oecologia 29:145-162.

Franey, J. R., M. V. Ivanov, and H. Rhodhe. 1983. The sulphur cycle. Ch. 2.5 in B. Bolin and R. B. Cook, eds. The Major Biogeochemical Cycles and Their Interactions. Scientific Committee on Problems of the Environment Publ. 21. John Wiley & Sons, Chichester, Eng.

Frank, P. W. 1968. Life histories and community stability. Ecology 49:355-356.

Frankel, O. H., and M. E. Soulé. 1981. Conservation and Evolution. Cambridge University Press, New York.

Franklin, I. R. 1980. Evolutionary change in small populations. Pp. 135-149 in M. Soulé and B. Wilcox, eds. Conservation Biology: An Evolutionary-Ecological Perspective. Sinauer Associates, Sunderland, Mass.

Frenkiel, F. N., and D. W. Goodall, eds. 1978. Simulation Modelling of Environmental Problems. SCOPE Report 9. International Council of Scientific Unions-Scientific Committee on Problems of the Environment. John Wiley & Sons, Chichester, Eng.

Fretwell, S. D. 1972. Populations in a Seasonal Environment. Princeton University Press, Princeton, N.J.

Fritz, E. S., P. J. Rago, and I. D. Murarka. 1980. Strategy for Assessing Impacts of Power Plants on Fish and Shellfish Populations. FWS/OBS-80/34. National Power Plant Team, Fish and Wildlife Service, U.S. Department of the Interior, Ann Arbor, Mich.

Furniss, R. L., and V. M. Carolin. 1980. Western Forest Insects. Miscellaneous Publication 1339. Forest Service, U.S. Department of Agriculture, Washington, D.C.

Galli, A. E., C. F. Leck, and R. T. T. Forman. 1976. Avian distribution patterns in forest islands of different sizes in central New Jersey. Auk 93:356-364.

Gause, G. F. 1934. The Struggle for Existence. Williams and Wilkins, Baltimore, Md.

Gavereski, C. A. 1976. Relation of park size and vegetation to urban bird populations in Seattle, Washington. Condor 78:375-382.

Georghiou, G. P., A. Lagunes, and J. D. Baker. 1983. Effect of insecticide rotations on the evolution of insecticide resistance. Pp. 183-189 in J. Miyamoto, ed. Pesticide Chemistry: Human Welfare and the Environment. Pergamon Press, New York.

Geppert, R. R., C. W. Lorenz, and A. G. Larson. 1984. Cumulative Effects of Forest Practices on the Environment: A State of the Knowledge. Ecosystems, Inc., Olympia, Wash.

Giddings, J. M. 1980. Field experiments. Pp. 315-331 in F. S. Sanders, S. M. Adams, L. W. Barnthouse, J. M. Giddings, E. E. Huber, K. D. Kumar, D. Lee, B. Murphy, G. W. Suter, and W. Van Winkle, eds. Strategies for Ecological Effects Assessment at DOE Energy Activity Sites. Environmental Sciences Division Publ. 1639. ORNL/TM-6783. Oak Ridge National Laboratory, Oak Ridge, Tenn.

Gilman, A. P., D. B. Peakall, D. J. Hallett, G. A. Fox, and R. J. Norstrom. 1979. Animals as Monitors of Environmental Pollutants. National Academy of Sciences, Washington, D.C.

Gilpin, M. E. 1975. Limit cycles in competition communities. Am. Nat. 109:51-60.

Gilpin, M. E., and J. M. Diamond. 1982. Factors contributing to non-randomness in species co-occurrences on islands. Oecologia 52:75-84.

Gleason, H. A. 1922. On the relation between species and area. Ecology 3:158-162.

Godron, M., and R. T. T. Forman. 1983. Landscape modification and changing ecological characteristics. Pp. 12-28 in H. A. Mooney and M. Godron, eds. Disturbance and Ecosystems. Springer, New York.

Goldberg, E. D., V. T. Bowen, J. W. Farrington, G. Harvey, J. H. Martin, P. L. Parker, R. W. Riseborough, W. Robertson, E. Schneider, and E. Gamble. 1978. The Mussel Watch. Environ. Conserv. 5:101-125.

Gomez-Pompa, A. C., C. Vasquez-Yaues, and S. Guerarra. 1972. The tropical rainforest: A nonrenewable resource. Science 177:762-765.

Goodman, D. 1975. The theory of diversity-stability relationships in ecology. Q. Rev. Biol. 50:237-266.

Grassle, J. F., H. L. Sanders, R. R. Hessler, G. T. Rowe, and T. McLellan. 1975. Pattern and zonation: A study of the bathyal megafauna using the research submersible Alvin. Deep-Sea Res. 22:457-481.

Green, R. H. 1979. Sampling Design and Statistical Methods for Environmental Biologists. John Wiley & Sons, Toronto.

Greenland, D. J., and R. Lal. 1977. Soil Conservation and Management in the Humid Tropics. John Wiley & Sons, Chichester, Eng.

Gross, M. R. 1984. Sunfish, salmon, and the evolution of alternative reproductive strategies and tactics in fishes. Pp. 55-75 in G. W. Potts and R. J. Wootton, eds. Fish Reproduction. Academic Press, New York.

Gross, M. R. 1985. Disruptive selection for alternative life histories in salmon. Nature 313:47-48.

Hairston, N. G., F. E. Smith, and L. B. Slobodkin. 1960. Community structure, population control and competition. Am. Nat. 94:421-425.

Handford, P., G. Bell, and T. Reimchen. 1977. A gillnet fishery considered as an experiment in artificial selection. J. Fish. Res. Bd. Can. 34:954-961.

Harcourt, D. G., and E. J. Leroux. 1967. Population regulation in insects and man. Am. Sci. 55:400-415.

Hare, J. D. 1983. Manipulation of host suitability for herbivore pest management. Pp. 655-680 in R. F. Denno and M. S. McClure, eds. Variable Plants and Herbivores in Natural and Managed Ecosystems. Academic Press, New York.

Harper, J. L. 1969. The role of predation in vegetational diversity. Pp. 48-62 in Diversity and Stability in Ecological Systems. Brookhaven Symposia in Biology 22. Brookhaven National Laboratory, Upton, N.Y.

Harper, J. L., P. H. Lovell, and K. G. Moore. 1970. The shapes and sizes of seeds. Annu. Rev. Ecol. Syst. 1:327-356.

Harris, G. P. 1980. Temporal and spatial scales in phytoplankton ecology: Mechanisms, models and management. Can. J. Fish. Aquat. Sci. 37:877-900.

Harrison, W. G. 1980. Nutrient regeneration and primary production in the sea. Pp. 433-460 in P. G. Falkowski, ed. Primary Productivity in the Sea. Plenum, New York.

Harshbarger, J. C., and J. J. Black. In press. A strategy for using fish bioassays and surveys to identify and eliminate point source environmental carcinogens. In Towards a Transboundary Monitoring Network: Proceedings of a Workshop. International Joint Commission, Washington, D.C., and Ottawa, Ont.

Hartzbank, D. J., and A. McCusker. 1979. Establishing criteria for offshore sampling design. Pp. 59-78 in Proceedings of the Ecological Damage Assessment Conference. Society of Petroleum Industry Biologists, Los Angeles.

Hasler, A. D. 1947. Eutrophication of lakes by domestic sewage. Ecology 28:383-395.

Hassell, M. P. 1985. Insect natural enemies as regulating factors. J. Anim. Ecol. 54:323-334.

Haukioja, E., and T. Hakala. 1975. Herbivore cycles and periodic outbreaks. Formulation of a general hypothesis. Rep. Kevo Subarctic Res. Stn. 12:1-9.

Haukioja, E., and P. Niemelä. 1979. Birch leaves as a resource for herbivores: Seasonal occurrence of increased resistance in foliage after mechanical damage of adjacent leaves. Oecologia 39:151-159.

Hawksworth, D. 1971. Lichens as litmus for air pollution: A historical review. Int. J. Environ. Stud. 1:281-296.

Heath, R. T. 1979. Holistic study of an aquatic microcosm: Theoretical and practical implications. Int. J. Environ. Stud. 13:87-93.

Hecky, R. E. 1984. Thermal and optical characteristics of Southern Indian Lake before,

124   *KINDS OF ECOLOGICAL KNOWLEDGE AND THEIR APPLICATIONS*

during, and after impoundment and Churchill River diversion. Can. J. Fish. Aquat. Sci. 41:579-590.

Hecky, R. E., R. W. Newbury, R. A. Bodaly, K. Patalas, and D. M. Rosenberg. 1984. Environmental impact prediction and assessment: The Southern Indian Lake experience. Can. J. Fish. Aquat. Sci. 41:720-732.

Heinrich, B. 1979. Resource heterogeneity and problems of movement in foraging bees. Oecologia 40:235-245.

Hilborn, R. 1979. Some failures and successes in applying systems analysis to ecological systems. J. Appl. Syst. Anal. 6:25-31.

Hilborn, R., and C. Walters. 1981. Some pitfalls of environmental baseline and process studies. Report 3. Cooperative Fisheries Research Unit, University of British Columbia, Vancouver.

Hilborn, R., C. S. Holling, and C. J. Walters. 1980. Managing the unknown: Approaches to ecological policy design. Pp. 103-113 in Biological Evaluation of Environmental Impacts. FWS/OBS-80/26. Council on Environmental Quality and Fish and Wildlife Service, U.S. Department of the Interior, Washington, D.C.

Hirsch, A. 1980. The baseline study as a tool in environmental impact assessment. Pp. 84-93 in Biological Evaluation of Environmental Impacts. FWS/OBS-80/26. Council on Environmental Quality and Fish and Wildlife Service, U.S. Department of the Interior, Washington, D.C.

Holdridge, L. R. 1967. Life Zone Ecology. Tropical Science Center, San Jose, Costa Rica.

Holling, C. S. 1973. Resilience and stability of ecological systems. Annu. Rev. Ecol. Syst. 4:1-23.

Holling, C. S., ed. 1978. Adaptive Environmental Assessment and Management. Int. Ser. on Applied Systems Analysis 3, International Institute for Applied Systems Analysis. John Wiley & Sons, Toronto.

Horn, H. S. 1974. The ecology of secondary succession. Annu. Rev. Ecol. Syst. 5:25-37.

Horn, H. 1976. Succession. Pp. 187-204 in R. M. May, ed. Theoretical Ecology: Principles and Applications. Blackwell Scientific, Oxford, Eng.

Hornbeck, J. W., G. E. Likens, R. S. Pierce, and F. H. Bormann. 1975. Strip cutting as a means of protecting site and streamflow quality when clearcutting northern hardwoods. Pp. 209-225 in B. Bernier and C. H. Winget, eds. Forest Soils and Forest Land Management. Proceedings of the Fourth North American Forest Soils Conference. Les Presses de l'Université Laval, Quebec, Que.

Huston, M. 1979. A general hypothesis of species diversity. Am. Nat. 113:81-101.

Hutchinson, G. E. 1959. Homage to Santa Rosalia, or why are there so many kinds of animals? Am. Nat. 93:145-159.

Istock, C. A. 1967. The evolution of complex life cycle phenomena: An ecological perspective. Evolution 21:592-605.

Jewell, P. A., and S. Holt, eds. 1981. Problems in Management of Locally Abundant Wild Mammals. Academic Press, New York.

Jonasson, P. M. 1972. Ecology and production of the profundal benthos in relation to phytoplankton in Lake Estrom. Oikos 14(Suppl.):1-148.

Jordan, C. F. 1971. A world pattern in plant energetics. Am. Sci. 59:425-433.

Juday, C., and E. A. Birge. 1931. A second report on the phosphorus content of Wisconsin lake waters. Trans. Wis. Acad. Sci. Arts Lett. 26:353-382.

Juday, C., E. A. Birge, G. I. Kemmerer, and R. J. Robinson. 1928. Phosphorus content of lake waters of northeastern Wisconsin. Trans. Wis. Acad. Sci. Arts Lett. 23:233-248.

Karr, J. K. 1982. Avian extinction on Barro Colorado Island, Panama: A reassessment. Am. Nat. 119:228-239.

Karr, J. R., and R. R. Roth. 1971. Vegetation structure and avian diversity in several New World areas. Am. Nat. 105:423-435.

Kettlewell, H. B. D. 1973. The Evolution of Melanism. Clarendon Press, Oxford, Eng.

Kimball, K. D., and S. A. Levin. 1985. Limitations of laboratory bioassays: The need for ecosystem-level testing. BioScience 35:165-171.

Kodric-Brown, A., and J. H. Brown. 1979. Competition between distantly related taxa in the coevolution of plants and pollinators. Am. Zool. 19:1115-1127.

Kozlowski, T. T., and C. E. Ahlgren, eds. 1974. Fire and Ecosystems. Academic Press, New York.

Krebs, C. J. 1985. Ecology: The Experimental Analysis of Distribution and Abundance. 3rd ed. Harper and Row, New York.

Kubicki, B. 1969a. Investigations on sex determination in cucumber (*Cucumis sativus*). Parts 3-8. Genet. Pol. 10:5-144.

Kubicki, B. 1969b. Sex determination in muskmelon (*Cucumis melo*). Genet. Pol. 10:145-166.

Kubicki, B. 1969c. Comparative studies on sex determination in cucumber (*Cucumis sativus*) and muskmelon (*Cucumis melo*). Genet. Pol. 10:167-184.

Kumar, K. D. 1980. Statistical considerations. Pp. 333-348 in F. S. Sanders, S. M. Adams, L. W. Barnthouse, J. M. Giddings, E. E. Huber, K. D. Kumar, D. Lee, B. Murphy, G. W. Suter, and W. Van Winkle, eds. Strategies for Ecological Effects Assessment at DOE Energy Activity Sites. Environmental Sciences Division Publ. 1639. ORNL/TM-6783. Oak Ridge National Laboratory, Oak Ridge, Tenn.

Lack, D. 1954. The Natural Regulation of Animal Numbers. Clarendon Press, Oxford, Eng.

Larkin, P. A. 1978. Fisheries management—An essay for ecologists. Annu. Rev. Ecol. Syst. 9:57-73.

Larkin, P. A. 1984. A commentary on environmental impact assessment for large projects affecting lakes and streams. Can. J. Fish. Aquat. Sci. 41:1121-1127.

Law, R. 1979. Optimal life histories under age-specific predation. Am. Nat. 114:399-417.

Lawrey, J. D., and M. E. Hale, Jr. 1979. Lichen growth responses to stress induced by automobile exhaust pollution. Science 204:423-424.

Lawton, J. H., and M. P. Hassell. 1981. Asymmetrical competition in insects. Nature 289:793-795.

Leonard, D. E. 1974. Recent developments in ecology and control of the gypsy moth. Annu. Rev. Entomol. 19:197-229.

Levin, B. R., and R. E. Lenski. 1983. Coevolution in bacteria and their viruses and plasmids. Pp. 99-127 in D. J. Futuyma and M. Slatkin, eds. Coevolution. Sinauer Associates, Sunderland, Mass.

Levins, R., and R. H. MacArthur. 1969. An hypothesis to explain the incidence of monophagy. Ecology 15:910-911.

Lewis, R. A., N. Stein, and C. W. Lewis, eds. 1984. Environmental Specimen Banking and Monitoring as Related to Banking. Martinus Nijhoff, Boston.

Lewontin, R. C. 1965. Selection for colonizing ability. Pp. 77-94 in H. G. Baker and G. L. Stebbins, eds. The Genetics of Colonizing Species. Academic Press, New York.

Likens, G. E., ed. 1972. Nutrients and eutrophication: The limiting nutrients controversy. Am. Soc. Limnol. Oceanogr. Spec. Symp. 1:1-328.

Likens, G. E., and F. H. Bormann. 1972. Nutrient cycling in ecosystems. Pp. 25-67 in

J. A. Wiens, ed. Ecosystem Structure and Function. Oregon State University Press, Corvallis.

Lovejoy, T. E. 1979. A projection of species extinctions. Pp. 328-329 in Council on Environmental Quality and Department of State. The Global 2000 Report to the President: Entering the Twenty-First Century. Council on Environmental Quality, Washington, D.C.

Lucas, H. L. 1976. Some statistical aspects of assessing environmental impact. Pp. 295-306 in R. K. Sharma, J. D. Buffington, and J. T. McFadden, eds. Proceedings of a Workshop on the Biological Significance of Environmental Impacts. NR-CONF-002. U.S. Nuclear Regulatory Commission, Washington, D.C.

Ludwig, D., and R. Hilborn. 1983. Adaptive probing strategies for age-structured fish stocks. Can. J. Fish. Aquat. Sci. 40:559-569.

MacArthur, R. H. 1955. Fluctuations of animal populations, and a measure of community stability. Ecology 36:533-536.

MacArthur, R. H. 1965. Patterns of species diversity. Biol. Rev. 40:510-533.

MacArthur, R. H. 1972. Geographical Ecology. Harper and Row, New York.

MacArthur, R. H., and J. W. MacArthur. 1961. On bird species diversity. Ecology 42:594-598.

MacArthur, R. H., and E. O. Wilson. 1967. The Theory of Island Biogeography. Princeton University Press, Princeton, N.J.

MacClintock, L., R. F. Whitcomb, and D. L. Whitcomb. 1977. Island biogeography and "habitat islands" of eastern forest. II. Evidence for the value of corridors and minimization of isolation in preservation of biotic diversity. Am. Birds 32:6-16.

Mack, R. N. In press. Alien plant invasion into the Intermountain West: A case history. In H. A. Mooney, ed. Ecology of Biological Invasions in North America and Hawaii. Springer, New York.

Mack, R. N., and J. N. Thompson. 1982. Evolution in steppe with few large, hooved mammals. Am. Nat. 119:757-773.

MacLeod, J. M. 1977. Discontinuous Stability in a Sawfly Life System and Its Relevance to Pest Management. Working Paper W-22, Institute of Animal Resource Ecology, University of British Columbia, Vancouver, B.C.

Mann, K. H., and P. A. Breen. 1972. The relation between lobster abundance, sea urchins, and kelp beds. J. Fish. Res. Bd. Can. 29:603-605.

Maryland Department of Natural Resources. 1984. Power Plant Cumulative Environmental Impact Report. PPSP-CEIR-4. Maryland Department of Natural Resources, Annapolis, Md.

Massachusetts Department of Environmental Management. 1981. Gypsy Moth Control. Final Environmental Impact Report. Commonwealth of Massachusetts, Boston.

Mathisen, O. A. 1962. The effect of altered sex ratios on the spawning of red salmon. Pp. 137-245 in T. Y. Koo, ed. Studies of Alaska Red Salmon. University of Washington Press, Seattle.

Maxwell, F. G., and P. R. Jennings, eds. 1980. Breeding Plant Resistance to Insects. John Wiley & Sons, New York.

May, R. M. 1973. Stability and Complexity in Model Ecosystems. Princeton University Press, Princeton, N.J.

May, R. M. 1977. Food lost to pests. Nature 267:669-670.

May, R. M. 1980. Mathematical models in whaling and fisheries management. Am. Math. Soc. Lect. Math. Life Sci. 13:1-64.

May, R. M. 1981. Models for single populations. Pp. 5-29 in R. M. May, ed. Theoretical Ecology. Sinauer Associates, Sunderland, Mass.

May, R. M. 1983. The structure of food webs. Nature 301:566-568.

May, R. M., ed. 1984. Exploitation of Marine Communities. Dahlem Konferenzen. Springer, Berlin.

May, R. M. 1985. Evolution of pesticide resistance. Nature 315:12-13.

May, R. M., and A. P. Dobson. In press. Population dynamics and the rate of evolution of pesticide resistance. In National Research Council. Management of Resistance to Pesticides: Strategies, Tactics, and Research Needs. National Academy Press, Washington, D.C.

McNab, B. 1963. Bioenergetics and the determination of home range size. Am. Nat. 97:133-140.

McNamee, P. J., J. M. McLeod, and C. S. Holling. 1981. The Structure and Behavior of Defoliating Insect/Forest Systems. Research Publication R-25. Institute of Animal Resource Ecology, University of British Columbia, Vancouver.

Menge, B. A., and J. P. Sutherland. 1976. Species diversity gradients: Synthesis of the roles of predation, competition, and temporal heterogeneity. Am. Nat. 110:351-369.

Meyer, S. L. 1975. Data Analysis for Scientists and Engineers. John Wiley & Sons, New York.

Middleton, J. T. 1956. Response of plants to air pollution. J. Air Pollut. Control Assoc. 6:7-9.

Mindinger, P. C., and S. Hope. 1982. Expansion of the range of the house finch. Am. Birds 36:347-353.

Mineau, P. G., G. A. Fox, R. J. Norstrom, D. V. Weseloh, D. J. Hallett, and J. A. Ellenton. 1984. Using the herring gull to monitor levels and effects of organochlorine contaminants in the Canadian great lakes. Pp. 425-452 in J. O. Nriagu and M. S. Simmons, eds. Contaminants in the Great Lakes. John Wiley & Sons, Toronto.

Moav, R., T. Brody, and G. Hulata. 1978. Genetic improvement of wild fish populations. Science 201:1090-1094.

Moore, F. R. 1978. Interspecific aggression: Toward whom should it be addressed? Behav. Ecol. Sociobiol. 3:173-176.

Mortimer, A. M. 1984. Population ecology and weed science. Pp. 363-388 in R. Dirzo and J. Sarukhan, eds. Perspectives on Plant Population Ecology. Sinauer Associates, Sunderland, Mass.

Munn, R. E., ed. 1979. Environmental Impact Assessment: Principles and Procedures. SCOPE 5. John Wiley & Sons, Chichester, Eng.

Murphy, G. I. 1977. Clupeoids. Pp. 283-308 in J. Gulland, ed. Fish Population Dynamics. John Wiley & Sons, New York.

Myerberg, A. A., Jr., and R. E. Thresher. 1974. Interspecific aggression and its relevance to the concept of territoriality in reef fishes. Am. Zool. 14:81-96.

Myers, N. 1979. The Sinking Ark. Pergamon Press, Oxford, Eng.

Myers, N. 1980. Conversion of Tropical Moist Forests. National Academy of Sciences, Washington, D.C.

Myers, N. 1984. The Primary Source. Norton, New York.

National Research Council. 1969. Eutrophication: Causes, Consequences, Correctives. National Academy of Sciences, Washington, D.C.

National Research Council. 1981. Testing for Effects of Chemicals on Ecosystems. National Academy Press, Washington, D.C.

National Research Council. 1982. Assessment of Multichemical Contamination. Proceedings of an International Workshop, Milan, April 28-30, 1981. National Academy Press, Washington, D.C.

National Research Council. 1983a. Acid Deposition in Eastern North America. National Academy Press, Washington, D.C.

National Research Council. 1983b. Effects of atmospheric transformations of polycyclic aromatic hydrocarbons. Pp. 3-1 to 3-48 in Polycyclic Aromatic Hydrocarbons: Evaluation of Sources and Effects. National Academy Press, Washington, D.C.

National Research Council. 1983c. Changing Climate. Report of the Carbon Dioxide Assessment Committee. National Academy Press, Washington, D.C.

National Research Council. In press. Monitoring and Assessment of Trends in Acid Deposition. National Academy Press, Washington, D.C.

Niering, W. A., and F. E. Egler. 1955. A shrub community of *Viburnum lentagon*, stable for twenty-five years. Ecology 36:356-360.

O'Donald, P. 1967. A general model of sexual and natural selection. Heredity 22:499-518.

Odum, E. P. 1969. The strategy of ecosystem development. Science 164:262-270.

Odum, E. P., J. T. Finn, and E. H. Franz. 1979. Perturbation theory and the subsidy-stress gradient. BioScience 29:349-352.

Odum, W. E. 1982. Environmental degradation and the tyranny of small decisions. BioScience 32:728-729.

Olson, J. S. 1958. Rates of succession and soil changes on southern Lake Michigan sand dunes. Bot. Gaz. 119:125-170.

Orians, G. H. 1969. The number of bird species in some tropical forests. Ecology 50:783-801.

Orians, G. H. 1975. Diversity, stability and maturity in natural ecosystems. Pp. 139-150 in W. H. Van Dobben and R. H. Lowe-McConnell, eds. Unifying Concepts in Ecology. W. Junk, The Hague.

Orians, G. H. 1980. Some Adaptations of Marsh-nesting Blackbirds. Princeton University Press, Princeton, N.J.

Orians, G. H. In press. Site characteristics favoring invasions. In H. A. Mooney, ed. Ecology of Biological Invasions in North America and Hawaii. Springer, New York.

Orians, G. H., and O. T. Solbrig, eds. 1977. Convergent Evolution in Warm Deserts. Dowden, Hutchinson & Ross, Stroudsburg, Pa.

Orians, G. H., and M. F. Willson. 1964. Interspecific territories of birds. Ecology 17:736-745.

Paine, R. T. 1966. Food web complexity and species diversity. Am. Nat. 100:65-75.

Paine, R. T. 1969. The *Pisaster-Tegula* interaction: Prey patches, predator food preference, and intertidal community structure. Ecology 50:950-961.

Paine, R. T. 1974. Intertidal community structure: Experimental studies on the relationship between a dominant competitor and its principal predator. Oecologia 15:93-120.

Paine, R. T. 1980. Food webs: Linkage, interaction strength and community infrastructure. J. Anim. Ecol. 49:667-685.

Paine, R. T. 1981. Truth in ecology. Bull. Ecol. Soc. Am. 62:256-258.

Paine, R. T. 1984. Ecological determinism in the competition for space. Ecology 65:1339-1348.

Paine, R. T., and S. A. Levin. 1981. Intertidal landscapes: Disturbance and the dynamics of pattern. Ecol. Monogr. 51:145-198.

Park, T. 1948. Experimental studies of interspecies competition. I. Competition between populations of the flour beetles *Tribolium confusum* Duval and *Tribolium castaneum* Herbst. Ecol. Monogr. 18:265-308.

Parrish, R. H., and A. D. MacCall. 1978. Climatic Variation and Exploitation in the Pacific Mackerel Fishery. Fish Bull. 167. California Department of Fish and Game, La Jolla.

Partridge, L. 1978. Habitat selection. Pp. 351-376 in J. R. Krebs and N. B. Davies, eds. Behavioral Ecology. Sinauer Associates, Sunderland, Mass.

Patrick, R., and R. R. Kiry. 1976. Estuarine surveys, biomonitoring and bioassays. Department of Limnology of the Academy of Natural Sciences, Philadelphia.

Patrick, R., B. Crum, and J. Coles. 1967. Temperature and manganese as determining factors in the presence of diatom or blue-green algal floras in streams. Proc. Natl. Acad. Sci. U.S.A. 64:472-478.

Peakall, D. B. 1975. Physiological effects of chlorinated hydrocarbons on avian species. Pp. 343-360 in R. Hauge and V. Freed, eds. Environmental Dynamics of Pesticides. Plenum, New York.

Perkins, J. H. 1982. Insects, Experts, and the Insecticide Crisis: The Quest for New Pest Management Strategies. Plenum, New York.

Peterman, R. M. 1978. Testing for density-dependent marine survival in marine salmonids. J. Fish. Res. Bd. Can. 35:1434-1450.

Peterman, R. M. 1984. Interaction among sockeye salmon in the Gulf of Alaska. Pp. 187-199 in W. G. Pearcy, ed. The Influence of Ocean Conditions on the Production of Sockeye Salmonids. Oregon State University Sea Grant Communications, Corvallis.

Peters, W. 1984. New answers through chemotherapy? Experientia 40:1351-1357.

Peters, W. In press. Resistance to antiparasitic drugs and its prevention. Parasitology.

Pianka, E. R. 1966. Latitudinal gradients in species diversity. Am. Nat. 100:33-46.

Pickett, S. T. A., and J. N. Thompson. 1978. Patch dynamics and the design of nature reserves. Biol. Conserv. 13:27-37.

Pimentel, D., D. Andow, R. Dyson-Hudson, D. Gallahan, S. Jacobson, M. Irish, S. Kroop, A. Moss, I. Schreiner, M. Shepard, T. Thompson, and B. Vinzant. 1980. Environmental and social costs of pesticides: A preliminary assessment. Oikos 34:126-140.

Pimm, S. L. 1979. The structure of food webs. Theor. Pop. Biol. 16:144-158.

Pimm, S. L. 1980. Food web design and the effect of species deletion. Oikos 35:139-147.

Pimm, S. L. 1982. Food webs. Chapman and Hall, London.

Pimm, S. L. 1984. The complexity and stability of ecosystems. Nature 307:321-326.

Policansky, D. 1983. Size, age and demography of metamorphosis and sexual maturation in fishes. Am. Zool. 23:57-63.

Porter, J. W., J. D. Woodley, G. J. Smith, J. E. Neigel, J. F. Battey, and D. G. Dallmeyer. 1981. Population trends among Jamaican reef corals. Nature 294:240-250.

Preston, F. W. 1960. Time and space and the variation of species. Ecology 29:254-253.

Price, P. W. 1980. Evolutionary Biology of Parasites. Princeton University Press, Princeton, N.J.

Pyke, G. H., H. R. Pulliam, and E. L. Charnov. 1977. Optimal foraging: A selective review of theory and tests. Q. Rev. Biol. 52:137-154.

Ratcliffe, P. R. 1984. Population dynamics of red deer (*Cervus elephas* L.) in Scottish commercial forests. Proc. R. Soc. Edinburgh 82B:291-302.

Recher, H. F. 1969. Bird species diversity and habitat diversity in Australia and North America. Am. Nat. 103:75-80.

Reckhow, K. H. 1979. Empirical lake models for phosphorus: Development, applications, limitations and uncertainty. Pp. 193-221 in D. Scavia and A. Robertson, eds. Perspectives in Lake Ecosystem Modeling. Ann Arbor Sciences, Ann Arbor, Mich.

Reif, A. E. 1984. Synergism in carcinogenesis. J. Natl. Cancer Inst. 73:25-39.

Rhoades, D. F. 1979. Evolution of plant chemical defense against herbivores. Pp. 3-54 in G. A. Rosenthal and D. N. Janzen, eds. Herbivores: Their Interaction with Secondary Plant Metabolites. Academic Press, New York.

Rhoades, D. F. 1982. Response of alder and willow to attack by tent caterpillars and fall webworms: Evidence for pheromonal sensitivity of willows. Pp. 55-68 in P. Hedin, ed. Plant Resistance to Insects. ACS Symposium Series 208. American Chemical Society, Washington, D.C.

Rhoades, D. F. 1983. Offensive-defensive interactions between herbivores and plants: Their relevance in herbivore population dynamics and ecological theory. Am. Nat. 125:205-238.

Rhoades, D. F., and R. G. Cates. 1976. Toward a general theory of plant antiherbivore chemistry. Rec. Adv. Phytochem. 10:168-213.

Rice, E. L. 1974. Allelopathy. Academic Press, New York.

Ricker, W. E. 1963. Big effects from small causes: Two examples from fish population dynamics. J. Fish. Res. Bd. Can. 20:257-264.

Ricker, W. E. 1981. Changes in the average size and average age of Pacific salmon. Can. J. Fish. Aquat. Sci. 38:1636-1656.

Ricklefs, R. E. 1979. Ecology. 2nd ed. Chiron Press, New York.

Ricklefs, R. E., and K. O'Rourke. 1975. Aspect diversity in moths: A temperate-tropical comparison. Evolution 29:313-324.

Robinson, R. W, H. M. Munger, T. W. Whitaker, and G. W. Bohn. 1976. Genes of the Cucurbitaceae. Hortoscience 11:554-568.

Romesburg, H. C. 1981. Wildlife sciences: Gaining reliable knowledge. J. Wildl. Manage. 45:293-313.

Rosenberg, D. M., V. H. Resh, *et al.* 1981. Recent trends in environmental impact assessment. Can. J. Fish. Aquat. Sci. 38:591-624.

Rosenzweig, M. L. 1972. Stability of enriched aquatic ecosystems. Science 175:562-565.

Roth, R. R. 1976. Spatial heterogeneity and bird species diversity. Ecology 57:773-782.

Ryan, C. A., and T. R. Green. 1974. Proteinase inhibitors in natural plant protection. Rec. Adv. Phytochem. 8:123-140.

Ryman, N., R. Baccus, C. Reuterwall, and M. Smith. 1981. Effective population size, genetic interval, and potential loss of genetic variability in game species under various hunting regimes. Oikos 26:257-266.

Salisbury, F. B. 1942. The Reproductive Capacities of Plants. George Bell & Sons, Ltd., London.

Sanders, F. S., S. M. Adams, L. W. Barnthouse, J. M. Giddings, E. E. Huber, K. D. Kumar, D. Lee, B. Murphy, G. W. Suter, and W. Van Winkle, eds. 1980. Strategies for Ecological Effects Assessment at DOE Energy Activity Sites. Environmental Sciences Division Publ. 1639. ORNL/TM-6783. Oak Ridge National Laboratory, Oak Ridge, Tenn.

Schindler, D. W. 1974. Eutrophication and recovery in experimental lakes: Implications for lake management. Science 184:897-898.

Schindler, D. W., E. J. Fee, and T. Ruszczynski. 1978. Phosphorus input and its consequences for phytoplankton standing crop and production in the Experimental Lakes Area and in similar lakes. J. Fish. Res. Bd. Can. 35:190-196.

Schoener, T. W. 1968. Sizes of feeding territories among birds. Ecology 49:123-141.

Schoener, T. W. 1971. Theory of feeding strategies. Annu. Rev. Ecol. Syst. 2:369-404.

Schoener, T. W. 1983. Field experiments on interspecific competition. Am. Nat. 122:240-285.

Schonewald-Cox, C. 1983. Conclusions: Guidelines to management: A beginning attempt. Ch. 25 in C. Schonewald-Cox, S. M. Chambers, B. MacBryde, and L. Thomas, eds. Genetics and Conservation: A Reference for Managing Wild Animal and Plant Populations. Benjamin/Cummings, Menlo Park, Calif.

Schonewald-Cox, C., S. M. Chambers, B. MacBryde, and L. Thomas, eds. 1983. Genetics and Conservation: A Reference for Managing Wild Animal and Plant Populations. Benjamin/Cummings, Menlo Park, Calif.

Schultz, J. C. 1983a. Habitat selection and foraging tactics of caterpillars in heterogeneous trees. Pp. 61-90 in R. F. Denno and M. S. McClure, eds. Variable Plants and Herbivores in Natural and Managed Systems. Academic Press, New York.

Schultz, J. C. 1983b. Impact of variable plant defensive chemistry on susceptibility of insects to natural enemies. Pp. 37-54 in P. A. Hedin, ed. Plant Resistance to Insects. ACS Symposium Series 208. American Chemical Society, Washington, D.C.

Schultz, J. C., and I. T. Baldwin. 1982. Oak leaf quality declines in response to defoliation by gypsy moth larvae. Science 217:149-151.

Schwabe, C. 1984. Animals as monitors of the environment. Pp. 562-578 in Veterinary Medicine and Human Health. 3rd ed. Williams and Wilkins, Baltimore, Md.

Scruggs, C. G. 1975. The Peaceful Atom and the Deadly Fly. Jenkins, Austin, Tex.

Searcy, W. A. 1982. The evolutionary effects of mate selection. Annu. Rev. Ecol. Syst. 13:57-85.

Seber, G. A. F. 1982. The Estimation of Animal Abundance and Related Parameters. Griffin, London.

Selander, R. K. 1983. Evolutionary consequences of inbreeding. Pp. 201-215 in S. C. Schonewald-Cox, S. M. Chambers, B. MacBryde, and L. Thomas, eds. 1983. Genetics and Conservation: A Reference for Managing Wild Animal and Plant Populations. Benjamin/Cummings, Menlo Park, Calif.

Shaffer, M. L. 1981. Minimum population sizes for species conservation. BioScience 31:131-134.

Sharma, R. K., J. D. Buffington, and J. T. McFadden, eds. 1976. Proceedings of a Workshop on the Biological Significance of Environmental Impacts. NR-CONF-002. U.S. Nuclear Regulatory Commission, Washington, D.C.

Sharp, J. M., S. G. Appan, M. E. Bender, T. L. Linton, D. J. Reisch, and C. H. Ward. 1979. Natural variability of biological community structure as a quantitative basis for ecological impact assessment. Pp. 257-284 in Proceedings of the Ecological Damage Assessment Conference. Society of Petroleum Industry Biologists, Los Angeles.

Shepherd, J. G. 1984. The availability and information content of fisheries data. Pp. 95-109 in R. M. May, ed. Exploitation of Marine Communities. Dahlem Konferenzen. Springer, Berlin.

Shugart, H. H., and D. C. West. 1980. Forest succession models. BioScience 30:308-313.

Silliman, R. P. 1975. Selective and unselective exploitation of experimental populations of *Tilapia mossambica*. Fish. Bull. 73:495-507.

Simberloff, D. In press. Ecological communities. In McGraw-Hill Encyclopedia of Science and Technology. 6th ed. McGraw-Hill, New York.

Simberloff, D., and L. G. Abele. 1976. Island biogeography theory and conservation practice. Science 191:285-286.

Simenstad, C. A., J. A. Estes, and K. W. Kenyon. 1978. Aleuts, sea otters, and alternate stable-state communities. Science 200:403-411.

Sissenwine, M. P. 1984. Why do fish populations vary? Pp. 59-94 in R. M. May, ed. Exploitation of Marine Communities. Dahlem Konferenzen. Springer, Berlin.

Skutch, M. M., and R. T. N. Flowerdeu. 1976. Measurement techniques in environmental impact assessment. Environ. Conserv. 3:209-217.

Slawson, G. C., Jr., and B. C. Marcy, Jr. 1976. Evaluation of effects of multiple power plants on a river ecosystem. Ch. 17-3 in International Conference on Environmental

Sensing and Assessment. Vol. 1. Institute of Electrical and Electronics Engineers, New York.

Smith, C. E. G. 1982. Practical problems in the control of infectious diseases. Pp. 177-190 in R. M. Anderson and R. M. May, eds. Population Biology of Infectious Diseases. Dahlem Konferenzen. Springer, Berlin.

Snaydon, R. W. 1984. Plant demography in an agricultural context. Pp. 369-407 in R. Dirzo and J. Sarukhan, eds. Perspectives on Plant Population Ecology. Sinauer Associates, Sunderland, Mass.

Soulé, M. E. 1980. Thresholds for survival: Maintaining fitness and evolutionary potential. Pp. 151-169 in M. E. Soulé and B. A. Wilcox, eds. Conservation Biology. Sinauer Associates, Sunderland, Mass.

Soulé, M. E., and B. A. Wilcox, eds. 1980. Conservation Biology. Sinauer Associates, Sunderland, Mass.

Sousa, W. P. 1984. The role of disturbance in natural communities. Annu. Rev. Ecol. Syst. 15:353-391.

Southwood, T. R. E., R. M. May, M. P. Hassell, and G. R. Conway. 1974. Ecological strategies and population parameters. Am. Nat. 108:791-804.

States, J. B., P. T. Haug, T. G. Shoemaker, L. W. Reed, and E. B. Reed. 1978. A Systems Approach to Ecological Baseline Studies. FWS/OBS-78/21. Fish and Wildlife Service, U.S. Department of the Interior, Fort Collins, Colo.

Stearns, S. C. 1976. Life-history tactics: A review of the ideas. Q. Rev. Biol. 51:3-47.

Steele, J. H. 1984. Kinds of variability and uncertainty affecting fisheries. Pp. 245-262 in R. M. May, ed. Exploitation of Marine Communities. Dahlem Konferenzen. Springer, Berlin.

Steele, J. H., and E. W. Henderson. 1984. Modeling long-term fluctuations in fish stocks. Science 224:985-987.

Strong, D. R. Jr. 1984. Exorcising the ghost of competition past: Phytophagous insects. Ch. 11 in D. R. Strong, Jr., D. Simberloff, L. G. Abele, and A. B. Thistle, eds. Ecological Communities: Conceptual Issues and the Evidence. Princeton University Press, Princeton, N.J.

Strong, D. R., Jr., L. A. Szyska, and D. S. Simberloff. 1979. Tests of community-wide character displacement against null hypotheses. Evolution 33:897-913.

Strong, D. R., Jr., D. Simberloff, L. G. Abele, and A. B. Thistle, eds. 1984. Ecological Communities: Conceptual Issues and the Evidence. Princeton University Press, Princeton, N.J.

Suchanek, T. H. 1979. The *Mytilus californianus* Community: Studies on the Composition, Structure, Organization, and Dynamics of a Mussel Bed. Ph.D. thesis, University of Washington, Seattle.

Suter, G. W., II. 1981. Commentary: Ecosystem theory and NEPA assessment. Bull. Ecol. Soc. Am. 62:186-192.

Suter, G. W., II. 1982. Terrestrial perturbation experiments for environmental assessment. Environ. Manage. 6:43-54.

Thomas, J. W., ed. 1979. Wildlife Habitats in Managed Forests: The Blue Mountains of Oregon and Washington. Agriculture Handbook 553. Forest Service, U.S. Department of Agriculture, Washington, D.C.

Thompson, J. N. 1982. Interaction and Coevolution. John Wiley & Sons, New York.

Tilman, D. 1982. Resource Competition and Community Structure. Princeton University Press, Princeton, N.J.

Tinbergen, J. M. 1981. Foraging decisions in starlings (*Sturnus vulgaris* L.). Ardea 69:1-67.

Turner, R. M., S. M. Alcom, and S. Hope. 1969. Mortality of transplanted saguaro seedlings. Ecology 50:835-844.

Underwood, A. J., and E. J. Denley. 1984. Paradigms, explanations, and generalizations in models for the structure of intertidal communities on rocky shores. Pp. 151-180 in D. R. Strong, Jr., D. Simberloff, L. G. Abele, and A. B. Thistle, eds. Ecological Communities: Conceptual Issues and the Evidence. Princeton University Press, Princeton, N.J.

U.S. Environmental Protection Agency. 1980. Acid Rain. U.S. EPA Report 600/9-79-036. Office of Research and Development, U.S. Environmental Protection Agency, Washington, D.C.

U.S. Fish and Wildlife Service. 1984. Endangered classification proposed for four fishes in Southeast and Utah. Endang. Species Tech. Bull. 9(8):1, 10-11.

Van Horne, B. 1983. Density as a misleading indicator of habitat quality. J. Wildl. Manage. 47:893-901.

Velich, I., and L. Satyko. 1974. Possibilities of increasing earliness in melon by means of the genetics of sex and floral biology. Agrartudomanyi Kozlemenyek 33:459-472.

Vollenweider, R. A. 1969. Möglichkeiten und Grenzen elementärer Modelle der Stoffbilanz von Seen. Arch. Hydrobiol. 66:1-36.

Vollenweider, R. A. 1975. Input-output models with special reference to the phosphorus loading concept in limnology. Schweiz. Z. Hydrol. 37:53-84.

Vollenweider, R. A. 1976. Advances in defining critical loading levels for phosphorus in lake eutrophication. Mem. Ist. Ital. Idrobiol. 33:53-83.

Walters, C. J. 1975. An interdisciplinary approach to development of watershed simulation models. J. Fish. Res. Bd. Can. 32:177-195.

Walters, C. J. 1984. Managing fisheries under biological uncertainty. Pp. 263-274 in R. M. May, ed. Exploitation of Marine Communities. Dahlem Konferenzen. Springer, Berlin.

Walters, C. J. In press. Adaptive Management of Renewable Resources. Macmillan, New York.

Walters, C. J., and R. Hilborn. 1976. Adaptive control of fishing systems. J. Fish. Res. Bd. Can. 33:145-159

Ward, D. V. 1978. Biological Environmental Impact Studies: Theory and Methods. Academic Press, New York.

Watt, K. E. F. 1964. Comments on fluctuations of animal populations and measures of community stability. Can. Entomol. 96:1434-1442.

Weinstein, L., and D. McCune. 1970. Field surveys, vegetation monitoring. Pp. G1-4 in J. S. Jacobson and A. C. Hill, eds. Recognition of Air Pollution Injury to Vegetation: a Pictorial Atlas. Air Pollut. Control Assoc., Pittsburgh.

Wells, P. V. 1965. Scarp woodlands, transplanted grassland soils, and the concept of grassland climate in the Great Plains region. Science 148:246-249.

Werner, G. E., and D. J. Hall. 1974. Optimal foraging and the size selection of prey by the bluegill sunfish (*Lepomis macrochirus*). Ecology 55:1042-1052.

Westoby, M. 1978. What are the biological bases of varied diets? Am. Nat. 112:627-631.

Wheelwright, N. T. 1983. Fruits and the ecology of resplendent quetzals. Auk 100:286-301.

Whitham, T. G., A. G. Williams, and A. M. Robinson. 1984. The variation principle: Individual plants as temporal and spatial mosaics of resistance to rapidly evolving pests. Pp. 15-52 in P. W. Price, C. N. Slobodchikoff, and W. S. Gaud, eds. A New Ecology. Wiley-Interscience, New York.

Wilcove, D. S. 1985. Nest-predation in forest tracts and the decline of songbirds. Ecology 66:1211-1214.

Willis, E. O. 1974. Populations and local extinctions of birds on Barro Colorado Island, Panama. Ecol. Monogr. 44:153-169.

Wooster, W. S., ed. 1983. From Year to Year: Interannual Variability of the Environment and Fisheries in the Gulf of Alaska and the Eastern Bering Sea. University of Washington, Washington Sea Grant, Seattle.

Zar, J. H. 1976. Statistical significance and biological significance of environmental impacts. Pp. 285-293 in R. K. Sharma, J. D. Buffington, and J. T. McFadden, eds. Proceedings of a Workshop on the Biological Significance of Environmental Impacts. NR-CONF-002. U.S. Nuclear Regulatory Commission, Washington, D.C.

Zaret, T. M. 1980. Predation and Freshwater Communities. Yale University Press, New Haven, Conn.

II

Selected Case Studies

The case studies in this part are presented as examples of how ecological knowledge has been used in planning and carrying out problem-solving efforts, including how the knowledge has been limited and how it has been adapted to specific problems. The case studies presented here are referred to frequently in Part I. Their selection and preparation are described in the Introduction. The Committee is responsible for the introductory statements and comments.

12

North Pacific Halibut Fishery Management

The goal of management of a commercially important resource is to resolve the conflict between maintenance and exploitation. Unfettered exploitation often leads to disappearance of the resource, as in the cases of whales, passenger pigeons, and buffaloes. But management that is too conservative leads to inefficient use of the resource. One approach to the conflict in fishery management has been to use the idea of some maximal sustainable yield that can be taken from the fishery. Unfortunately, owing largely to unpredictable variations in the environment, maximal yields are usually not sustainable for very long. The management of the Pacific halibut by the International Pacific Halibut Commission (IPHC) is an example of the responsive (adaptive) approach to management. Additionally, in this case there has been strong commitment to an understanding of the biology of the species, and the activities of IPHC have been superbly documented, as has the biological information obtained.

Case Study

DAVID POLICANSKY, National Research Council, Washington, D.C.

INTRODUCTION

The International Pacific Halibut Commission (IPHC) was established by a convention between the United States and Canada in 1923, which was revised in 1930, 1937, and 1953 (Bell, 1969) and again in 1983. (The commission was originally named the International Fisheries Commission and renamed in 1953; I use the abbreviation IPHC throughout.) The purpose of the Commission was to provide a mechanism for joint management of the Pacific halibut (*Hippoglossus stenolepis* Schmidt), whose abundance had been declining up to 1923. The management objective, originally the maximization of sustainable yield, was changed in 1983 to the optimization of sustainable yield (R. B. Deriso, personal communication); the idea of "optimal" sustained yield includes the "quality of the fishery," as well as the weight of the harvest (Roedel, 1975). The primary ecological problem has been stock assessment; some related issues have also been important.

The Commission originally had four members and now has six, drawn from industry and government. Half the members are Canadian and half are from the United States. The Commission is supported by a scientific staff, headquartered in Seattle, with a full-time director appointed by the commissioners; a Conference Board composed of fishermen and vessel owners, which makes recommendations to the Commission with respect to regulations; and an Advisory Group composed of 14 fishermen, dealers, and fish-processors, half of whom are selected by the Conference Board and half by the Halibut Association of North America. Members of the Board participate in the Commission's meetings as observers.

The history of IPHC can be traced in its many thorough reports and in other sources; the following account relies especially on Thompson and Freeman (1930) and Bell (1981). Until 1888, the Indians had conducted an important halibut fishery, with a catch probably exceeding 3 million pounds a year. The Northern Pacific Railroad was completed in 1888, and the Canadian Pacific Railroad in 1892. These new railroads profoundly affected the development of the halibut fishery by providing ready access to large markets for halibut in the East. Attracted by the new profitability of the Pacific halibut fishery, men and boats from the eastern fishery arrived, and the annual catch increased rapidly. The next 30 years saw depletion of known halibut banks, maintenance of yield by discovery and exploitation of new banks, and technical innovations. By the time of World War I, it was obvious

138

that, from an economic point of view, the banks were being overexploited. For this reason and because winter fishing was dangerous and expensive, a winter closed season was desirable to the fishermen.

Thus, economic pressure was the impetus for the birth of IPHC. The institution of a closed season required international regulation, which required a treaty. Many attempts to enact fishery treaties between the United States and Canada had failed, because they included both conservation and unrelated economic considerations (such as reciprocal port privileges). However, when the halibut treaty was finally ratified in 1924, it contained only a conservation measure (a closed season) and provision for the establishment of IPHC.

An IPHC report to the two governments in 1928 offered specific recommendations for the development of the fishery and the conservation of the resource. The report detailed the decline in abundance of halibut in all the areas where they were exploited. It recommended establishment of management areas in each of which the total catch of halibut could be reduced until the yield was stable, with the amount of the reduction being responsive to the catch per unit of effort (CPUE); closure of the nursery grounds; prohibition of the use of destructive gear; extension of the closed season; provision for future modifications of the closed season; and licensing of all vessels for statistical and other purposes. These recommendations, based on scientific activities of IPHC under the direction of W. F. Thompson (including tagging experiments, analyses of catch statistics, and hydrographic studies), resulted in the Halibut Convention of 1930, which gave IPHC broad regulatory authority.

BIOLOGICAL BASIS OF MANAGEMENT

The overriding issue here is the ecological problem of stock assessment, i.e., knowing how many fish are in the sea. If that is known and the catch is known, then the effect of fishing on the fish stocks can be determined. Other aspects of the biology of the halibut have also been studied and are discussed below.

Assessment of Stock

Knowledge of stock abundance is desirable if a fishery is to be monitored. The assessment of fish stocks is perhaps the major fishery problem and is often intractable. IPHC has relied heavily on CPUE as an index of stock abundance (Skud, 1978); but the catch has been measured in biomass (weight), rather than numbers of fish. Number of fish is also important, because declining numbers can be masked by increased growth rate if only information on biomass is used (Schmitt and Skud, 1978). In part for that reason, mark-and-recapture (tagging) experiments have been done.

In addition to the estimates just mentioned, a number of models have been used (they are discussed below). In general, these models have provided estimates of abundance relative to that at some specified time. As long as the relativity coefficients vary by only small amounts, knowledge of absolute abundance is not needed for successful fishery management, although it would be intellectually satisfying.

Relationship Between Stock and Recruitment

The relationship between stock and recruitment is difficult to quantify—the density-related form described by Ricker's equation (Ricker, 1954) is seldom accurate enough to be used in management. The reason for this is probably that recruitment is determined largely by larval survival, and larval survival seems to be very much affected by environmental conditions (for recent reviews, see Beverton, 1984; Hunter, 1983; Sissenwine, 1984; and Wooster, 1983). Although there have been attempts to relate stock and recruitment in halibut (e.g., Schmitt and Skud, 1978), the relationship is not directly used in managing the halibut fishery. It appears as a term in some of the models discussed below, but they usually assume that recruitment is independent of stock size.

Catchability

The catchability of fish with a given type of fishing gear is a measure of the efficiency of that gear. If the catchability coefficient were a constant for a particular gear, then catch would depend on fishing effort and fish abundance only. But catchability varies with fish behavior, as well as with gear type, and must be known if comparisons of CPUEs with different types of gear are to be valid. Such comparisons are required for tracking the history of a fishery over periods in which different types of gear were used; in other words, they are necessary for an understanding of the effect of fishing on stock abundances. Some of the difficulties associated with catchability estimates are discussed below.

The commonest gear in the halibut fishery is the setline, and the functional unit of this longline gear is the "skate"—an 1,800-foot length of line, gangions (branch lines), and hooks. The hooks, regularly spaced on a given line, are 9-26 feet apart. The catch not only is proportional to the number of hooks, but varies with their spacing; therefore, neither the skate nor a number of hooks is a good unit of fishing effort. Changes in these variables complicate analyses of historical trends in CPUE (Skud, 1978). Hook shape also influences catchability, and a much more efficient hook has recently been introduced into the halibut fishery (Deriso, personal communication). Other variables influence catchability as well (Hoag and

Deriso, 1984; IPHC, 1982; Myhre and Quinn, 1984). All these variables affect estimates of CPUE and hence of stock abundance, and considerable effort has been devoted to the understanding of their effects (Hamley and Skud, 1978; Myhre and Quinn, 1984; Skud, 1975, 1978).

Migration Patterns

The halibut fishery extends over a very large area, from the Oregon coast to the Bering Sea. For intelligent management, it is obviously important to know whether this area is occupied by one large, intermingling stock or by a number of independent stocks. The fishery is divided into management areas; the largest catches come from area 2 (Cape Spencer, in the middle of the Alaska panhandle, south to Oregon) and area 3 (Cape Spencer to the western end of the Aleutian Islands). Tagging experiments before the mid-1950s suggested that adult halibut from the two areas did not mix, and studies of the distributions of eggs and larvae coupled with the use of drift bottles suggested that there was not much interchange of those stages either.

Skud (1977a) re-examined the data in the early life stages and considered more recent data on later stages. He concluded that, although there is little drift of the early stages from area 3 to area 2, there is substantial drift northward from area 2 to area 3, much migration of juvenile stages between the two areas, and seasonal movement of adults.

The longline fishery currently operates only in the summer and preferentially takes fish that return to the same grounds each summer. Because of this, Skud concluded that the areas could reasonably be managed separately, if the fishery continued to operate only in the summer. He did not comment on the implications of the intermingling of the earlier life stages, but noted that a trawl fishery operates year round and preferentially takes juvenile halibut, which migrate extensively. For these reasons, the areas should not be managed independently (Skud, 1977a).

Natural and Fishing Mortality

It is extremely difficult in practice to separate causes of mortality in a fished population. The estimates of natural and fishing mortality are closely related and are important in models of stock abundance (as discussed later).

Growth Rates

Growth rates have been studied by IPHC for a variety of management purposes (e.g., Schmitt and Skud, 1978), most recently for use in mathematical models of stock abundances. Because catch in this fishery is measured in biomass, rather than numbers of fish, a knowledge of growth rates is essential to an understanding of the population dynamics of the resource.

Growth rates have been estimated by weighing fish and determining their ages by otolith counts. The growth rates of halibut have varied with location and over time, with marked increases since the 1950s (Schmitt and Skud, 1978). Both natural environmental changes and density-dependent responses to fishing appear to have influenced growth rates (Schmitt and Skud, 1978; Southward, 1967).

GENERAL APPROACH TO MANAGEMENT

The general approach, as indicated earlier, has been that an understanding of the population dynamics of halibut is required for successful management of the fishery. The Pacific halibut was studied before the existence of IPHC, to a large extent by W. F. Thompson, who became its first director. The 1923 convention required the Commission to "make a thorough investigation into the life history of the Pacific halibut." The aim of the Commission "has been to establish beyond doubt the actual condition of the fishery" by "close adherence to the facts and avoidance of unsupported theory" (Babcock *et al.*, 1931). Thompson was familiar with the already large European literature on fishery biology and was probably guided in his studies by that literature.

The specific objective of IPHC's program, stock assessment, has not yet been reached, although *relative* abundances are used successfully for management.

Boundaries

The boundaries, originally set by the range of the boats and the demand for fish, represented the range of fishing activities. Fishing is now limited in the south by the abundance of halibut, and in the north and west by deep water and low abundance. The setting of the jurisdictional boundaries by the formation of IPHC required many years of difficult negotiations (Bell, 1981; Thompson and Freeman, 1930).

Monitoring

The nature of this case has made monitoring an essential part of the scientific program from its inception. There is continuous monitoring of many biological parameters, including numbers, biomass, growth rate, fecundity, and catchability. Information from the commercial catch and results of research sampling of the fish are both used. The information derived from this monitoring is applied in regulating the fishery.

Cumulative Effects

The population has changed in a number of ways since 1930. The fish grow faster, fecundity increased up to about 1960 and then declined, and abundance declined and recently increased. The sex ratio might also have become more female-biased (Schmitt and Skud, 1978). These changes could reflect the results of fishing, which is selective (e.g., Myhre, 1969), and they might therefore have a genetic basis, at least in part. If the changes are the result of evolution under fishing pressure, then fishing has had an important cumulative effect.

SOURCES OF KNOWLEDGE

Ecological Facts

All the ecological facts that IPHC used in managing the fishery were obtained through study of the halibut fishery itself and from persons involved with it. No useful facts were obtained from the published literature, data banks and files, or other stored information. However, general knowledge in fishery biology includes many ecological theories, principles, and models that can be found in textbooks of ecology or fishery biology, and some were tested on the halibut fishery.

A great deal of interesting and new ecological information has been gathered by IPHC, but it appears that only total catch, CPUE, catch-at-age, growth rates, and migration information from tagging were actually used in managing the fishery. Other facts—such as egg size, fecundity, and age and size at metamorphosis—could not be used, because there was no practical basis for incorporating them. Although other available data were not used and there has been at least some success in managing the fishery, additional ecological information would probably improve management. But the additional information would probably come from improvement of the types of information already used, e.g., stock numbers and biomass, recruitment, and catchability.

Theory and General Principles

Of the theoretical aspects of fishery management that are actually used in management of the halibut fishery, the law of accumulated stock is perhaps the most fundamental. The crux of this idea is that yield is some function of stock size. If the stock is too small to provide adequate yield, and "if a greater yield is desired, the only possible way to obtain it is not by increasing the amount of gear fished but by decreasing it" (Thompson, 1937). Other theories and principles, some of them implicit, are discussed in the next section.

Specific Models of Stock Abundance

Several models have been used in the attempt to assess stock abundances. The more important include the use of CPUE as a measure of stock abundance; constant or random stock-recruitment models; von Bertalanffy and other growth-rate models; models of mortality rate ranging from constant mortality to variable, age-specific (usually fishing) mortalities; population age-structure models; and such complex models as cohort analysis, delay-difference models, and nonlinear catch models. Some of these models were developed in connection with the halibut fishery, but most were taken from the general fisheries literature or developed with no specific fishery in mind. Some of the important ones are briefly discussed below.

• *CPUE.* This method uses CPUE as an index of biomass (not numbers) each year and assumes that catchability is constant. Because gear can become saturated (there is a limit to the proportion of hooks that will catch fish, however abundant the population is), CPUE can fail to increase even if the population increases. In addition, catchability varies, as discussed earlier, and gear does not catch halibut at random, but selects for different sizes of fish (Myhre, 1969). Thus, the relationship between biomass and numbers changes when the age composition of the population or the growth rate changes. Because increased growth can counteract decreased abundance, the use of biomass instead of numbers can be misleading (Schmitt and Skud, 1978).

• *Cohort Analysis* (Pope, 1972). In this method, very similar to virtual population analysis (Gulland, 1965), one starts with the number of living fish in the oldest year class and follows the year class back through preceding years, estimating abundances by using the natural mortality (M_i) and fishing mortality (F_i) for each year. The information needed includes an estimate of the ages of the fish caught in the preceding years. In general, natural mortality is assumed to be constant at around 0.2, and fishing mortality is estimated from catch information (Hoag and McNaughton, 1978). If natural mortality is constant from year to year, then fluctuations in CPUE are due to changes in abundance.

• *Delay-Difference Population Model* (Deriso, 1980). This model was developed as a general fishery model, although it has been specifically applied to halibut. It is in effect a simulation model that takes into account starting biomass, a growth coefficient, catch, natural survival (1 − natural mortality), and a spawner-recruit function. The model is run with available information; the measurable variables include starting relative abundance or starting biomass, growth rates, and catch. As in cohort analysis, natural mortality has to be estimated (by tagging or other means) or assumed to be constant. The stock-recruitment relationship has defied elucidation and is assumed to be independent of density. The model is attractive, because it demonstrates the validity of using biomass of sexually mature fish, rather than numbers; the rationale is that fecundity is more closely proportional to the biomass of sexually mature (female) fish than to their numbers. The model also uses parameters whose values can be estimated independently of CPUE data.

● *Nonlinear Catch-Age (Catch at Age) Analysis.* Developed by Pope (1974) and Doubleday (1976), this method estimates abundance by estimating relative year-class strength and cumulative fishing and natural mortalities. It takes into account the differential selectivity of gear for different age classes and the inconstancy of catchability. But it assumes that age selectivity is constant over time; if growth rates change, that assumption is not valid.

A comparison of the four models, which are not completely independent, indicates fairly close agreement among them in their estimates of stock abundance; the most reliable are cohort analysis and the delay-difference population model (Quinn *et al.*, 1984).

● *Annual Surplus Production (ASP).* ASP is defined as the amount of fish that can be taken in a year without changing the biomass of the stock over the course of the year. It is estimated by adding the catch in a given year to the annual change in stock biomass as estimated by one or more of the four methods previously described. Although ASP has been referred to as "equilibrium yield" (Schaefer, 1954), it is not necessary to view it that way, because neither the population size nor the catch need be at equilibrium.

Project as Experiment

The adaptive management approach to the halibut fishery implies an element of experimentalism of the form "let's try X and see if it works," where X is based on the scientific activities of IPHC. In fact, in the 1970s IPHC deliberately increased the catch to test estimates of maximal sustainable yield to demonstrate that the Bering Sea stock was fully utilized and therefore qualified for exemption from Japanese fishing. However, there were no controls, and the great number of variables involved makes it impossible to evaluate the results with certainty.

ROLE OF ECOLOGICAL KNOWLEDGE

The main role of ecological knowledge in this case has been in management, and the most important piece of information has been the relative abundance of the stock. IPHC has based its regulations on stock assessments, derived at first only from CPUE and total catch and later from additional estimates. When the estimated stock size declined, the total allowable catch was decreased; when the estimated stock size increased, the total catch was allowed to increase.

To this day, it is uncertain how much of the decline in abundance up to 1923 was due to fishing mortality and how much the increase after 1930 was due to the regulations imposed by the Commission, although fishing and regulations probably had at least some effect (Skud, 1975). The primary uncertainty concerns estimates of stock abundance and CPUE, because the effect of changes in fishing gear is unknown. An alternative hypothesis to fishing mortality is the one of environmentally caused fluctuations advanced by Burkenroad (see Skud, 1975, for review).

Management areas were established in recognition of the differences in the fishery between different places: fish grow at different rates, spawn at different ages, and respond differently to fishing pressure in the different areas. Although more recent work has revealed much migration between the areas (Skud, 1977a), there is some ecological justification for the establishment of the management areas, and certainly a management justification.

Estimating fishing mortality from sources other than the directed longline fishery for halibut posed an assessment problem. At first, IPHC had to deal only with the effects of the directed fishery on the halibut stock. The closed seasons imposed were in winter, when very little other fishing took place. However, trawl fisheries and directed line fisheries for blackcod, *Anoplopoma fimbria* (Pallas), became more common and resulted in incidental catches of halibut. "Incidental" catches were also made by foreign vessels from the late 1950s on.

Detection of unreported catches is essential to any rational fishery management. At first, IPHC reduced the legal catch, to maintain CPUE at an acceptable level without estimating the illegal catch. That is, legal fishing was reduced to counterbalance illegal, unreported fishing. In effect, the resource was being preserved for the outlaws. It thus became necessary for IPHC to estimate unreported catches, and it resorted to persuasion and a number of clever stratagems, including the use of undercover agents and monitoring of the sales of halibut livers for comparison with the reported catch. Little biological sophistication is required; halibuts have one liver apiece, and there is a statistically valid ratio of liver weight to halibut weight. Skud (1977b) has described IPHC's regulations fully.

IPHC has used monitoring, biological information, and models in management of the halibut fishery. Regulations are based on biological information and on models using that information, and the population is monitored to ascertain their effects. The results of this monitoring lead to annual, or even more frequent, revisions of the regulations.

IPHC is aware of impediments to management that it is powerless to deal with (Skud, 1976). It has no control of such matters as limiting entry to the fishery, enforcement, price of the catch, and efficiency of harvest. However, these are not biological issues, although they exert strong influences on the biological issues.

ACKNOWLEDGMENTS

The International Pacific Halibut Commission staff, particularly R. B. Deriso, has been extremely generous with information and advice. R. B. Deriso, R. M. May, and other members of the Committee on Applications of Ecological Theory to Environmental Problems have made many helpful comments on this paper.

REFERENCES

Babcock, J. P., W. A. Found, M. Freeman, and H. O'Malley. 1931. Report of the International Fisheries Commission. International Fisheries Commission, Seattle, Wash.

Bell, F. H. 1969. Agreements, Conventions and Treaties between Canada and the United States of America with Respect to the Pacific Halibut Fishery. Int. Pac. Halibut Comm. Rep. 50. International Pacific Halibut Commission, Seattle, Wash.

Bell, F. H. 1981. The Pacific Halibut: The Resource and the Fishery. Alaska Northwest Publishing Co., Anchorage, Alaska.

Beverton, R. J. H., rapporteur. 1984. Dynamics of single species. Group report. Pp. 13-58 in R. M. May, ed. Exploitation of Marine Communities. Dahlem Konferenzen. Springer, Berlin.

Deriso, R. B. 1980. Harvesting strategies and parameter estimation for an age-structured model. Can. J. Fish. Aquat. Sci. 37:268-282.

Doubleday, W. G. 1976. A least squares approach to analyzing catch at age data. Int. Comm. Northwest Atl. Fish. Res. Bull. 12:69-81.

Gulland, J. A. 1965. Manual of Methods for Fish Stock Assessment. FAO Manuals in Fisheries Science 4. U.N. Food and Agriculture Organization, Rome.

Hamley, J. M., and B. E. Skud. 1978. Factors affecting longline catch and effort: II. Hook-spacing. Pp. 16-24 in Int. Pac. Halibut Comm. Sci. Rep. 64. International Pacific Halibut Commission, Seattle, Wash.

Hoag, S. H., and R. B. Deriso. 1984. Recent Changes in Halibut CPUE: Studies on Area

Differences in Setline Catchability. Int. Pac. Halibut Comm. Sci. Rep. 71. International Pacific Halibut Commission, Seattle, Wash.

Hoag, S. H., and R. J. McNaughton. 1978. Abundance and Fishing Mortality of Pacific Halibut, Cohort Analysis, 1935-1976. Int. Pac. Halibut Comm. Sci. Rep. 65. International Pacific Halibut Commission, Seattle, Wash.

Hunter, J. R. 1983. Commentary: On the determinants of stock abundance. Pp. 11-16 in W. S. Wooster, ed. From Year to Year: Interannual Variability of the Environment and Fisheries of the Gulf of Alaska and the Eastern Bering Sea. Washington Sea Grant, University of Washington, Seattle.

IPHC (International Pacific Halibut Commission). 1982. The Pacific Halibut: Biology, Fishery, and Management. Int. Pac. Halibut Comm. Tech. Rep. 16. International Pacific Halibut Commission, Seattle, Wash.

Myhre, R. J. 1969. Gear Selection and Pacific Halibut. Int. Pac. Halibut Comm. Rep. 51. International Pacific Halibut Commission, Seattle, Wash.

Myhre, R. J., and T. J. Quinn II. 1984. Comparison of Efficiency of Snap Gear to Fixed-Hook Setline Gear for Catching Pacific Halibut. Int. Pac. Halibut Comm. Sci. Rep. 69. International Pacific Halibut Commission, Seattle, Wash.

Pope, J. G. 1972. An investigation of the accuracy of virtual population analysis using cohort analysis. Int. Comm. Northwest Atl. Fish. Res. Bull. 9:65-74.

Pope, J. G. 1974. A Possible Alternative Method to Virtual Population Analysis for the Calculation of Fishing Mortality from Catch at Age Data. Int. Comm. Northwest Atl. Fish. Res. Doc. 74/20. International Commission of the Northwest Atlantic Fisheries, Dartmouth, N.S.

Quinn, T. J., II, R. B. Deriso, S. H. Hoag, and R. J. Myhre. 1984. A summary of methods of estimating annual surplus production for the Pacific halibut fishery. Int. North Pac. Fish. Comm. Bull. 42:73-81.

Ricker, W. E. 1954. Stock and recruitment. J. Fish. Res. Bd. Can. 11:559-623.

Roedel, P. M., ed. 1975. Optimum Sustained Yield as a Concept in Fisheries Management. Proceedings of a Symposium, September 1974. Special Pub. 9. American Fisheries Society, Washington, D.C.

Schaefer, M. B. 1954. Some aspects of the dynamics of populations important to the management of commercial marine fisheries. Int. Am. Trop. Tuna Comm. Bull. 1:27-56.

Schmitt, C. C., and B. E. Skud. 1978. Relation of Fecundity to Long-Term Changes in Growth, Abundance and Recruitment. Int. Pac. Halibut Comm. Sci. Rep. 66. International Pacific Halibut Commission, Seattle, Wash.

Sissenwine, M. P. 1984. Why do fish populations vary? Pp. 59-94 in R. M. May, ed. Exploitation of Marine Communities. Dahlem Konferenzen. Springer, Berlin.

Skud, B. E. 1975. Revised Estimates of Halibut Abundance and the Thompson-Burkenroad Debate. Int. Pac. Halibut Comm. Sci. Rep. 56. International Pacific Halibut Commission, Seattle, Wash.

Skud, B. E. 1976. Jurisdictional and Administrative Limitations Affecting Management of the Halibut Fishery. Int. Pac. Halibut Comm. Sci. Rep. 59. International Pacific Halibut Commission, Seattle, Wash.

Skud, B. E. 1977a. Drift, Migration, and Intermingling of Pacific Halibut Stocks. Int. Pac. Halibut Comm. Sci. Rep. 63. International Pacific Halibut Commission, Seattle, Wash.

Skud, B. E. 1977b. Regulations of the Pacific Halibut Fishery, 1924-1976. Int. Pac. Halibut Comm. Tech. Rep. 15. International Pacific Halibut Commission, Seattle, Wash.

Skud, B. E. 1978. Factors affecting longline catch and effort: I. General review. Pp. 5-

14 in Int. Pac. Halibut Comm. Sci. Rep. 64. International Pacific Halibut Commission, Seattle, Wash.

Southward, G. M. 1967. Growth of Pacific Halibut. Int. Pac. Halibut Comm. Rep. 43. International Pacific Halibut Commission, Seattle, Wash.

Thompson, W. F. 1937. Theory of the Effect of Fishing on the Stock of Halibut. Int. Fish. Comm. Rep. 12. International Fisheries Commission, Seattle, Wash.

Thompson, W. F., and N. L. Freeman. 1930. History of the Pacific Halibut Fishery. Int. Fish. Comm. Rep. 5. International Fisheries Commission, Seattle, Wash.

Wooster, W. S. 1983. On the determinants of stock abundance. Pp. 1-10 in W. S. Wooster, ed. From Year to Year: Interannual Variability of the Environment and Fisheries of the Gulf of Alaska and the Eastern Bering Sea. Washington Sea Grant, University of Washington, Seattle.

Committee Comment

The management of marine fisheries has not been distinguished by many great successes. In the United States, striped bass (*Roccus saxatilis*), haddock (*Melanogrammus aeglefinus*), and Atlantic halibut (*Hippoglossus hippoglossus*) were once abundant on the East Coast (Smith, 1833), as were Pacific sardines (*Sardinops sagax*) and chub mackerel (*Scomber japonicus*) in California (Murphy, 1966; Parrish and MacCall, 1978). They are now much reduced, although overfishing cannot be given all the blame (see Chapter 8). Fifty-five years after establishment of IPHC, and in the face of continuous, intensive harvesting, the Pacific halibut fishery is still viable and valuable.

It is not clear how much the survival of the halibut fishery is due to the activities of the Commission and how much is due to luck, but the intelligent approach to management taken by IPHC seems to have contributed. The idea of adaptive fishery management, which seems at least partly responsible for the success of IPHC, is not peculiar to that organization; it has been recommended specifically (e.g., Parrish and MacCall, 1978) and as a general approach (e.g., Walters and Hilborn, 1976).

From the beginning, the Commission's primary focus has been on the stock (population) size of the fishery. Regulations have been based on relative stock sizes; when the stock seems to have decreased, fishing pressure is reduced, and when the stock seems to have increased, fishing pressure is allowed to increase. There has been no serious attempt to manage the fishery on any equilibrium or static basis, and that is a major strength of the management program. Because a responsive (adaptive) approach is used, management is not tied to an estimation procedure that could be wrong, and errors are not compounded from year to year. This is strikingly different from the management of striped bass on the East

Coast of the United States, where all the regulations, strict as some are, are static and not directly tied to stock sizes.

Beginning with Thompson, the first director of IPHC, there has been a commitment to an understanding of the biology of the halibut. This is based on the premise that better understanding of the biology of the halibut can lead to better management. As a result of this commitment, there is an enormously detailed, long-term, and intelligently collected set of data on halibut biology. Although it is possible to manage a fishery for short periods with minimal biological information (Shepherd, 1984), the halibut data have been put to good use. They have allowed the development of sophisticated models of the fishery, have made it possible to analyze the long-term effects of previous management practices (i.e., there has been excellent monitoring), and can be used to acquire new biological knowledge (i.e., the project has become a well-designed and well-executed research project). The documentation also allows this case to inform other attempts to manage fisheries.

The case study describes an excellent solution to the problem of managing in more than one jurisdiction, which makes many fish stocks particularly difficult to manage. The Fishery Conservation and Management Act of 1976 resulted in part from a recognition by the United States that the sharing of its marine fisheries by many diverse fishing nations made it very difficult to manage them. One of the Act's major provisions places almost all fisheries within 200 miles of the coast under the sole jurisdiction of the United States, and this appears to have resulted in an increase in the stocks of many fisheries (Finch, 1985). Similar legislation is now common elsewhere. The case also illustrates how difficult it is to achieve agreements and how strong the incentive for agreement must be.

References

Finch, R. 1985. Fishery management under the Magnuson Act. Mar. Policy 9:170-179.

Murphy, G. I. 1966. Population biology of the Pacific sardine (*Sardinops caerulea*). Proc. Calif. Acad. Sci. 34:1-84.

Parrish, R. H., and A. D. MacCall. 1978. Climatic Variation and Exploitation in the Pacific Mackerel Fishery. Fish. Bull. 167. Calif. Department of Fish and Game, La Jolla, California.

Shepherd, J. 1984. The availability and information content of fisheries data. Pp. 95-109 in R. M. May, ed. Exploitation of Marine Communities. Dahlem Konferenzen. Springer, Berlin.

Smith, J. V. C. 1833. A Natural History of the Fishes of Massachusetts; Embracing a Practical Essay on Angling. Allen and Ticknor, Boston.

Walters, C. J., and R. Hilborn. 1976. Adaptive control of fishing systems. J. Fish. Res. Bd. Can. 33:145-159.

13

Vampire Bat Control
in Latin America

To find an effective control agent that affects only the species of concern normally requires knowledge of the life history of the target species. The current control of vampire bats in Latin America rests on the following aspects of the biology of the bats: they are much more susceptible than cattle to the action of anticoagulants, they roost extremely close to each other, they groom each other, their rate of reproduction is low, they do not migrate, and they forage only in the absence of moonlight. These pieces of information were acquired in a search for control methods, and they yielded methods that are effective and that have minimal or no effect on nontarget species. The approach in this case was to find and use weak links in the life history of the animal in question. It also reflects an attempt to save money by comparing the cost and availabilities of various chemicals.

Case Study

G. CLAY MITCHELL, Denver Wildlife Research Center, Denver, Colorado

INTRODUCTION

When the Conquistadors arrived in Mexico in 1527, their horses were immediately attacked by the common vampire bat, *Desmodus rotundus* (Molina Solis, 1896). Over four centuries later, Bernardo Villa (1969) made the following statement after reviewing the vampire bat problem in Brazil: "No notable success has been achieved in controlling bats of the family Desmodontidae, particularly the species *Desmodus rotundus*, in any of the countries of tropical America."

Of the three species of vampire bats, only the common vampire, the subject of this review, is economically damaging. The others, *Diphylla ecaudata* and *Diaemus youngii*, have specialized feeding habits, preferring to feed on the blood of birds (Uieda, 1982; Villa, 1966). Vampire bats range from tropical Mexico to northern Argentina and northern Chile, and they cost the Latin American livestock industry $350 million a year: $100 million in direct losses to rabies and $250 million in secondary losses, such as reduction in milk production and secondary infections (Kverno and Mitchell, 1976). In addition, several humans die each year of rabies transmitted by vampire bats (Baer, 1975; Irons *et al.*, 1957; Venters *et al.*, 1954).

Previous attempts to reduce vampire bat populations in Latin America have been ineffective, dangerous, destructive, impractical, too localized, or too expensive. Methods tried included the use of flame throwers in Trinidad (Greenhall, 1970), the dynamiting of several thousand caves in the State of Grande do Sul, Brazil (Villa, 1969), the placement of a strychnine-syrup mixture at old bite sites (Greenhall, 1963), the gassing of thousands of caves in Latin America (Arteche, 1969), the use of Japanese mist nets (Dalquest, 1954; Greenhall, 1963), and the placement of traps at the entrances of caves (Constantine, 1969). Bats have also been killed with clubs or firearms in stables and dwellings where they attack their prey (Constantine, 1970).

In 1967, under the authority of the Foreign Assistance Act, the U.S. Agency for International Development (AID) asked the U.S. Department of the Interior's Fish and Wildlife Service to conduct research on vertebrate pests (rodents, pest birds, and vampire bats) in developing countries. In

June 1968, the Service's Denver Wildlife Research Center (DWRC) initiated a program of research on the vampire bat problem under the auspices of AID. The program consisted of two phases: (1) research conducted by a laboratory team at DWRC and a field team at the Instituto Nacional de Investigaciones Pecuarias in Palo Alto, Mexico; and (2) a utilization and training phase, developed after species-specific control methods were available.

THE ENVIRONMENTAL PROBLEMS

The main objective of the program was to increase livestock production in Latin America by reducing vampire bat populations. The main environmental problems were how to accomplish this objective without destroying habitat, contaminating the environment with pesticides, destroying desirable species, and adversely affecting an endangered species, the white-winged vampire (*Diaemus youngii*). It was most important to develop species-specific control methods that were effective, inexpensive, and safe and required little training to apply.

APPROACHES TO CONTROLLING VAMPIRE BATS

When this project began, the morphology, physiology, and distribution of vampires (primarily *Desmodus*) were reasonably well known, but little was known of their social structure, behavior, movement, population dynamics, and interspecific relationships, and no techniques were available for determining their relative or absolute abundance (Linhart, 1975). Hence, the proposed initial research was aimed at acquiring some knowledge of the behavior and ecology of vampires.

The following studies were undertaken during the research phase from 1969 to 1973:

- Pharmacological evaluation of chemical agents, such as toxicants and substances that alter physiology or behavior, for controlling vampire bats.
- Laboratory evaluation of techniques for applying control agents, including feeding and grooming behavior of captive vampires, carrier compounds or liquid vehicles for application of control agents to livestock, and the application of control agents directly to livestock.
- Field evaluation of techniques for applying control, including devices for applying carrier agents and their persistence on livestock, feeding behavior of free-ranging vampires, and treatment of cattle with control agents.

• Estimation of actual numbers and repopulation or recovery rates of natural populations of vampire bats.

Additional studies focused on vampire bat movements, sensory mechanisms associated with locating prey, and pharmacological hazards associated with the application of control agents to livestock.

Ecology of Vampire Bats

Research identified two potential weak links in vampires. First, *Desmodus* produce only one or possibly two young each year (Burns, 1970), although each has a potential longevity of 13-14 years (Linhart, 1973; Trapido; 1946). Therefore, population recovery after reduction by control is slow. Second, vampire populations are concentrated near herds of livestock to which they fly nightly (Constantine, 1970).

Vampires do not migrate, although they can move locally during the breeding season (Burns and Flores Crespo, 1975). They forage every night, except when there is a full moon, but only during the darkest part of the night, either before the moon rises or after it sets (Flores Crespo *et al.*, 1972). Finally, vampire bats form mobile communities that use multiple roosts; the number of vampires in a colony might remain the same, but there can be nightly exchange of individuals among roosts (Mitchell *et al.*, 1973). The most practical time to capture bats in mist nets set around corralled cattle is during the week after a full moon, when the period of complete darkness is still short.

Development of Control Methods

Two species-specific control methods for reducing vampire bat populations were developed in this project: the systemic method, in which cattle are treated with a chemical; and the topical method, in which captured vampire bats are treated with the same chemical.

Systemic Method. The idea of treating cattle with systemic toxicants originated at a Communicable Disease Center (CDC) laboratory working on bats and rabies in Las Cruces, New Mexico. Denny G. Constantine, a CDC veterinarian, was told that a Bayer systemic insecticide, Neguvon, would kill vampire bats. He treated cattle with Neguvon, but no vampires that fed on the treated cattle died. Although Neguvon did not kill vampires in Constantine's experiments, other systemic insecticides recommended by the U.S. Department of Agriculture (USDA) Livestock Insects Investigations Laboratory were tested. The recommendations on agents and dosages were based on margins of safety for cattle. These compounds were tested in Mexico, and one of them, Famophos, gave positive results. In the laboratory, cattle were treated with a 13.2% pour-on formulation of Famophos recommended for the control of screwworm larvae and sucking flies. All vampire bats that fed on these treated cattle died (Mitchell *et al.*, 1970, 1971). However, research with Famophos was discontinued, because it was not available in Latin America and it was prohibitively expensive.

When it was shown that vampires could be killed with a systemic insecticide, DWRC scientists began looking at other substances that could be administered to cattle, especially anticoagulants. Because they bind to blood proteins, anticoagulants are excreted by the cow very slowly and are available longer than most other chemicals for the bat to ingest with a blood meal. Bats ingest about 50% of their body weight in blood at each feeding and concentrate it by excreting water (Breidenstein, 1982), so finding a dose of anticoagulant lethal to the bat but not harmful to cattle appeared feasible.

Treating cattle with the anticoagulant actually selected, diphenadione, at a concentration of 1.0 mg/kg gave a wide margin of safety for the cattle and killed all the vampires that fed on blood taken from the treated cattle up to 72 hours after treatment. Diphenadione was not retained in the liver of treated animals, as was another effective anticoagulant, chlorophaci-none (Bullard *et al.*, 1970, 1971).

In 1971, diphenadione was administered with a balling gun to cattle in Mexico. Results were good at Rancho Huichi (90.2% reduction in fresh bites) and poor at Rancho Don Tomas (20% reduction in fresh bites) (Mitchell *et al.*, 1971). The poor results at Rancho Don Tomas were attributed to weather that reduced the normal foraging activities of the bats. In 1972, diphenadione was again field-tested in Mexico, but was administered intraruminally with a syringe; this test reduced biting by 93% (Thompson *et al.*, 1972).

Topical Method.   The idea of topical treatment was based on the like-lihood that bats would ingest a toxicant while grooming. Similar behavior is a basis of common methods of rat control. Greenhall (1965) reported that vampires are thorough groomers and spend a considerable amount of time in this activity. Captive vampires were observed to spend some 2 hours per day grooming (Flores Crespo *et al.*, 1971a)—enough time for a control compound on the fur to be ingested.

To determine whether there would be sufficient contact between the vampire and its prey for the vampire to receive the chemical from a cow's skin and whether the vampire has a preferred biting area, 49 feedings were observed in a bat-proof corral under laboratory conditions. Of the 49 bites, 27 were made in the hoof region while the cattle were standing, and 22 on various parts of the body while they were lying down; moreover, each feeding entailed minimal contact between the vampire and its prey (Flores Crespo *et al.*, 1971b). As a followup to the laboratory study, vampires were observed with Starlight night-vision telescopes, which electronically magnify available light, while they fed under natural conditions. Vampires were observed while feeding under normal conditions on three races of cattle—Brahma, Charolais, and Holstein. Approximately one-third of the bats landed on the Brahma before biting and feeding, and 31% on the Charolais, but 85% landed directly on the Holstein. Only on the Holstein was there a preferred area for biting: 74.5% of the bites were on the neck (Flores Crespo *et al.*, 1974).

Almost all cattle in the geographical range of vampire bats are Brahma. Because the Brahma are on open range and because only 21% of the vampires that fed on Brahma cattle fed on their necks, the idea of applying a control compound to an appropriate area of the prey animal was discarded as impractical.

An alternative topical method would be to apply the toxicant to the bat itself. While writing a manuscript on grooming, DWRC biologist Samuel B. Linhart recognized that vampires are extensive groomers and roost in compact groups separate from beneficial bats in the same cave. Hence, a substance placed on the fur of a captured bat might be passed to other bats in the colony once the treated bat returned to the roost. These ideas led Linhart to develop the topical method now in use.

On the basis of the recommendations of DWRC specialists, four chemicals were tested for toxicity to vampire bats. Chlorophacinone, an anticoagulant, was the most toxic. Ten carriers were evaluated by mixing them with a dye and applying them to the dorsal surface of vampire bats. Bats were checked several times a day to determine the persistence of the carriers. Three carriers were selected for additional study: a saturated solution of acetone and an acrylic polyester fluorescent paint pigment; petroleum jelly; and a mixture of abalyn (a resin), mineral oil, and polyethylene (Epoline-10). For each candidate carrier, one vampire was treated with a mixture of carrier and a dye and then introduced into a simulated roost containing 19 other vampires. The roost was checked several times a day to determine the degree of transfer among colony members by grooming. Petroleum jelly was superior to the other carriers.

In the final test with the simulated roost, 50 mg of chlorophacinone was mixed with 1.5 ml of petroleum jelly and applied to the dorsal surface of one vampire. This vampire was put into the roost with 19 other bats. The carrier bat was found dead on the morning after reintroduction into the roost, and 18 of the 19 other bats died 5-19 days after reintroduction (Linhart, 1970).

Several field tests confirmed these results. Vampire bats captured in two caves were treated with a mixture of 50 mg of chlorophacinone and 1.5 ml of petroleum jelly. Six vampires captured in one cave were treated and released; later, 94 dead vampires were found—a ratio of 15-16 vampires killed to each one treated. After 2 weeks, the caves contained only one live vampire (Linhart *et al.*, 1972). In another test, bats were captured in mist nets as they flew in to feed on corralled cattle on two ranches, were treated with the control mixture, and were released. Two weeks later, vampire bites on the same cattle had decreased by 95% (Linhart *et al.*, 1972).

At first, all vampires captured were treated with the chlorophacinone-petroleum jelly mixture. Later, it was determined that one treated bat would kill 19 others in the roost. There is a high correlation between the number of fresh bites on members of a herd and the number of vampires feeding on the herd. If a herd had 100 fresh bites, the vampire population was estimated to be 100 and only five bats were treated.

There were two reasons for changing the control chemical from chlorophacinone to diphenadione. First, other anticoagulants were as effective as chlorophacinone in controlling vampire bat populations with the topical method. Second, diphenadione was available for manufacture, because the patent had expired. The use patent is held by the U.S. government, which gives permission to anyone to use the chemical as a vampiricidal agent.

Although the topical method is more difficult to use than the systemic method, because it requires night work and the proper identification and treatment of only vampires, it is sometimes the preferred method. Injection of the anticoagulant into the rumen is the easiest way to administer it; but, of the animals commonly attacked by vampires, only cattle have a rumen and can be treated this way. If other livestock or humans are being attacked, the topical method is recommended.

Another control method, application of anticoagulants to roost walls, was evaluated in a pilot study in Mexico. Diphenadione in petroleum jelly was placed at vampire roost sites in a tunnel and in a mine (Mitchell *et al.*, 1971, 1972). In both cases, all the vampires were killed. Four non-hematophagous bats also died in the tunnel. In view of these results, control of vampire bats by treating roosts was not recommended.

Control Techniques

Systemic Treatment of Cattle with Diphenadione. Diphenadione is injected into cattle intraruminally at 1.0 mg/kg of body weight. The drug is absorbed and circulates in the blood. Any vampire bat that feeds on a properly treated animal within 72 hours after treatment receives a lethal dose of the drug. Treating all cattle in a herd reduces biting by 90-95% (Thompson et al., 1972).

Topical Treatment of Vampire Bats with Diphenadione. Vampire bats are captured with mist nets that are set around corralled cattle or at cave entrances. Approximately 1.5 ml of a diphenadione-petroleum jelly mixture is placed on the dorsal surface of each captured bat, and the bat is released. The bats return to their roost, and, because they live in compact colonies, pass the chemical from one to another. The bats die after ingesting the chemical during grooming (Linhart et al., 1972). For every treated bat, approximately 20 vampires die at the roost.

Vaccines

Although topical or systemic use of anticoagulants can reduce the number of bites by up to 96%, and thus reduce secondary losses by a similar amount, rabies can still occur among unvaccinated cattle even when the bat population is reduced. Therefore, both cattle vaccination and bat population control are recommended if the aim is to eliminate the disease, rather than only to reduce cattle losses to an acceptable point (Lord, 1980).

Piccinini (1977) used both vaccines and anticoagulants in a 500-km^2 area in the State of Pernambuco, Brazil. In the 2 years before control, 17,870 cattle were in the study area, 2,342 were not protected by vaccines, and 140 died of rabies. After application of both control methods, the vampire population was reduced by 96.1%. In the 2 years after control, 17,431 cattle were in the area, 4,848 were not vaccinated, and only 2 died of rabies. This shows that reduction of the vampire population gives cattle some protection from rabies without the use of vaccines. It should be noted that the treatments in Piccinini's study were made by scientists; the results might be less impressive under normal operational field conditions.

KINDS OF ECOLOGICAL KNOWLEDGE USED

Ecological Facts

Ecological and physiological facts were obtained from published literature and from the observations of experienced people in the field. Experiments were done in simulated and natural situations and yielded much knowledge of the life history and behavior of vampire bats. The ecological observations provided the basis of the control programs.

Pilot Studies

The studies of topical anticoagulants in simulated roosts constitute an example of the successful use of analog studies, and the pilot tests of both systemic and topical anticoagulants were important in the development of practical control programs. The studies did not all lead directly to successful control, but the ones that "failed" enabled workers to avoid expensive and ineffective methods.

Project as Experiment

Experience gained in the utilization phase led to improvements in the procedures. During that phase, one problem was encountered in Nicaragua, where 9 of 14 calves died after treatment with diphenadione (1.0 mg/kg). These calves were less than 3 months old and lacked a fully functional rumen. Further tests showed that calves without a functioning rumen should not be treated (Elias *et al.*, 1978).

UTILIZATION AND TRAINING

From 1974 to 1978, AID funded DWRC biologists to develop a utilization or training phase for vampire bat control. It wanted DWRC to take the control methods to Latin American cattlemen and to continue adaptive research to answer questions that could arise. Initially, DWRC biologists conducted training seminars in Latin America, often involving teams of veterinarians organized to vaccinate cattle. The basic training tool was a pamphlet, *Chemical Control of Vampire Bats*, published in Spanish, English, and Portuguese (Mitchell and Burns, 1973a,b, 1978). Later, the Peace Corps in Belize (Mitchell *et al.*, 1975, 1976) and the Pan American Health Organization in Venezuela, Trinidad, and Surinam (Mitchell *et al.*, 1974) became involved. In-country training also occurred. For example, 125 veterinarians from 19 states were trained in Brazil (Mitchell *et al.*, 1976). In 1976, Rodrigo Gonzales presented results of the campaign in Nicaragua at an international conference (Gonzales and Mitchell, 1976), and the Nicaraguan government hosted a seminar on vampire bat control, attended by 46 people from 11 countries (Mitchell *et al.*, 1976).

EFFECTIVENESS OF THE PROGRAM

After the publication of our control methods, new questions arose. The possibility of hazards to other bats from use of the topical method was suggested (Turner, 1975). Killing of nonhematophagous bats where vampires have been controlled topically has not been reported, and none was killed in the many field tests conducted. Brazilian workers controlled vampires in 3,062 caves and never found a nonhematophagous bat killed by the topical method (Mitchell *et al.*, 1976).

The topical method was field-tested in every Latin American country exposed to vampire predation. Results were consistent: 90-95% reduction in the vampire population. The systemic method of treating cattle with diphenadione is even more specific in that it kills only the vampires that are attacking cattle. If cattle are treated with the recommended dosage, 1.0 mg/kg, no chemical is passed in the milk, and residues in tissue are negligible (Bullard and Thompson, 1977; Bullard *et al.*, 1976, 1977). With sound management and husbandry practices in Ecuador, the killing of vampire bats had no effect on milk production. In Nicaragua, under more tropical conditions, milk production increased by 16% (Thompson *et al.*, 1977).

These methods of vampire bat control are very effective even when applied by nonprofessionals. For example, in Nicaragua, 20 high-school graduates were trained to apply the control methods over 4 years. They examined 270,825 cattle on 2,124 ranches and found 123,376 fresh vampire bites. They treated 148,142 of the cattle and 2,696 vampires and reduced fresh bites by 90% to about 12,000. The annual benefit to farmers from the vampire bat control program in Nicaragua was US$2,414,158, and annual costs, $129,750, for a very favorable benefit-to-cost ratio of about 18.6:1 (Badger and Schmidt, 1979).

ACKNOWLEDGMENTS

Although I prepared this case study, the solution to the problem is based on an accumulation of many scientific experiments conducted by a host of scientists. Much of the work was done by DWRC scientists in an array of disciplines, including wildlife biology, pharmacology, animal psychology, statistics, physiology, electronics, chemistry, nutrition, and ecology. Undoubtedly, this multidisciplinary approach was influential in solving the problem.

Thanks go to Bernardo Villa, Master Zoologist at the University of Mexico, for sharing his vast knowledge of bats.

I also wish to give special thanks to Charles Ladenheim (AID retired), who had the insight to support the transition from a research project to an operational program. Research cannot be considered accomplished or successful until its results are made known and implemented—in this case, by the small producers of livestock in Latin America.

Funds for this research were provided to DWRC by AID under Participating Agency Service Agreements RA (ID) 01-67 and ID-TAB-000-10-76.

Reference to trade names in this review does not imply U.S. government endorsement.

REFERENCES

Arteche, E. 1969. Rabia en Rio Grande do Sul, Brazil. Zoonosis 11:34-35.

Badger, D. D., and K. M. Schmidt. 1979. Evaluation of the Vampire Bat Control Program. Department of Agricultural Economics, Oklahoma State University, Stillwater, Okla. Available from U.S. Agency for International Development, Washington, D.C.

Baer, G. M. 1975. Rabies in nonhematophagous bats. Pp. 79-98 in G. M. Baer, ed. The Natural History of Rabies. Vol. 2. Academic Press, New York.

Breidenstein, C. P. 1982. Digestion and assimilation of bovine blood by a vampire bat (*Desmodus rotundus*). J. Mammal. 63:482-484.

Bullard, R. W., and R. D. Thompson. 1977. Efficacy and safety of the systemic method of vampire bat control. Interciencia 2:149-152.

Bullard, R. W., S. R. Kilburn, and G. Holguin. 1970. Effectiveness of chlorophacinone (DRC-3776) as a livestock systemic for control of vampire bats. Pp. 33-34 in 1970 Annual Report, Denver Wildlife Research Center, Denver, Colo.

Bullard, R. W., S. R. Kilburn, and G. Holguin. 1971. Livestock systemic technique for control of vampire bats (*Desmodus rotundus*). Pp. 36-41 in 1971 Annual Report, Denver Wildlife Research Center, Denver, Colo.

Bullard, R. W., R. D. Thompson, and G. Holguin. 1976. Diphenadione residues in tissues of cattle. J. Agric. Food Chem. 24:261-263.

Bullard, R. W., R. D. Thompson, and S. R. Kilburn. 1977. Diphenadione residues in milk of cattle. J. Agric. Food Chem. 25:79-81.

Burns, R. J. 1970. Twin vampire bats born in captivity. J. Mammal. 51:391-392.

Burns, R. J., and R. Flores Crespo. 1975. Local movement and reproduction of common vampire bats in Colima, Mexico. Southwest. Nat. 19:446-449.

Constantine, D. G. 1969. Trampa portatil para vampiros usada en programas de campana antirrabica. Bol. Of. Sanit. Panam. 67:39-42.

Constantine, D. G. 1970. Bats in relation to health, welfare, and economy of man. Pp. 319-449 in W. A. Wimsatt, ed. Biology of Bats. Vol. 2. Academic Press, New York.

Dalquest, W. W. 1954. Netting bats in tropical Mexico. Trans. Kans. Acad. Sci. 57:1-10.

Elias, D. J., R. D. Thompson, and P. J. Savarie. 1978. Effects of the anticoagulant diphenadione on suckling calves. Bull. Environ. Contam. Toxicol. 20:71-78.

Flores Crespo, R., S. B. Linhart, and R. J. Burns. 1971a. Comportamiento del vampiro (*Desmodus rotundus*) en cautiverio. Southwest. Nat. 17:139-143.

Flores Crespo, R., R. J. Burns, and S. B. Linhart. 1971b. Comportamiento del vampiro (*Desmodus rotundus*) durante su alimentacion en ganado bovino en cautiverio. Tecnica Pecuaria en Mexico 18:40-44.

Flores Crespo, R., S. B. Linhart, R. J. Burns, and G. C. Mitchell. 1972. Foraging behavior of the common vampire bat related to moonlight. J. Mammal. 53:366-368.

Flores Crespo, R., S. Said Fernandez, R. J. Burns, and G. C. Mitchell. 1974. Observaciones sobre el comportamiento del vampiro comun (*Desmodus rotundus*) al alimentarse en condiciones naturales. Tecnica Pecuaria en Mexico 27:39-45.

Gonzales, R., and G. C. Mitchell. 1976. Vampire bat control programs in Latin America. Proc. Vertebr. Pest Control Conf. 7:254-257.

Greenhall, A. M. 1963. Use of mist nets and strychnine for vampire bat control in Trinidad. J. Mammal. 44:396-399.

Greenhall, A. M. 1965. Notes on behavior of captive vampire bats. Mammalia 29:441-451.

Greenhall, A. M. 1970. Vampire bat control: A review and proposed research programme for Latin America. Proc. Vertebr. Pest Control Conf. 4:41-54.

Irons, G. V., R. B. Eads, J. E. Grimes, and A. Conklin. 1957. The public health importance of bats. Tex. Rep. Biol. Med. 15:292-298.

Kverno, N. B., and G. C. Mitchell. 1976. Vampire bats and their effect on cattle production in Latin America. World Anim. Rev. 17:1-7.

Linhart, S. B. 1970. Work Unit DF-101.1: Basic Research in Mammal and Bird Damage Control: Chemicals. 1970 Annual Progress Report, Denver Wildlife Research Center, Denver, Colo.

Linhart, S. B. 1973. Age determination and occurrence of incremental growth lines in the dental cementum of the common vampire bat (*Desmodus rotundus*). J. Mammal. 54:493-496.

Linhart, S. B. 1975. The biology and control of vampire bats. Pp. 221-241 in G. M. Baer, ed. The Natural History of Rabies. Vol. 2. Academic Press, New York.

Linhart, S. B., R. Flores Crespo, and G. C. Mitchell. 1972. Control of vampire bats by topical application of an anticoagulant, chlorophacinone. Bol. Of. Sanit. Panam. 6:31-38.

Lord, R. D. 1980. An ecological strategy for controlling bovine rabies through elimination of vampire bats. Proc. Vertebr. Pest Control Conf. 9:170-175.

Mitchell, G. C., and R. J. Burns. 1973a. Combate Quimico de los Murcielagos Vampiros. Regional Technical Assistance Center, U.S. Embassy, Mexico City, Mex.

Mitchell, G. C., and R. J. Burns. 1973b. Chemical Control of Vampire Bats. Denver Wildlife Research Center, Denver, Colo. (mimeo)

Mitchell, G. C., and R. J. Burns. 1978. Combate Quimico aos Morcegos Hematofagos. Ministry of Agriculture, EMBRAPA, Coronel Pacheco, Braz.

Mitchell, G. C., R. Flores Crespo, R. J. Burns, S. Said Fernandez, and S. B. Linhart. 1970. Vampire Bats: Rabies Transmission and Livestock Production in Latin America. 1970 Annual Report of the Denver Wildlife Research Center's Mexican Field Station, Palo Alto, Mex.

Mitchell, G. C., R. Flores Crespo, R. J. Burns, and S. Said Fernandez. 1971. Vampire Bats: Rabies Transmission and Livestock Production in Latin America. 1971 Annual Report of the Denver Wildlife Research Center's Mexican Field Station, Palo Alto, Mex.

Mitchell, G. C., R. Flores Crespo, R. J. Burns, and S. Said Fernandez. 1972. Vampire Bats: Rabies Transmission and Livestock Production in Latin America. 1972 Annual Report of the Denver Wildlife Research Center's Mexican Field Station, Palo Alto, Mex.

Mitchell, G. C., R. J. Burns, and A. L. Kolz. 1973. Rastreo del comportamiento nocturno de los murcielagos vampiros por radio-telemetria. Tecnica Pecuaria en Mexico 24:47-56.

Mitchell, G. C., R. J. Burns, J. W. De Grazio, and D. J. Elias. 1974. Vampire Bats:

Rabies Transmission and Livestock Production in Latin America. 1974 Annual Report. Denver Wildlife Research Center, Denver, Colo.

Mitchell, G. C., R. D. Thompson, D. J. Elias, C. E. Shuart, R. W. Bullard, P. J. Savarie, and R. J. Burns. 1975. Vampire Bats: Rabies Transmission and Livestock Production in Latin America. 1975 Annual Report. Denver Wildlife Research Center, Denver, Colo.

Mitchell, G. C., R. D. Thompson, D. J. Elias, C. E. Shuart, and R. W. Bullard. 1976. Vampire Bats: Rabies Transmission and Livestock Production in Latin America. 1976 Annual Report. Denver Wildlife Research Center, Denver, Colo.

Molina Solis, J. F. 1896. Historia del Descubrimiento y Conquista de Yucatan, con una Resena de la Historia Antigua de la Peninsula. R. Caballero, Merida, Mex.

Piccinini, R. S. 1977. The Use of Diphenadione (2-Diphenylacetyl-1,3 indandione) for Vampire Bat Control in Endemic Rabies Areas, Northeastern Brazil. Master's thesis, University of California, Davis.

Thompson, R. D., G. C. Mitchell, and R. J. Burns. 1972. Vampire bat control by systemic treatment of livestock with an anticoagulant. Science 177:806-808.

Thompson, R. D., D. J. Elias, and G. C. Mitchell. 1977. Effects of vampire bat control on bovine milk production. J. Wildl. Manage. 41:736-739.

Trapido, H. 1946. Observations on the vampire bat with special reference to longevity in captivity. J. Mammal. 27:217-219.

Turner, D. C. 1975. The Vampire Bat. John Hopkins University Press, Baltimore.

Uieda, W. 1982. Aspectos do Comportamento Alimentar das Tres Especies do Morcegos Hematofagos (Chiroptera, Phyllostomidae). Master's thesis, Instituto de Biologia da Universidade Estadual de Campinas, Braz.

Venters, H. D., W. R. Hoffert, J. E. Scatterday, and A. V. Hardy. 1954. Rabies in bats in Florida. Am. J. Public Health 44:182-185.

Villa, R. B. 1966. Los Murcielagos de Mexico. Institute of Biology, National Autonomous University, Mexico City, Mex.

Villa, R. B. 1969. The ecology and biology of vampire bats and their relationship to paralytic rabies. Report to the Government of Brazil. Rep. No. TA 2656. U.N. Development Programme/Food and Agriculture Organization. Rome.

Committee Comment

The control of pests is a common theme in humans' attempts to live in a world populated with animals and plants, and that theme is reflected in several of the cases discussed in this report (Chapters 14, 15, and 24). Many serious pests are controlled only poorly, despite great expense; those controlled successfully usually have idiosyncrasies of life history that make them vulnerable, and successful control depends on detailed knowledge of those idiosyncrasies.

Early attempts to control vampires did not rely on knowledge of the animals' natural history; they were ineffective and destructive. The resulting careful and thorough research into the natural history of vampires by the Denver Wildlife Research Center (DWRC) exemplifies an approach that is usually needed for pest control; the idiosyncrasies discovered represent a certain amount of the luck that so often seems to accompany successful cases.

DWRC biologists studied the behavior, dispersal, physiology, and ecology of vampires, using field and laboratory experiments and observations. These studies and literature review revealed important aspects of vampires' natural history that made them susceptible to control. Two of these aspects commonly help in pest control: the bats have a low reproductive rate, which meant that less powerful control would be needed to keep their numbers down, and they do not migrate.

Successful control also depended on several idiosyncrasies. The bats, because of their diet of blood, are much more susceptible to anticoagulants than cattle are. They forage only on dark nights, so when the moon rises shortly after sunset or sets shortly before sunrise, their activity is confined to short periods, and that makes them easier to trap. They roost extremely close to each other and they groom each other—both characteristics that favor transfer of vampiricides between individuals and that make it unnecessary to treat more than a few members of each roost.

Thus, successful control in this case depended on thorough research into the pests' natural history, which is nearly always needed, and on the vulnerabilities that the research revealed. The vulnerable points might not exist, but in the absence of thorough research they will almost surely not be found.

14

Biological Control of
California Red Scale

Interest in biological control of agricultural pests has been stimulated by the evolution of resistance of the target organisms to pesticides, by secondary outbreaks caused by losses of natural predators and parasites, and by the increasing seriousness of unwanted side effects of pesticides on many species, including humans. However, biological control is not yet a highly predictive branch of ecology. Most biological control efforts fail, either because the control agent does not survive or because, if it survives, it does not result in satisfactory control. The California red scale study illustrates features common to many instances of biological control, but it also shows how imaginative use of some new elements of ecological theory, such as those related to foraging and sex allocation, led to changes in project organization that illuminated the processes underlying effective biological control.

Case Study

ROBERT F. LUCK, Department of Entomology, University of California, Riverside

INTRODUCTION

California red scale, *Aonidiella aurantii* (Maskell) (Diaspididae: Homoptera), is an insect pest of citrus in arid and semiarid regions (Bodenheimer, 1951; Ebeling, 1959). It inhabits all aboveground parts of a citrus plant. At moderate densities, the scale infests the fruit, which might then be culled in accordance with industry standards for appearance. At higher densities, it inhibits fruit production and kills branches. In California, it is one of three principal arthropod pests of citrus and infests all four major cultivars grown in the state: grapefruit, lemons, and two orange varieties.

California red scale was introduced into southern California between 1868 and 1875 on shipments of citrus nursery stock from Australia (Quayle, 1938). *Lindorus lophanthae* (Blaisd.) (Coccinellidae: Coleoptera), a predatory ladybird beetle introduced from Australia, was the first of 52 predators and parasites (36 and 16, respectively) introduced against California red scale. Of the eight that became established, three were predators—*L. lophanthae*, introduced in 1889; *Orcus chalybeus* (Bvdl.), in 1892; and *Chilocorus similis*, in 1924-1925—and five were parasitoids—*Comperiella bifasciata* Howard, in 1940; *Habrolepis rouxi* Compere, in 1937-1939; *Aphytis lingnanensis* Compere, in 1948; *Encarsia* (= *Prospaltella*) *perniciosi* Tower, in 1949; and *A. melinus* DeBach, in 1956-1957 (Rosen and DeBach, 1978). *Aphytis chrysomphali* (Mercet), a thelytokous (parthenogenetic, mostly female-producing) parasitoid, was unknowingly introduced around 1900, probably in parasitized scale on infested nursery stock from the Mediterranean Basin (Rosen and DeBach, 1978, 1979).

Before the introduction of the wasp *Aphytis lingnanensis* (Aphelinidae: Hymenoptera), in 1947, California red scale was economically controlled by *A. chrysomphali* in the Santa Barbara, California, area (DeBach and Sisojevic, 1960), perhaps with some help from ladybird beetles, so long as ants, dust, and insecticide interference were minimized. The scale remained a problem, however, in inland coastal valleys and other coastal areas, even though *A. chrysomphali* was found throughout these regions. When *A. lingnanensis* was introduced, biological control was achieved in

166

most coastal areas. But only with the introduction of *A. melinus* in 1956-1957 was biological control achieved in inland valleys (Rosen and DeBach, 1979).

In contrast with the situation in southern California, economical biological control of California red scale does not exist in the San Joaquin Valley or the desert, even though both *Comperiella bifasciata* (Encyrtidae: Hymenoptera) and *A. melinus* are present. Remaining natural enemy species, both introduced and native, are scarce. Moreover, citrus production in the San Joaquin Valley is assuming increasing importance as that in southern California declines with increasing urbanization. In the absence of economic biological control of scale, growers have relied on synthetic organic insecticides (Flaherty *et al.*, 1973).

THE ENVIRONMENTAL PROBLEM

Synthetic organic insecticides have several drawbacks. Their initial effectiveness, convenience, and low cost all promote their use (Luck *et al.*, 1977). But these materials are general in their effects, so they also kill native predators and parasites of the target pests, as well as the pests themselves, and sometimes cause outbreaks of the target pests and secondary pests (phytophagous arthropods normally held at bay by their natural enemies). Pesticide applications increase the nutritional value of plants for some prey species (Dittrich *et al.*, 1974; Jones and Parrella, 1984; Maggi and Leigh, 1983). Pest resurgence and secondary-pest outbreaks promote further pesticide use, and the grower becomes enmeshed in a pesticide treadmill in which still further pesticide use increases control costs and selects for resistance in primary or secondary pest species (Luck *et al.*, 1977).

Citrus is one of many crops that have followed this scenario (Luck *et al.*, 1977). California red scale populations in South Africa, Israel, and Lebanon have evolved resistance to one or more of the synthetic organic scalicides (Georghiou and Taylor, 1977). California populations have evolved resistance to hydrogen cyanide gas, used as a fumigant before World War II. Secondary-pest outbreaks have also occurred in citrus. Since the 1960s, several Lepidoptera species and citrus red mite, *Panonychus citri* McG. (Tetranichidae), have become increasingly frequent pests in California citrus because of insecticide use (Luck *et al.*, 1977). Citrus red mite has also evolved resistance to many acaricides used against it (Georghiou and Taylor, 1977).

ECOLOGICAL APPROACH TO THE
ENVIRONMENTAL PROBLEM

We have had two objectives in our studies of California red scale, its host plants, and its natural enemies: the specific objective of improving the biological control of the scale, especially in the San Joaquin Valley, and the general objective of understanding how biological control of red scale works and what limits its success. The latter objective presupposes that understanding the reasons for success and failure and the associated processes and mechanisms in a specific case will improve our ability to predict the outcome both of this particular natural enemy-pest interaction and of such interactions in general.

The specific objective motivates a continuing biological control project, whose key components are the obtaining and manipulation of natural enemies of the scale, the importation of exotic predator and parasitoid species reared from California red scale or related species around the world, and the mass rearing and release of endemic and exotic natural enemies. We seek by these means to develop practical and economical ways of augmenting endemic natural enemy populations by releasing insectary-reared individuals. The duration of this project depends on the discovery of new species of natural enemies, on political conditions in the geographical region where we wish to collect natural enemies, and on the degree of biological control achieved by introducing enemies. Importation and release of natural enemies would cease once effective biological control were achieved throughout California.

We have pursued the general objective of using California red scale as a model system through a combination of field and laboratory studies to identify attributes and processes that characterize a successful biological control project. Laboratory studies have sought to understand the host selection and foraging behavior of the several natural enemies of California red scale. The results suggested hypotheses to be tested with field populations of the scale. In addition to field tests of hypotheses (e.g., augmentative releases), comparative demographic studies of red scale populations were begun in three of California's four citrus regions; these studies sought to evaluate the effect of natural enemies on the dynamics of red scale populations (cf. Rosen and DeBach, 1978) and to provide a baseline for assessing new introductions.

USES OF KNOWLEDGE AND UNDERSTANDING

Ecosystem Definition

The ecosystem components of interest in these studies are several citrus cultivars grown in commercial and dooryard plantings in the state, honeydew-producing insects and ant species that tend them, natural enemies that attack California red scale and its primary natural enemies, California red scale, and climates in the three citrus regions. Also of interest are associated phytophagous arthropods that receive frequent or occasional insecticide treatments, whose consequences can affect red scale or its interaction with its natural enemies. Chemical control of red scale might also affect densities of other phytophagous species or their natural enemies, including the citrus thrips, *Scirtothrips citri* (Thripidae: Thysanoptera), citrus red mite, and several lepidopterans (Luck, 1984). With increasing information and understanding, we might also need to include other components of the ecosystem, such as citrus nematode, phytophthora root rot, or *Tristeza* (seedling yellows), because they and their treatment influence the physiology of the tree or otherwise affect the interaction between red scale and its entomophages. The ecosystem components named earlier were identified because they were economically important or because previous research with citrus, with other agricultural ecosystems, or with natural ecosystems suggested that these components were biologically important.

Significance of Impact

If biological control of California red scale is achieved, chemical control will be required only infrequently when weather patterns, dust, insecticides (drifting from adjacent crops from treatments of an associated pest), or ants interfere with the scale's natural enemies and thus induce increases in local scale populations. Biological control should reduce red scale population densities, pest-control costs, and pesticide-related environmental and health problems. These expectations are based on experience with biological control of pests in general and of red scale in both southern California (Rosen and DeBach, 1978, 1979) and other citrus-growing regions, such as Greece (DeBach and Argyriou, 1967), Australia (Furness et al., 1983), South Africa (Bedford and Georgala, 1978), and Cyprus (Orphanides, 1984).

Study Strategy and Monitoring

Our first specific goal associated with the second objective was to understand why *Aphytis melinus* replaced *A. lingnanensis* in southern California and what limits the degree of parasitism achieved by *A. lingnanensis*. *A. lingnanensis* was first introduced into California in 1947. It was not the first *Aphytis* present on red scale in California, however; *A. chrysomphali* had apparently been introduced before 1900 (Rosen and DeBach, 1979). *A. lingnanensis* replaced *A. chrysomphali*, probably because it was better adapted to the climate of California's citrus districts (DeBach and Sisojevic, 1960). However, *A. lingnanensis* failed to achieve economic scale control in the more interior citrus districts, so additional natural enemies were sought. One of these, *A. melinus*, was established in the interior in 1956-1957 and quickly spread throughout the area, displacing *A. lingnanensis*. *A melinus* provided economic scale control in much of southern California. In the late 1970s, the three species were distributed as follows: *A. melinus* in the interior citrus areas, *A. lingnanensis* along the coast and inland approximately 10 km, and *A. chrysomphali* as a relict in a few coastal enclaves (Rosen and DeBach, 1979).

DeBach and his co-workers monitored establishment and spread of these introduced *Aphytis* species by sampling red scale populations in many places throughout southern California (DeBach, 1965). DeBach noticed *A. melinus* displacing *A. lingnanensis*. In some groves, the displacement occurred even when third-stage female scale, the putative host for the two *Aphytis* species, was abundant. DeBach concluded that hosts for the parasitoids were not limiting and that competition for scarce hosts could not explain the displacement (DeBach, 1966; DeBach and Sundby, 1963; Rosen and DeBach, 1979). Laboratory experiments designed to test this hypothesis, however, contradicted the field observation: *A. lingnanensis* displaced *A. melinus* in the laboratory—the opposite of what happened in the field (DeBach and Sundby, 1963). DeBach then suggested that *A. melinus* performed better in the field because it was both a better searcher and better adapted to California's climate than *A. lingnanensis*.

Established Boundaries

For demographic studies, we used the citrus tree as a sampling universe. Such a universe is reasonable for characterizing an interaction between red scale and its natural enemies because red scale is mobile only during the crawler stage (first instar) and, in the case of male scale, briefly (less than 24 hours) as an adult while seeking mates. Between-tree movement depends on wind dispersal of crawlers and sometimes on crawlers that

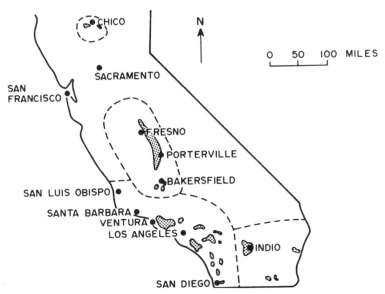

FIGURE 1   Climatic zones in which California citrus is grown.

walk between contiguous trees. Adult natural enemies can easily fly between trees, but we understand little of their behavior or patterns of movement. However, because we wanted to know the effects of natural enemies on the red scale population of a tree, it seemed reasonable, at least initially, to consider the natural enemies on a within-tree basis also.

For our studies of economic effects, we again used the tree as a universe, but we were also concerned with the economic effects of biological control. Thus, we chose "blocks" (management units of citrus groves) of about 2.5 hectares as our economic universe.

We limited our study to California citrus groves and dooryard citrus. We divided the state's citrus growing areas into zones (Figure 1) on the basis of climate and industry experience. We used the differences between zones as manipulations with which to compare changes in the scale's age structure caused by seasonal temperature patterns and the influence of these changes on control of the scale.

Developing and Implementing a Study Strategy

We first tested DeBach's assertion that *A. melinus* and *A. lingnanensis* used the same host resource (scale stage and size) (DeBach, 1966; DeBach and Sundby, 1963; Rosen and DeBach, 1979) by characterizing the oviposition behaviors of the two species. We used these behaviors to see what the parasitoids recognized as hosts and how many eggs of what gender they chose to allocate to those hosts (male, female, or mixed gender). An *Aphytis* egg is laid on the host's body beneath a scale cover where it cannot be seen, but the laying of an egg is reflected in obvious oviposition behavior (Luck *et al.*, 1982) (Figure 2).

We next determined what size and stage of scale each wasp species used as hosts for its progeny. Most hymenopterans, including *A. melinus* and *A. lingnanensis*, have diploid females and haploid males. Thus, the parental female determines the gender of her offspring by "deciding" whether or not to fertilize the egg. Some hymenopterans lay male eggs on small hosts and female eggs on large hosts (Charnov, 1982; Charnov *et al.*, 1981; Chewyreuv, 1913; Clausen, 1939). This behavior characterizes both *A. melinus* (Figure 3) and *A. lingnanensis* (Figure 4), although *A. melinus* allocates sons and daughters to smaller scale than does *A. lingnanensis* (Luck and Podoler, 1985). We hypothesized that *A. melinus* displaced *A. lingnanensis* because it usurped a substantial proportion of the scale population before scale attained the size that *A. lingnanensis* required for its offspring—especially the size needed to produce daughters. Both *Aphytis* species paralyze the scale before they deposit eggs on it, thus preventing further scale growth. As a scale grows, it attains the size used by *A. melinus* first. Use of the scale by *A. melinus* then eliminates the scale as a potential resource for *A. lingnanensis*. Each species tends to avoid depositing eggs on scale previously parasitized by the other (Luck and Podoler, 1985). Thus, the initial laboratory studies on host selection by *Aphytis* led us to propose a hypothesis to explain how *A. melinus* displaced *A. lingnanensis* in southern California. Some of the predictions derived from this hypothesis were testable in the field.

Specific Predictions and Hypotheses

If our hypothesis is correct, *A. melinus* should eventually eliminate *A. lingnanensis* from southern California. Rosen and DeBach (1979) found that in 1965 *A. melinus* occurred in the interior coastal valleys and that *A. lingnanensis* occurred along the coast to a point about 10 km inland. By 1972, *A. melinus* had spread coastally while the area populated by *A. lingnanensis* had contracted (Luck and Podoler, 1985). In 1983, *A. lingnanensis* was restricted to a few locations, all within a kilometer or so of the coast (R. F. Luck and S. Warner, unpublished data).

The age structure of California red scale varies seasonally with location in California. Lower winter temperatures inhibit crawler production by

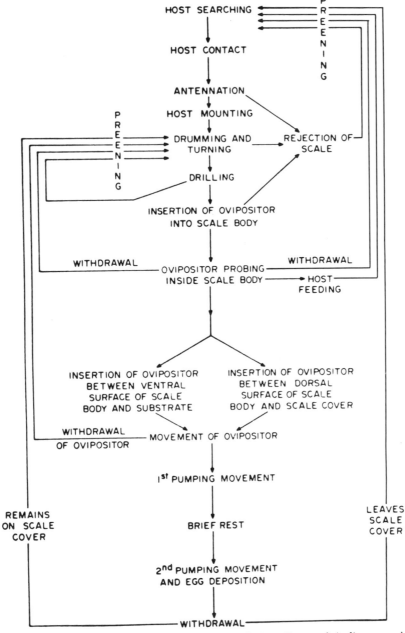

FIGURE 2 Behavioral sequence exhibited by *Aphytis melinus* and *A. lingnanensis* during oviposition.

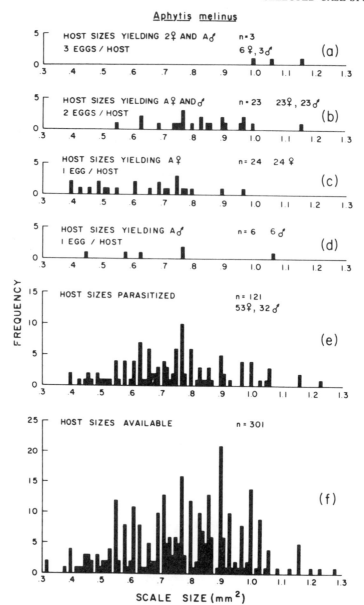

FIGURE 3   a-d, size distribution of third-stage California red scale on which *Aphytis melinus* laid indicated numbers and sexes of eggs. e-f, size distribution of scale chosen by and offered to *A. melinus* for parasitization.

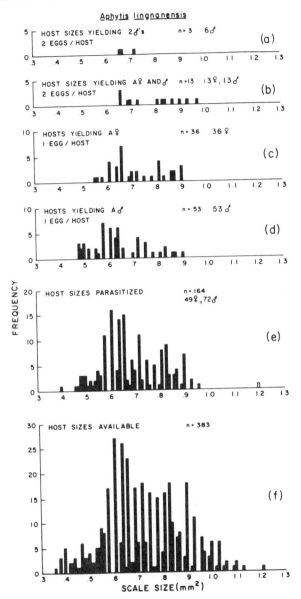

FIGURE 4   a-d, size distribution of third-stage California red scale on which *Aphytis lingnanensis* laid indicated numbers and sexes of eggs. e-f, size distribution of scale chosen by and offered to *A. lingnanensis* for parasitization.

gravid and parturient female red scale. They also inhibit male scale emergence; hence, virgin (third-instar) female scale remain uninseminated. First-instar and first-molt scale (those molting to second instar) are easily killed by lower winter temperatures (under 8°C) (Abdelrahman, 1974). These stages are expected to die during the winter if below-average winter temperatures prevail in the interior coastal valleys or if normal winter temperatures prevail in the San Joaquin Valley. Finally, normal winters are warm enough to allow some development of second-stage scale. They usually attain the third instar, but the lack of males prevents them from developing further as they remain uninseminated. Development of the male scale is arrested at the unemerged imago or pupal stage. The degree to which the scale's age structure is dominated by one or two age classes thus depends on temperature and is correlated with the citrus zone: the more interior the zone, the more variable the seasonal temperature pattern and, thus, the more the scale's age structure is dominated by the older scale stages (virgin and gravid females).

With the higher spring temperatures, males inseminate the virgin females, and crawlers appear after a gestation period. The more extreme the scale's age structure (dominated by one or two scale classes), the more synchronized the pulse of crawlers that initiates the spring generation. This pulse can be observed passing through the successive scale stages (ages). It becomes less distinct with each successive generation and is usually unrecognizable by autumn (third generation). Such a pattern produces a host resource for the *Aphytis* species that is at first abundant and then scarce. The abundant third-stage (virgin female) scale present in the spring escapes parasitization by *Aphytis*, because the low temperatures (under 18°C) inhibit searching by *Aphytis*. We therefore expected *Aphytis* to be ineffective as a biological control agent where the early-season scale population (spring generation) is dominated by one or two age classes. Thus, a hypothesis that evolved initially to explain the competitive displacement of *A. lingnanensis* by *A. melinus* also suggested a testable explanation for the absence of economic biological control in the San Joaquin Valley. These expectations provided the rationale for the experimental design that compared the scale's and parasitoid's demographic characteristics in the different citrus zones.

Developing and Implementing a Monitoring Program

The comparative demographic studies of *Aphytis* species involved a sampling approach aimed at monitoring seasonal patterns in the scale's age structure, the scale size (stage) from which an *Aphytis* species emerges, the sex of the emerging wasp, the geographical distribution of the scale's natural enemies, and the diurnal temperature pattern during the season. In addition to the work already performed, a survey will be conducted every 10 years to determine the distribution of natural enemies of California red scale. However, if a new natural enemy becomes established, an extensive sampling program will be conducted more frequently.

Parasitized scale encountered in the samples were individually reared to determine the gender and size of the emerging parasitoid. Second-stage and older scale were measured, as were parasitized scale that retained their shape (*Aphytis* species are ectoparasitoids; hence, as they feed they cause the scale to shrivel). Natural-enemy field densities and female parasitoid sizes were also monitored. We expect to modify these sampling schemes as we gain experience and as our hypotheses change.

We monitored diurnal temperatures hourly in each study grove. Scale phenology was also monitored. Because adult male scale live less than 24 hours in the field, males captured on pheromone traps mirror the temporal pattern of virgin third-stage female scale, i.e., the scale's age structure relevant to female progeny production by *A. melinus* and *A. lingnanensis*.

Future Goals

Future research is planned along several lines. First, why are some parasitoid species that exist on red scale elsewhere unable to exist on the scale in California, and, if they are present there, why are their densities and distributions limited there? Numerous parasitoid species have been introduced into southern California to control red scale, but most have failed to become established. Others have become established, but remain scarce or geographically limited. For example, *Aphytis africanus*, a common parasitoid of California red scale in South Africa, has been introduced several times, but has not become established. *Encarsia* (= *Prospaltella*) *perniciosi* (red scale strain) is limited to coastal citrus and immediately adjacent areas; it is infrequently encountered in the interior coastal valleys. Similarly, *Comperiella bifasciata*, a parasitoid introduced into southern California in 1939, is rare in the interior coastal valleys and coastal citrus zones, but is common in the San Joaquin Valley.

Second, what causes size variations in California red scale? Scale are largest on fruit, of intermediate size on leaves, and smallest on wood. Scale on the twigs (above the fourth internode) are smaller than those on the leaves. High laboratory temperatures (above 27°C) produce smaller scale (D. Sicki-Yu and R. F. Luck, unpublished data), and our field data on scale size suggest that smaller scale are present during the hottest months. Scale size influences vulnerability to *Aphytis melinus* and whether the scale receives a female or a male egg. Biological control is less effective on lemon and grapefruit cultivars; we suspect that the difference is associated with scale size and the relative abundance of the substrate supporting the smaller scale. We suspect that these substrates offer red scale a refuge from *A. melinus*: the scale are too small to be hosts for female wasp progeny, so the wasps do not spend much time searching these substrates.

Third, how does a parasitoid recognize a scale as a potential host? The parasitoids usually hesitate on encountering a scale, and they drum and turn if the scale is initially accepted (Luck *et al.*, 1982). However, not all apparently suitable individual scale are parasitized. Parasitoids walk over some without hesitating. Host recognition is based, in part, on chemical cues called kairomones—transspecific chemical messengers that favor the recipient, rather than the emitter (Brown *et al.*, 1970)—according to Luck and Uygun (in press) and Quednau and Hubsch (1964). Kairomones and their recognition might also depend partly on the chemistry of the scale's host plant and on the parasitoid's experience, associative learning.

SOURCES OF KNOWLEDGE

Generally Accepted Ecological Facts

Much is known about the natural history of California red scale, its predators, and its parasitoids, especially the parasitoid genus *Aphytis* (Bodenheimer, 1951; Ebeling, 1959; Rosen and DeBach, 1978, 1979). Less is known about processes that govern the scale's biological control. Information is also available on the red scale's pheromone and its use as a predictive tool (Moreno and Kennett, 1985). All these sources were consulted when this research project was designed and were initially summarized in a population model (Luck *et al.*, 1980). The only body of literature we ignored was that associated with chemical control of the scale.

General Theory and General Principles of Ecology

Four ecological theories provided the structure for our California red scale research—those dealing with predator-prey relations (e.g., Hassell, 1978, 1980), competition (e.g., Pianka, 1981), foraging (e.g., Pyke *et al.*, 1977), and sex allocation (e.g., Charnov, 1982). All these theories are relevant and helped to direct the development of the experimental designs. Predator-prey relations are at the heart of biological control. As stated earlier, biological control is predicated on the notion that host or prey densities are limited by natural enemies, because these agents compete intraspecifically and interspecifically for hosts or prey. Thus, the agents are themselves limited by the densities of their prey or hosts in a reciprocal interaction. However, DeBach and colleagues (DeBach, 1966; DeBach and Sundby, 1963; Rosen and DeBach, 1979) concluded, on the basis of laboratory and field experiments, that the two principal parasitoids responsible for biological control of California red scale were not limited by host density. But if hosts were not limiting for the parasitoids, how did *Aphytis melinus* control California red scale biologically? We used the principles of foraging theory to identify what scale types the parasitoids chose as hosts, to rank the scale for quality as hosts in the laboratory, and to test this ranking in the field. Because *Aphytis* allocates sons to small hosts and daughters to large hosts, sex-allocation theory became an extension of foraging theory. Thus, the hypothesis we tested was an amalgam of these principles—that resources are indeed limiting and that the limitation is evinced as a male-biased sex ratio in the field population of *A. melinus*. As scale grow, they reach the size used by *Aphytis* to produce sons before they can attain the size used for daughters. Thus, a male-biased sex ratio results from a scarcity of hosts large enough to produce daughters. This pattern might be episodic, depending on the dynamics of the scale's age structure.

A second body of theory used in the design of the experiments concerns the effect of temperature on the developmental rate of insects (Wagner *et al.*, 1984). An insect's rate of development increases with increasing temperature (to a maximum), but the relative duration spent in a developmental stage at a given temperature and the developmental rate of that stage often differ with developmental stage. These effects, coupled with the influence of temperature on reproduction, lead to variations in an insect's age structure as a consequence of temperature variations in the field. We hypothesized that climatically induced variation in the scale's age structure, coupled with the requirement of *A. melinus* and *A. lingnanensis* for a specific size or large scale on which to produce daughters, leads to bottlenecks in the availability of suitable hosts and that displacement of A. *lingnanensis* by A. *melinus* occurred during such bottlenecks.

Specific Models

First, we used a simulation model of the life history of California red scale and its parasitoids that organized information about these organisms (Luck *et al.*, 1980) and included a modified host-parasitoid model similar to that of Hassell (1978). This model, which embodied the verbal model specific to biological control, was inadequate, because biological knowledge was insufficient. Thus, we used the verbal model to predict the field host-parasitoid interaction. This model specifies that natural enemies reduce their host or prey populations to densities at which intraspecific competition and interspecific competition among the predators and parasitoids for prey or hosts reciprocally limit the natural enemies' densities (DeBach, 1974; Huffaker and Laing, 1972). Because it became clear in the course of the research that, from the parasitoid's perspective, the scale was not a homogeneous resource, we used foraging and sex-allocation theory to define, in the laboratory, both host resource requirements for *A. melinus* and *A. lingnanensis* progeny production (especially daughters) and the value of a host to a parental female in terms of the quality of the daughters (size = fecundity) that the scale produced. Finally, we used a day-degree model—developmental rate of the scale and parasitoid as a function of temperature (Wagner *et al.*, 1984)—to evaluate the effects of diurnal and seasonal temperature patterns on age structure of California red scale.

Analog Studies

No other evaluations of the effectiveness of a natural enemy appear to have used a similar conceptual approach. Most of the available studies (CIBC, 1971; Clausen, 1978; Greathead, 1976; Greathead *et al.*, 1971; McGuggan and Coppell, 1962; McLeod, 1962; Rao *et al.*, 1971; Turnbull and Chant, 1961; Wilson, 1960) at best simply documented success or failure. Important exceptions were the studies of Wellington (1960, 1964, 1965) and Whitham (1978, 1979, 1980).

Project as Experiment

The California red scale research project was designed to improve the biological control of California red scale and to test conventional wisdom as to the attributes that characterize an effective biological control agent, in this case a parasitoid. Huffaker *et al.* (1977), Hassell (1978), and Waage and Hassell (1982) proposed the following attributes as characterizing an effective biological control agent: high search rates, ability to aggregate in patches of high host density, close synchrony with the host population, high degree of host specificity, and sufficient reproductive rate to overtake and suppress the host population. Because we knew that *A. melinus* and *A. lingnanensis* were successful biological control agents in part of their ranges, we began testing them to see whether they had these attributes and whether *A. melinus* was better endowed with them.

We quickly faced the problem of how to measure a high rate of search or determine whether a parasitoid has a sufficiently high reproductive rate to overtake its host in the field. Moreover, on the basis of a reproductive-rate criterion, *A. lingnanensis* would be judged the better of the two parasitoids, because it is the more fecund (DeBach and Sundby, 1963). Unfortunately, *A. lingnanensis* was unable to realize its reproductive advantage in the form of daughters, because it was more restrictive in its host (red scale) requirements for daughters than was *A. melinus*: it required larger scale. Similar problems arose with several other attributes. Thus, after our initial effort to measure some of the attributes for *A. lingnanensis* and *A. melinus* in the laboratory (Kfir and Luck, 1979, 1984; Luck *et al.*, 1980; Podoler, 1981), we changed our approach and asked what the specific host requirements were for these two parasitoids with respect to host stage (instar) and size and the pattern of host availability in the field. We used the parasitoids' behavior and the size of the daughters to find the value of a host to the female wasp. On the basis of the results of these laboratory experiments, we are continuing to test the pattern of resource use in the field. We are nearing the end of this phase of our research in late 1985. Thus, we have used a flexible research strategy of hypothesis formation and testing that incorporates laboratory and field experiments mixed with computer simulation models of aspects of California red scale and parasitoid biology.

Expert Judgment

Clearly, a research project of this sort is not undertaken by a single researcher. First, several colleagues have participated in the evolution and conduct of the research. Rami Kfir, Haggai Podoler, Devin Carrol, Dani Blumberg, Dicky Sicki-Yu, Daniel Moreno, W. W. Murdoch, Charles Kennett, and John Reeves—all associated with academic or research institutions—have been involved in various aspects of hypothesis formation, experimental design, experimental conduct, analysis, and interpretation. Several persons—Harry Griffith, Jim Stewart, Jim Gorden, and Stanley Warner—with many years of experience in citrus pest management contributed as participants in the research and as critics of our interpretations of the results. Finally, in response to seminars, in conversations with colleagues—especially F. Galis, J. M. M. van Alphen, J. C. van Lentern, P. DeBach, L. E. M. Vet, L. Nunny, and D. R. Strong—or in peer reviews of submitted papers, questions or comments often called attention to an aspect of the problem of which we were unaware or stimulated us to see the problem from a different perspective. Thus, the project represents a collage of contributions from a variety of sources.

CONTRIBUTION OF ECOLOGICAL KNOWLEDGE
TO PROJECT RESULTS

Classical biological control has been very much a trial-and-error enterprise. It has been based on the notion that natural enemies regulate phytophagous insect populations and that an effective natural enemy is well adapted to and synchronized with its host or prey, is a good searcher, has a high reproductive rate, and aggregates at denser host or prey patches. The large number of successful biological control projects shows that natural enemies indeed limit the densities of their host or prey populations. Yet, if a measure of a successful theory is its ability to predict the outcome of a particular experiment, then predator-prey theory and biological control theory are inadequate. We cannot predict whether a given introduction will succeed or fail. More important, we cannot describe a natural enemy's attributes that characterize a successful biological control agent. Thus, predator-prey theory is to a great extent untested and, in its current formulation, not easily falsified. Therefore, we studied a biological control project in which both successful and unsuccessful parasitoids have been introduced, so that we could test several hypotheses arising from predator-prey theory and biological control theory and compare successful and unsuccessful parasitoids. Theory played a role both in formulating the hypotheses and in specifying the type of project to look at.

The California red scale biological control project was chosen for several reasons: the natural histories of the organisms involved were well known, the history of the project was well documented, the scale is controlled biologically in only a part of its California range, the scale and its natural enemies are easily reared in the laboratory, and the parasitoids responsible for the biological control of the scale are known. Initially, our experimental designs were guided by the Nicholson-Bailey (1935) model as modified by Hassell and Varley (Hassell, 1978). Although we obtained a modeled host-parasitoid interaction that persisted, it provided little insight into why *A. melinus* was so successful. The characteristics that defined the host-parasite interaction could not be measured in the field. Optimal-foraging theory became an alternative when evidence emerged that *A. melinus* and *A. lingnanensis* did not use the same host stages for offspring, as DeBach and co-workers had asserted (DeBach, 1966; DeBach and Sundby, 1963; Rosen and DeBach, 1979). Optimal-foraging theory provides ways of ranking hosts based on their value to the parasitoid, and we were able to use the size of the female progeny arising from a host as a surrogate measure of host quality. Thus, foraging theory played a role in reformulating the questions. More important, it led us to stop thinking of the host-parasitoid interaction as one in which hosts and parasitoids are homogeneous units (i.e., the old predator-prey theory) and to start thinking of hosts as resources of different value.

ACKNOWLEDGMENTS

The work described in this study was supported by grants from the National Science Foundation (BSR 84-03394) and the Binational Agricultural Research and Development Fund (1-138-79).

REFERENCES

Abdelrahman, I. 1974. Studies in the ovipositional behavior and control of sex in *Aphytis melinus* DeBach, a parasite of California red scale, *Aonidiella aurantii* (Mask.). Aust. J. Zool. 22:231-247.

Bedford, E. C. G., and M. B. Georgala. 1978. Citrus pests in the Republic of South Africa. Sci. Bull. Dept. Agric. Tech. Ser., Rep. S. Afr. 391:109-118.

Bodenheimer, F. S. 1951. Citrus Entomology in the Middle East. W. Junk, The Hague.

Brown, W. L., Jr., T. Eisner, and R. H. Whittaker. 1970. Allomones and kairomones: Transspecific chemical messengers. BioScience 20:21-22.

Charnov, E. L. 1982. The Theory of Sex Allocation. Princeton University Press, Princeton, N.J.

Charnov, E. L., R. L. Los-den Hartogh, W. T. Jones, and J. van den Assem. 1981. Sex ratio evolution in a variable environment. Nature 289:27-33.

Chewyreuv, I. 1913. Le role des femelles dans la determination du sexe et leur descendance dans le groupe des Ichneumonides. C. R. Soc. Biol. Paris 74:695-699.

CIBC (Commonwealth Institute of Biological Control). 1971. Biological Control Programmes Against Insects and Weeds in Canada, 1959-1968. Technical Communication 4. Commonwealth Agricultural Bureaux, Farnham Royal, Slough, Eng.

Clausen, C. P. 1939. The effects of host size upon the sex ratio of hymenopterous parasites and its relation to methods of rearing and colonization. J. N.Y. Entomol. Soc. 47:1-9.

Clausen, C. P. 1978. Introduced Parasites and Predators of Arthropod Pests and Weeds: A World Review. USDA/ARS Agric. Handbk. 480. Agricultural Research Service, U.S. Department of Agriculture, Washington, D.C.

DeBach, P. 1965. Some biological and ecological phenomena associated with colonizing entomophagous insects. Pp. 287-303 in H. G. Baker and G. L. Stebbins, eds. The Genetics of Colonizing Species. Academic Press, New York.

DeBach, P. 1966. The competitive displacement and coexistence principles. Annu. Rev. Entomol. 11:184-212.

DeBach, P. 1974. Biological Control by Natural Enemies. Cambridge University Press, New York.

DeBach, P., and L. C. Argyriou. 1967. The colonization and success in Greece of some imported species (Hym., Aphelinidae) parasitic on citrus scale insects (Hom., Diaspididae). Entomophaga 12:325-342.

DeBach, P., and P. Sisojevic. 1960. Some effects of temperature and competition on the distribution and relative abundance of *Aphytis lingnanensis* and *A. chrysomphali* (Hymenoptera: Aphelinidae). Ecology 41:153-160.

DeBach, P., and R. A. Sundby. 1963. Competitive displacement between ecological homologues. Hilgardia 34:105-166.

Dittrich, V. P., P. Streibert, and P. A. Bathe. 1974. An old case reopened: Mite stimulation by insecticide residues. Environ. Entomol. 3:534-539.

Ebeling, W. 1959. Subtropical Fruit Pests. University of California Division of Agricultural Science, Berkeley.

Flaherty, D. L., J. E. Pehrson, and C. E. Kennett. 1973. Citrus pest management studies in Tulare County. Calif. Agric. Nov., pp. 3-7.

Furness, G. O., G. A. Buchanan, R. S. Georgie, and N. L. Richardson. 1983. A history of the biological and integrated control of red scale, *Aonidiella aurantii*, on citrus in the lower Murry Valley of Australia. Entomophaga 28:199-212.

Georghiou, G. P., and C. E. Taylor. 1977. Pesticide resistance as an evolutionary phenomenon. Proc. Int. Congr. Entomol. 15:759-792.

Greathead, D. J., ed. 1976. A Review of Biological Control in Western and Southern Europe. Technical Communication 7. Commonwealth Agricultural Bureaux, Farnham Royal, Slough, Eng.

Greathead, D. J., J. F. G. Lionnet, N. Lodos, and J. A. Whellan. 1971. A Review of Biological Control in the Ethiopian Region. Technical Communication 5. Commonwealth Agricultural Bureaux, Farnham Royal, Slough, Eng.

Hassell, M. P. 1978. The Dynamics of Arthropod Predator-Prey Systems. Princeton University Press, Princeton, N.J.

Hassell, M. P. 1980. Foraging strategies, population models and biological control: A case study. J. Anim. Ecol. 49:603-628.

Huffaker, C. B., and J. E. Laing. 1972. "Competitive displacement" without a shortage of resources? Res. Pop. Ecol. 14:1-17.

Huffaker, C. B., R. F. Luck, and P. S. Messenger. 1977. The ecological basis of biological control. Proc. Int. Congr. Entomol. 15:560-586.

Jones, V. P., and M. P. Parrella. 1984. The sublethal effects of selected insecticides on life table parameters of *Panonychus citri* (Acari: Tetranychidae). Can. Entomol. 76:1178-1180.

Kfir, R., and R. F. Luck. 1979. Effects of constant and variable temperature extremes on sex ratio and progeny production by *Aphytis melinus* and *A. lingnanensis* (Hymenoptera: Aphelinidae). Ecol. Entomol. 4:335-344.

Kfir, R., and R. F. Luck. 1984. Effects of temperature and relative humidity on the developmental rate and adult life span of three *Aphytis* species (Hym., Aphelinidae) parasitizing California red scale. Z. Angew. Entomol. 97:314-320.

Luck, R. F. 1984. Integrated pest management in California citrus. Proc. Int. Soc. Citricul. 2:630-635.

Luck, R. F., and H. Podoler. 1985. The potential role of host size in the competitive exclusion of *Aphytis lingnanensis* by *A. melinus*. Ecology 66:904-913.

Luck, R. F., and N. Uygun. In press. Host recognition and selection by *Aphytis* species: Response to California red, *Aonidiella aurantii* (Mask.), oleander, *Aspidiotus nerii* Bouche, and cactus, *Diaspis echinocacti* (Bouche) scale cover extracts. Entomol. Exp. Appl.

Luck, R. F., R. van den Bosch, and R. Garcia. 1977. Chemical insect control—A troubled pest management strategy. BioScience 27:606-611.

Luck, R. F., J. C. Allen, and D. Baasch. 1980. A systems approach to research and decision making in the citrus ecosystem. Pp. 366-395 in C. B. Huffaker, ed. New Technology of Pest Control. John Wiley & Sons, New York.

Luck, R. F., H. Podoler, and R. Kfir. 1982. Host selection and egg allocation behaviour by *Aphytis melinus* and *A. lingnanensis*: A comparison of two facultatively gregarious parasitoids. Ecol. Entomol. 7:397-408.

Maggi, V. L., and T. F. Leigh. 1983. Fecundity response of the two-spotted spider mite to cotton treated with methyl parathion or phosphoric acid. J. Econ. Entomol. 75:616-619.

McGuggan, B. M., and H. C. Coppell. 1962. Biological control of forest insects, 1910-1958. Pp. 35-127 in A Review of the Biological Control Attempts Against Insects and Weeds in Canada. Technical Communication 2. Commonwealth Agricultural Bureaux, Farnham Royal, Slough, Eng.

McLeod, J. H. 1962. Biological control of pests of crops, fruit trees, ornamentals and weeds in Canada up to 1959. Pp. 1-33 in A Review of Biological Control Attempts Against Insects and Weeds in Canada. Technical Communication 2. Commonwealth Agricultural Bureaux, Farnham Royal, Slough, Eng.

Moreno, D., and C. E. Kennett. 1985. Predictive year-end California red scale (Homoptera: Diaspididae) orange fruit infestation based on male trap catches in the San Joaquin Valley, California. J. Econ. Entomol. 78:1-9.

Nicholson, A. J., and V. A. Bailey. 1935. The balance of animal populations. Proc. Zool. Soc. Lond. 3:551-598.

Orphanides, G. M. 1984. Competition displacement between *Aphytis* spp. [Hym.: Aphelinidae] parasites of the California red scale in Cyprus. Entomophaga 29:275-281.

Pianka, E. R. 1981. Competition and niche theory. Pp. 167-196 in R. M. May, ed. Theoretical Ecology: Principles and Applications. Sinauer Associates, Sunderland, Mass.

Podoler, H. 1981. Effects of variable temperatures on responses of *Aphytis melinus* and *A. lingnanensis* to host density. Phytoparasitica 9:179-190.

Pyke, G. H., H. R. Pulliam, and E. L. Charnov. 1977. Optimal foraging: A selective review of theory and tests. Q. Rev. Biol. 52:137-154.

Quayle, H. J. 1938. Insects of Citrus and Other Subtropical Fruits. Comstock Publ. Co., Ithaca, N.Y.

Quednau, F. W., and H. M. Hubsch. 1964. Factors influencing the host-finding and host acceptance pattern in some *Aphytis* species (Hymenoptera: Aphelinidae). S. Afr. J. Agric. Sci. 7:543-554.

Rao, V. P., M. A. Ghani, T. Sankaran, and K. C. Mathur. 1971. A Review of the Biological Control of Insects and Other Pests in South-east Asia and the Pacific Region. Commonw. Inst. Biol. Control Tech. Comm. 6. Commonwealth Agricultural Bureaux, Farnham Royal, Slough, Eng.

Rosen, D., and P. DeBach. 1978. Diaspididae. Pp. 78-128 in C. P. Clausen, ed. Introduced Parasites and Predators of Arthropod Pests and Weeds: A World Review. USDA/ARS Agriculture Handbook 480. Agricultural Research Service, U.S. Department of Agriculture, Washington, D.C.

Rosen, D., and P. DeBach. 1979. Species of *Aphytis* of the World (Hymenoptera: Aphelinidae). W. Junk, The Hague.

Turnbull, A. L., and D. A. Chant. 1961. The practice and theory of biological control of insects in Canada. Can. J. Zool. 39:697-753.

Waage, J. K., and M. P. Hassell. 1982. Parasitoids as biological control agents—A fundamental approach. Parasitology 84:241-268.

Wagner, T. L., H. Wu, P. J. H. Sharpe, and R. N. Coulson. 1984. Modeling distributions of insect development time: A literature review and application of the Weibull function. Ann. Entomol. Soc. Am. 77:475-483.

Wellington, W. G. 1960. Qualitative changes in natural populations during changes in abundance. Can. J. Zool. 38:289-314.

Wellington, W. G. 1964. Qualitative changes in populations in unstable environments. Can. Entomol. 96:436-451.

Wellington, W. G. 1965. Some maternal influences on progeny quality in the western tent caterpillar, *Malacosoma pluviale* (Dyar). Can. Entomol. 97:1-14.

Whitham, T. G. 1978. Habitat selection by *Pemphigus* aphids in response to resource limitation and competition. Ecology 59:1164-1176.

Whitham, T. G. 1979. Territorial behaviour of *Pemphigus* gall aphids. Nature 279:324-325.

Whitham, T. G. 1980. The theory of habitat selection: Examined and extended using *Pemphigus* aphids. Am. Nat. 115:449-466.

Wilson, F. 1960. A Review of the Biological Control of Insects and Weeds in Australia and Australian New Guinea. Commonw. Inst. Biol. Control Tech. Comm. 1. Commonwealth Agricultural Bureaux, Farnham Royal, Slough, Eng.

Committee Comment

The first point that emerges from the California red scale case is not treated in detail by Luck—that, even before the study that he describes was initiated, important partial control of red scale was achieved at little cost and with little attention to ecological theory (Clausen, 1978; DeBach, 1974). The key new knowledge deployed was taxonomic—until 1947, virtually all *Aphytis* reared overseas from California red scale were thought to be *A. chrysomphali*. Because this species had already been introduced and had usually not exerted very good control (partly because of pesticides and ants), specimens attributed to *A. chrysomphali* were not sent to the United States. Thus, in 1905-1909, a species that was probably *A. lingnanensis* was discovered in southern China, but not shipped. With the recognition of distinct species, *A. lingnanensis* was introduced in 1947 and very quickly exerted important control. The ecological knowledge used later to increase this control was rudimentary autecology. It appeared that *A. lingnanensis* was hindered by the harsh climate of interior southern California, so a search was initiated for other parasitoids in areas with similar climate. Success was achieved in northwestern India and Pakistan with the discovery of *A. melinus*. This approach is typical of biological control efforts; the first consideration in the search for natural enemies is usually that the climate of their native region approximate the climate where control is sought.

Luck's description is of the effort to extend control to the desert and the San Joaquin Valley, and his thrust is an interesting mixture of classical biological control (much like the introduction of the *Aphytis* species that are already present) and academic evolutionary ecology. One could simply have proceeded by trying more and more introductions from many parts of the world, perhaps focusing the search by looking for areas with citrus and with climates very similar to those of the desert and the San Joaquin Valley and hoping that some species would prove effective where the 52 already tried had not. Instead, Luck tried to understand why control operates differently in different regions and why *A. melinus* is replacing *A. lingnanensis*, despite the greater fecundity of the latter. The goal of this understanding is either to predict better what sort of parasitoid would succeed where *A. melinus* and *A. lingnanensis* have failed or to modify cultivation or other conditions so as to increase the effects of the *Aphytis* that are already present.

A number of apparently relevant ecological theories were examined and incorporated into this effort. For the most part, these proved inspirational or didactic, rather than directly applicable. The key to the entire biological control rationale is that the host (red scale) will ultimately be controlled by the parasitoids and will thus come to be scarce, so that parasitoid populations will ultimately be limited by intraspecific or interspecific competition; this is an entomological version of the "green earth" hypothesis of Hairston, Smith, and Slobodkin (1960). Luck observes that the degree of resolution afforded by this hypothesis did not allow answers to the most pertinent questions. He thus proceeded to the more specific traditional predation models of Nicholson and Bailey (1935) and Hassell and Varley (Hassell, 1978), but found that the model parameters were ambiguous and unmeasurable and so could not help to explain the geographic variation in parasitism rate or the apparent replacement of one parasitoid by another. The key to a more useful approach seems to have been the recognition that red scale individuals are not a uniform resource for the wasps, which led rather naturally to casting the problem in terms of optimal foraging. Although no particular existing optimal-foraging model was applicable, simply the formulation of the problem in these terms led to major insights on both geographic variation in parasitism and the superiority of *A. melinus* over *A. lingnanensis*. It is also noteworthy that Luck's familiarity with sex-allocation theory led him to examine sex ratio as an indicator of degree of competition for large hosts.

In sum, idiosyncrasies of key species in this study rendered even the most detailed theories inapplicable, but the theories, particularly optimal foraging, suggested new and fruitful ways of looking at the problem. Without the massive information on the biology of the organisms, one could not even have begun the study, but without familiarity with the theory, the beginnings of a solution might have been much longer in coming.

The entire project is conceived not only as an attempt to solve a particular environmental problem, but as an experiment to test hypotheses generated by various ecological theories and perhaps to aid in modifying the theories or hypotheses. For example, the intersection of optimal-foraging theory, competition theory, and sex-allocation theory in the explanation of why *A. melinus* outcompetes *A. lingnanensis* even when the latter would be successful by itself is more than just interesting entomology—it is an exciting theoretical advance that should spur further theoretical research. A quantitative generalization of this intersection does not exist, and the attention of theorists should be drawn to the problem. Whether the theoretical advance can ultimately help to solve the practical environmental problem—noninsecticidal control of red scale in the desert and the San Joaquin Valley—remains to be seen. However, the entire study is certainly an advance over the trial-and-error approach of much classical biological control, and the focus on precise mechanisms of species interaction seems to lend itself to more theoretical treatment.

It is interesting that the scoping of this project, as for all biological control (and, in fact, for most of applied ecology), is somewhat myopic. The myriad effects of introduced species are well documented in the ecological literature (e.g., Elton, 1958; Simberloff, 1981), and biological control rests fundamentally on introducing species, such as the *Aphytis* wasps used to control red scale. Although attention is paid to more than just the effects on red scale, it is restricted to effects that impinge on agricultural economy. For example, one wishes to avoid introducing a parasitoid or predator that attacks a beneficial species, such as a pollinator, and one screens potential imports to ensure that they are not carrying hyperparasites that would interfere with the parasites already active in controlling agricultural pests. However, little attention is paid to the possible effects of an introduced species on native species that are not agriculturally important.

Howarth (1983), for example, laments the reduction of native Hawaiian lepidopterans, partially by species introduced for biological control. He calls for a more narrowly focused release effort, rather than the hit-and-miss release of many potentially beneficial species that typifies some biological control efforts. The direction that the red scale study has taken certainly qualifies as a desirable narrowing of focus. Howarth also calls for more studious consideration of potential effects of control agents on nontarget species, just as one seeks such effects for pesticides. The three *Aphytis* species of most interest in red scale control—*A. chrysomphali*, *A. lingnanensis*, and *A. melinus*—all attack other insects; but, as far as is known, all the other hosts are themselves agricultural pests, such as yellow scale and dictyospermum scale (Clausen, 1978). Consequently, unexpected detrimental effects of their introduction are unlikely. However, the fact that a successful introduction is virtually impossible to eradicate (Sailer, 1983) and can easily spread from the site of introduction dictates a particularly broad scoping process.

References

Clausen, C. P. 1978. Introduced Parasites and Predators of Arthropod Pests and Weeds: A World Review. USDA/ARS Agriculture Handbook 480. Agricultural Research Service, U.S. Department of Agriculture, Washington, D.C.

DeBach, P. 1974. Biological Control by Natural Enemies. Cambridge University Press, New York.

Elton, C. S. 1958. The Ecology of Invasions by Animals and Plants. Methuen, London.

Hairston, N. G., F. E. Smith, and L. B. Slobodkin. 1960. Community structure, population control, and competition. Am. Nat. 94:421-425.

Hassell, M. P. 1978. The Dynamics of Arthropod Predator-Prey Systems. Princeton University Press, Princeton, N.J.

Howarth, F. G. 1983. Classical biocontrol: Panacea or Pandora's box? Proc. Hawaii. Entomol. Soc. 24:239-244.

Nicholson, A. J., and V. A. Bailey. 1935. The balance of animal populations. Proc. Zool. Soc. Lond. 3:551-598.

Sailer, R. I. 1983. History of insect introductions. Pp. 15-38 in C. Graham and C. Wilson, eds. Exotic Plant Pests and North American Agriculture. Academic Press, New York.

Simberloff, D. 1981. Community effects of introduced species. Pp. 53-81 in M. H. Nitecki, ed. Biotic Crises in Ecological and Evolutionary Time. Academic Press, New York.

15

Experimental Control of Malaria in West Africa

Epidemiology encompasses a sophisticated field of ecological knowledge, and complex mathematical models are available to describe and predict patterns of disease incidence in space and time. The Garki project in Nigeria began in 1970 to provide detailed information on all factors in the transmission of malaria and to test hypotheses about the dynamics of malaria. The project was based on epidemiological models of proven value, it was carefully planned, and it had appropriate controls. It was viewed as a prototype for more extensive malaria control projects elsewhere in Africa.

Although the project was carefully designed and executed in an integrated way over several years and data were gathered on the basis of a realistic mathematical model of malaria transmission, it failed to produce the predicted results, because incorrect assumptions were made about the biology of the mosquito vectors. In particular, it was assumed that all mosquitoes behave alike, whereas resting behavior actually varied in such a way that a large fraction of the mosquitoes did not land inside buildings, where they would have been exposed to the contact insecticide used. This case study highlights the importance of accurate and detailed natural history in the design of effective control strategies and the effect of initial assumptions on the results of using even sophisticated models.

Case Study

ROBERT M. MAY, Department of Biology, Princeton University

INTRODUCTION

Malaria might well be the most important public health problem in tropical Africa. The intensity of malaria transmission is so high that control is very difficult, and many malariologists have suggested that humans should not interfere with the established natural immunity in the population, inasmuch as intervention could increase the severity of the clinical manifestations of malaria and the mortality caused by it in older children and adults. Largely for that reason—and because the public health infrastructure was inadequate—theory predicted that available programs to control malaria would not work, and Africa was not included in the global malaria eradication program launched by the World Health Organization (WHO) in the 1950s.

However, spraying houses with DDT was locally effective in interrupting malaria transmission, and the results of later pilot projects in various African countries suggested that malaria transmission could be interrupted, particularly in forested areas, if insecticide coverage were total and were followed by full surveillance. But in none of these pilot projects were quantitative epidemiological data collected. In view of this background, WHO decided to begin a field research project, carefully designed to provide information on all the factors involved in malaria maintenance and transmission. The northern part of Nigeria, specifically the Garki district, was selected for this project.

The designers of the project decided to put more resources than usual into collection of baseline data and into evaluation of the impact of spraying houses with an effective residual insecticide (either alone or in combination with mass administration of drugs). As part of this effort to understand the dynamics of malaria transmission, a mathematical model was developed and tested against data. The model was developed in an attempt to simulate malaria transmission realistically, in the hopes of understanding the various factors involved and improving the planning of control programs. The Garki project was also designed to study an array of seroimmunological tests administered before, during, and after the application of control measures.

The project is described in detail in *The Garki Project*, edited by Molineaux and Gramiccia (1980).

191

THE ENVIRONMENTAL PROBLEMS

Malaria is caused by the multiplication of parasitic protozoa of the family Plasmodiidae in the blood cells or other tissues of the vertebrate host; the clinical symptoms in humans arise from multiplication of the blood stages of the parasite. The several genera of malarial parasites are associated with different hosts. Four main species of *Plasmodium* are found in humans: *P. falciparum*, *P. vivax*, *P. malariae*, and *P. ovale* (comparatively rare); *P. falciparum* is the most common in tropical regions and is of greatest concern in the Garki region, and *P. vivax* is most common in temperate zones.

Infection of a human host begins with the bite of a female anopheline mosquito and the injection of sporozoite stages of the parasite into the bloodstream. These stages are carried to the liver, where they develop in the parenchymal cells. After an incubation period of several days (or months, as some researchers have argued for *P. vivax*), these exoerythrocytic stages (i.e., from outside the red cells) grow, divide, and release merozoites into the bloodstream. The merozoites penetrate red cells, where they grow and subdivide to produce more merozoites, which rupture the host cells and invade other red cells. At some ill-understood point in this process, some of the merozoites develop into sexual stages, the gametocytes. Only the gametocytes are infective to the mosquito. When a vector mosquito bites a human and ingests male and female gametocytes, these are freed from the ingested blood cell, and the female gamete is fertilized and develops into an oocyst on the wall of the mosquito's gut. After about 10 days (the development time depends on temperature), immature sporozoites migrate from the ruptured oocyst to the mosquito's salivary glands and mature to infectivity, and the cycle is ready to repeat itself.

Thus, the overall cycle of the malaria parasite involves transmission both from mosquito to human and from human to mosquito. One method of control or eradication of malaria, therefore, is to reduce the mosquito population sufficiently to keep the parasite cycle from maintaining itself. After World War II, DDT was effective in reducing mosquito populations and the incidence of malaria in many parts of the world (more in Asia than in Africa, where transmission rates have always been very high). The pesticide chosen in the Garki project was propoxur, which is more specifically targeted against anopheline mosquitoes than are broad-spectrum insecticides, such as DDT.

Malaria is commonly believed to be responsible for more deaths than any other infectious disease, both today and throughout history. It is listed as the principal cause of morbidity and mortality by most African countries (notably by Nigeria, the most populous). In Latin America, respiratory tract infection tends to be listed as the most common cause of morbidity and mortality, followed by malaria and then measles; malaria ranks first in Brazil and Colombia. In Asia, gastrointestinal infection is the most common, followed by respiratory infection, with malaria ranking third as a source of morbidity; malaria ranks first in Malaysia. In short, malaria is a very serious public health problem throughout the tropical world (which is largely coincident with the developing world), and it is of especial importance in tropical Africa.

GENERAL APPROACH

The general approach of the Garki project was formulated on the basis of earlier pilot studies in Africa and elsewhere, all of which were qualitative, rather than quantitative. In contrast with the earlier studies, the Garki project had clearly formulated quantitative objectives. The main objectives were as follows (Molineaux and Gramiccia, 1980, p. 21):

● "To study the epidemiology of malaria in the lowland rural Sudan savanna. This means in particular a concentration of the study on the measurement of entomological, parasitological and seroimmunological variables and on their relationships. Also included were some meteorological, demographic and clinical variables, and the study of the prevalence of abnormal haemoglobins in the population."

● To measure the effect of specified activities directed toward interrupting transmission. Such intervention included spraying with a residual insecticide, propoxur, alone and in combination with mass administration of drugs.

● To undertake control measures in conjunction with the construction and testing of a mathematical model of malaria transmission. The general aim was to develop such a model into a planning tool that permitted comparison of the expected effects of different control strategies. Specifically, the model was developed to link entomological variables with parasitological variables (particularly the prevalence and transmissibility of *P. falciparum*) and to calculate the expected parasitological effect of defined changes in mosquito population density. The model also sought to permit evaluation of the effects of specific schemes for the mass administration of drugs, including the estimated role of immunity.

DEVELOPMENT OF THE PROJECT

Boundaries

In the preparatory phase, from September 1969 through September 1970, preliminary entomological and parasitological surveys were made, and specific areas and clusters of villages were chosen for detailed study. Protocols, forms, and operation manuals were drafted, and field and laboratory methods were tested. The boundaries of the study were, in effect, defined as specific villages in the Garki region.

Study Strategy and Monitoring

Drawing on experience gained in earlier, qualitative pilot projects in Africa and elsewhere, the designers of the Garki project produced a study strategy that divided the operation into three main phases of data collection and monitoring.

In the baseline phase, from October 1970 through March 1972 (which included a dry season, a wet season, and a second dry season), baseline epidemiological and parasitological data were collected in the clusters of villages that had been selected for followup. The preliminary insecticide and drug trials were also run.

In the intervention phase, from April 1972 through October 1973 (which included a wet season, a dry season, and a second wet season), the intervention strategies and the monitoring of epidemiological variables in the treated and untreated clusters of villages were implemented. The length of this phase of the project was not fixed in advance, but rather was left open to be guided by accumulated knowledge. In February 1972, a review of the results obtained up to then suggested that the additional information likely to be gained from a third year of intervention and monitoring would not justify the expense and the expected additional loss of population immunity (as discussed in the introduction of this case study).

In the postintervention phase, from November 1973 to the termination of the project in February 1976, selective administration of drugs in the villages covered during the intervention phase was continued, as was epidemiological monitoring in the villages treated with the most intensive strategies and in one cluster of untreated (control) villages.

The antimosquito insecticide propoxur was sprayed, primarily inside buildings, in 164 villages, distributed over roughly 1,000 km^2—a total of around 30,000 huts that housed a total population of some 50,000 persons. Propoxur is a contact chemical, so a mosquito must land on a sprayed surface if it is to pick up the insecticide. It remains active against mosquitoes on sprayed surfaces long after application. The coverage with insecticide, expressed as a proportion of huts sprayed, averaged about 99%. This coverage was estimated immediately after each spraying; the true coverage must have been somewhat lower, owing to building and repair activities between rounds of insecticide application.

A combination of the antimalarial drugs sulfalene and pyrimethamine was given to about 16,000 people in 60 of the sprayed villages. Infants were excluded, but visitors were included. The proportion of the total population given the drugs averaged about 85% (higher in the wet season, lower in the dry season).

USES OF KNOWLEDGE

The relevant ecological facts pertaining to the dynamics of malaria transmission are conveniently discussed under two headings: the mosquito and the malaria parasite. Human immunology, although extensively studied (Molineaux and Gramiccia, 1980), did not yield any findings particularly relevant to the overall design of control programs and is not treated here.

The Mosquito

The extent of contact between mosquitoes and humans in the Sudan savanna is very high indeed: the biting rates of *Anopheles gambiae sensu lato* and *A. funestus* were found to attain seasonal peaks of 174 and 94 bites per person per night, respectively (in studies that averaged more than 8 nights). The vectorial capacity, defined as the rate of contact between human hosts via the mosquito vectors, reached a seasonal peak of about 40, which is some 2,000 times the critical value for maintaining endemic malaria. The cumulative inoculation rate by mosquitoes reached a maximum of 145 sporozoite-positive bites per person in 1 year (of which 132 were in the wet season).

The mosquito density, and hence the extent of malaria transmission, in the Sudan savanna varies widely with season, year, and locale, according to Molineaux and Gramiccia (1980), who write:

In most villages, the vectorial capacity drops below its critical level for about half of the year (this does not necessarily prevent transmission, given the large reservoir of parasites), while in some it remains well above the critical level throughout the year. The variations from year to year are relatively important; over a period of 3 years, these variations followed the variations in total rainfall in the case of *A. funestus* but not in the case of *A. gambiae s.l.* Villages, even when not obviously different on inspection and located only a few kilometres apart, differ in vector density, anopheline fauna (*A. funestus* has a very uneven distribution) and probably in vector behaviour and karyotypes. Among 8 villages, the cumulative inoculation rate ranged from 18 to 145 sporozoite-positive bites in one year. The local and yearly variations stress the need for adequate baseline and comparison (control) data for a correct evaluation of the impact of control measures.

The Malaria Parasite

In general, the baseline prevalence of malaria parasitemia was high, and the age-specific curves of both prevalence and density of infection were typical of a high rate of transmission with a high level of acquired immunity. The findings were the same on each of the three parasites (*P. falciparum*, *P. malariae*, and *P. ovale*). There was relatively little variation among villages or between years.

The rate at which infants acquired malaria and the rate of onset of episodes of patent parasitemia (with fever, etc.) in the general population both testified to the high rate of transmission. The marked increase in the rate at which patent parasitemia disappeared with increasing age and the high ratio (nearly 1:1) of the entomological inoculation rate to the infant conversion rate (the ineffectiveness of sporozoite-positive bites) confirmed the high level of population immunity.

The combination of propoxur spraying and mass administration of sulfalene and pyrimethamine every 10 weeks reduced the prevalence of *P. falciparum* to a very low point in the dry season, but it did not significantly interrupt malaria transmission. Nor did it prevent an increase in prevalence in the wet season, when natural conditions favored mosquito breeding. A new equilibrium, with incidence oscillating between wet and dry seasons, was reached rapidly, and continuation of the intervention program would probably not have affected this result significantly. It was concluded that mobility of the human population, which was relatively pronounced, was unlikely to be the main cause of the maintenance of transmission. As discussed at length below, the main limiting factor was probably the exophily of some of the mosquitoes—i.e., their preference for resting outdoors.

When propoxur was sprayed and drugs administered more often (every 2 weeks in the wet season, every 10 weeks in the dry season), the prevalence of *P. falciparum* decreased to around 1% in the dry season and to 5% or less in the wet season. But transmission was still not interrupted greatly. As with the lower-frequency intervention, a new oscillating equilibrium was fairly rapidly attained, and it seems unlikely that continuation of the intervention would have modified the result. Again, it was concluded that exophily of the mosquitoes was the main cause, rather than population mobility.

As expected, mass application of the drugs at high frequencies for 1.5 years caused a temporary loss of immunity against *P. falciparum*, which was reflected in the resurgence of malaria in the postintervention phase of the project.

GENERAL THEORY

The basic model for malaria was first set out by Ross (1911, 1916) and later refined by MacDonald (1952, 1957, 1973), who included the latent, infected-but-not-yet-infectious period for the mosquito. If we divide the human host population into those who are susceptible and those who are infected and divide the female mosquito population into the susceptible and the infected, we can construct a pair of differential equations to describe the essentials of the dynamics of infection.

$$dx/dt = (abM/N)y(1 - x) - rx, \qquad (1)$$
$$dy/dt = ax(1 - y) - \mu y, \qquad (2)$$

where

x is the proportion of the human population infected,

y is the proportion of the female mosquito population infected,

N is the size of the human population,

M is the size of the female mosquito population,

a is the rate of biting on a human by a single mosquito (number of bites per unit time),

b is the proportion of infected bites on humans that produce an infection,

r is the per capita rate of recovery for humans ($1/r$ is the average duration of infection in a human host), and

μ is the per capita mortality rate for female mosquitoes ($1/\mu$ is the average lifetime of a mosquito).

In this simplest model, the total populations of both humans and female mosquitoes are assumed to be unchanging (N and M constant), so the dynamic variables are the proportions infected in the two populations (x and y). Equation 1 describes changes in the proportion of humans infected. New infections are acquired at a rate that depends on the number of mosquito bites per person per unit time (aM/N), on the probabilities that the biting mosquito is infected (y) and that a bitten human is not already infected ($1 - x$), and on the chance that an uninfected person bitten by an infected mosquito will become infected (b); and infections are lost by the return of infected people to the uninfected class, at a characteristic recovery rate (rx). Equation 2 describes changes in the proportion of mosquitoes infected. The gain term is proportional to the number of bites per mosquito per unit time (a) and to the probabilities that the biting mosquito is uninfected ($1 - y$) and that the bitten human is infected already (x); the loss term arises from the death of infected mosquitoes (y).

More formally, the loss terms for infected humans and for infected mosquitoes both involve death and recovery. But for human hosts the recovery rate is typically greater than the death rate (by 1-2 orders of magnitude), whereas for mosquitoes the opposite is typically the case; the above formulation is therefore a sensible approximation.

This model is, of course, highly simplified. One of its glaring flaws is the failure to disarticulate the "infected" categories of human and mosquito hosts to take account of the various developmental stages of the parasite. For instance, the model does not incorporate the incubation period of the parasite in the mosquito (during which no sporozoites are present in the salivary glands of the "infected" mosquitoes), even though this incubation period is comparable with the mean life span of the mosquito. We pursue this complication below. Likewise, the model does not distinguish between the pathological asexual merozoite blood stages and the infectious gametocyte sexual stages in the human. Encounters between biting mosquitoes and the humans they bite are assumed to be random.

Notwithstanding those shortcomings, the simple model defined by Equations 1 and 2 is useful in laying bare the essentials of the transmission process and in elucidating patterns in the diverse array of epidemiological data on different geographical regions. In particular, the model makes plain the significance of the "basic reproductive rate" of the parasite, in this context conventionally called z_0 (MacDonald, 1952, 1957). The basic reproductive rate is essentially the number of secondary infections generated by one infected individual in a population of susceptibles. If this number is, on the average, less than unity, the disease will be unable to maintain itself; if it equals or exceeds unity, the disease will be able to maintain itself. In general, the larger the basic reproductive rate, the greater the resistance of the disease to eradication. In this simple model,

$$z_0 = ma^2b/\mu r, \tag{3}$$

where m is the number of female mosquitoes per human host (M/N).

This result is intuitively understandable: the mosquito biting rate, a, enters twice in the cycle (hence, a^2); transmission is helped by large numbers of mosquitoes per human host (large m) and by large b; and transmission is hindered by a high mosquito death rate or by fast recovery (large μ and r, respectively).

To be more accurate, we should recognize that a latent period of duration T must elapse before an infected mosquito becomes infectious. This introduces an additional factor, $\exp(-\mu T)$, into Equation 3 to represent the probability that an infected mosquito will survive the latent period to become a transmitter of infection. The resulting expression for z_0 (as first derived by MacDonald, 1952, 1957) is then

$$z_0 = ma^2b \exp(-\mu T)/\mu r. \qquad (4)$$

A SPECIFIC MODEL

Dietz *et al.* (1974) developed a more realistic modification of the above model, which is presented in detail in Molineaux and Gramiccia (1980, Ch. 10). The essentials of the model are indicated schematically in Figure 1. There are two classes of people: one class has a low rate of recovery from malaria, and all infections can be detected; the other class has a high recovery rate, and infections have only a 70% chance of being detected (owing to the low densities of parasites). Members of both classes repeatedly are exposed, become infected, and recover, remaining within their own class except for a fixed rate of transition from the relatively susceptible class to the relatively immune class (the transition rate is determined by fitting the model to the data). The model thus takes into account the basic characteristics of immunity to malaria.

Within the framework of this model, it is possible to calculate the prevalence of *P. falciparum* as a function of the vectorial capacity and of its spontaneous and man-induced changes. It is also possible to calculate the effect of mass administration of drugs on the prevalence of malaria.

The model gave a good account of the basic dynamics of the interactions among humans, mosquitoes, and malaria. But the effects of insecticide

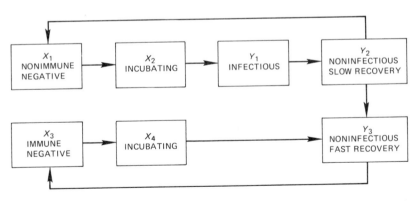

FIGURE 1   Schematic illustration of malaria model of Dietz *et al.* (1974). As discussed more fully in text, diagram shows various categories of individuals considered in model and possible transitions between categories.

on mosquito longevity were not accurately estimated by treating the mosquito population as homogeneous and as all resting indoors. Accurate prediction of the effects of insecticide on mosquito mortality, and on overall transmission dynamics, required an appreciation of the heterogeneity of the environment, which was not built into the original model.

The study concluded that the specific model described above does indeed simulate the epidemiology of *P. falciparum* infections with acceptable realism and that it can be used both for planning malaria control and for teaching the epidemiology and control of malaria. But important complications enter into the estimation of the parameters in the model, particularly the estimation of the effects of insecticides on mosquito mortality rates. These complications in many ways constitute the most interesting of the lessons learned in this case.

CONCLUSION: THE CASE STUDY AND ECOLOGICAL KNOWLEDGE

It will be seen that the Dietz *et al.* model does not discriminate among different classes of mosquito vectors. A substantial complication, however, arises from inhomogeneities in the effects of insecticides. In the Garki project, an insecticide was applied to interior surfaces of houses and was therefore more effective against mosquitoes that rest indoors after a blood meal. If insecticides do not affect all mosquitoes equally, measurements of the average biting rate and of average longevity might be severely distorted (Molineaux *et al.*, 1979). Suppose that after spraying there are in effect two mosquito populations: a minority (initially $p_1 = 0.2$) that rest outdoors (exophilous) and are consequently relatively unaffected by the spraying, and a majority (initially $p_2 = 0.8$) that rest indoors (endophilous) and suffer greater mortality. Let $a_1 = 1$ bite per day and $\mu_1 = 0.05$ per day for the exophilic mosquitoes, and $a_2 = 2$ and $\mu_2 = 0.5$ for the endophilic ones. Let the latent period, T, equal 10 days for all mosquitoes. The correct way to calculate the effective index, $a \exp(-\mu T)/\mu$, which determines the transmission rate from infected mosquitoes to human hosts and is therefore relevant to evaluating the effects of intervention, is to take the appropriate arithmetic average of the separate indexes:

$$p_1 a_1 \exp(-\mu_1 T)/\mu_1 + p_2 a_2 \exp(-\mu_2 T)/\mu_2 = 2.5.$$

But if the endophilic and exophilic categories are not properly distinguished, then the above index is likely to be estimated with average values of a and of μ: $a \exp(-\mu T)/\mu = 0.073$ (where $a = p_1 a_1 + p_2 a_2 = 1.8$

and $\mu = p_1\mu_1 + p_2\mu_2 = 0.41$). Molineaux *et al.* (1979) analyzed the consequences of this aggregation phenomenon in considerable detail and emphasized that aggregating the groups will always underestimate the biting capacity of the mosquito population and hence will always underestimate the basic reproductive rate, z_0, and overestimate the impact of insecticides. As the above example makes clear, these incorrect estimates can lead to seriously wrong conclusions.

The above discussion captures the essentials of the most important surprise found in the Garki project. It was known from the outset that the intensity of transmission of malaria in these parts of Africa was very high. Indeed, the study showed the basic reproductive rate of the infection to be about 1,000; in other words, the transmission rate, or "vectorial capacity of the mosquito," was about 1,000 times the critical value required for the maintenance of endemic malaria. This very high transmission rate put malaria in tropical Africa in a category of its own and constituted one reason that a control program like that tried in the Garki project was not attempted earlier. Against this background, however, it was not understood that heterogeneities in the environmental setting of the mosquitoes might invalidate simple estimates of the efficacy of insecticide application, which were based on assumptions that the mosquito population was effectively homogeneous and that most mosquitoes rested indoors much of the time.

Thus, it was found that spraying with the residual insecticide propoxur did not have the expected effect on the prevalence of malaria, although coverage was as nearly complete as possible and the insecticide was very effective against the mosquito vectors (it produced a high mortality rate even at the beginning of the third wet season after the last application). Immigration of vectors or humans from unsprayed villages did not appear to be a significant factor. The decisive factor appears to have been the exophily of a substantial fraction of the mosquito vectors. The combination of exophily with a high biting rate maintained a high transmission rate. Thus, although the insecticide program was supplemented by administration of drugs at high frequency and with high coverage and it reduced malaria to a very low incidence, it failed to interrupt transmission.

In more detail, it was found that the effect of residual spraying in reducing *A. gambiae s.l.* populations, and hence malaria transmission, varied significantly among villages. The variation was probably related not to variations in spraying coverage or to distance from unsprayed villages, but to variation in the amount of exophily. The amount of exophily appeared to be a relatively stable characteristic of particular villages, and it also appeared to be associated with genetic differences within species of the *A. gambiae* complex. The vector population attached to a village appeared to be relatively isolated genetically most of the time. This was also in accord with the observation that the effect of spraying was influenced little by the size of the sprayed area.

As to the possible genetic mechanisms underlying the behavioral differences between endophilic and exophilic mosquitoes, Molineaux and Gramiccia (1980) stated:

A. gambiae s.l. in the study area is composed of *A. gambiae s.s.* and *A. arabiensis.* The dominant species is usually *A. arabiensis* but the relative abundance of the 2 species varies between times and places in ways which are not explained. *A. gambiae s.s.* is the more anthropophilic and has higher sporozoite rates; no clear-cut difference was demonstrated regarding exophily or effectiveness of propoxur. The vectorial capacity was estimated as if *A. gambiae s.l.* were a single species; appropriate simulations show that this is unlikely to have introduced a large error in the estimate. The cytogenetic investigations of Coluzzi suggest that resting behaviour and exposure to propoxur are related less to the relative abundance of the 2 species than to the intraspecific frequency of certain chromosomal inversions, some of which may be associated with a relatively stable behaviour pattern of the individual.

In short, the essential conclusion of the Garki study, which was not anticipated in the early planning, was that,

if the resting behavior of a mosquito species is genetically determined, exophily will be a stable characteristic of individual vectors, and the usual method of interpreting the impact of residual insecticides on longevity, which tacitly assumes uniform behaviour, is overoptimistic.

From the parasitological point of view, the study was interesting, in that its longitudinal nature permitted it to demonstrate that almost everyone was infected early in life, not only by *P. falciparum,* but very probably also by *P. malariae* and even by *P. ovale,* which is commonly regarded as a "rare" parasite. In addition, early demonstration of the effect of parasitism on immunity confirmed that a degree of immunity is evoked by malarial infection and that continual reinfection is necessary to maintain such immunity over the long term. Other clinical studies showed interesting relationships between body temperature and parasitemia and showed a significant effect of malaria control on frequency of fever and on anthropometric indicators of the nutritional status of children.

In another conclusion, Molineaux and Gramiccia (1980) stated that "the new mathematical model, painstakingly tested against hard facts, allows much more realistic simulations of the epidemiology of malaria, both before and after the application of control measures, than was previously possible." Cohen and Singer (1979) developed the model further, incorporating additional elements of realism (such as infection with more than one species of malaria and a more accurate description of immune responses in human hosts) while retaining an essential understanding of the process of malaria transmission.

REFERENCES

Aron, J. L., and R. M. May. 1982. The population dynamics of malaria. Pp. 139-179 in R. M. Anderson, ed. Population Dynamics of Infectious Diseases. Chapman and Hall, London.

Cohen, J. E., and B. Singer. 1979. Malaria in Nigeria: Constrained continuous-time Markov models for discrete-time longitudinal data on human mixed-species infections. Pp. 69-133 in S. A. Levin, ed. Lectures on Mathematics in the Life Sciences. Vol. 12. Some Mathematical Questions in Biology. American Mathematical Society, Providence, R.I.

Dietz, K., L. Molineaux, and A. Thomas. 1974. A malaria model tested in the African savannah. Bull. WHO 50:347-357.

MacDonald, G. 1952. The analysis of equilibrium in malaria. Trop. Dis. Bull. 49:813-828.

MacDonald, G. 1957. The Epidemiology and Control of Malaria. Oxford University Press, London.

Macdonald, G. 1973. Dynamics of Tropical Diseases. Oxford University Press, London.

Molineaux, L., and G. Gramiccia, eds. 1980. The Garki Project. World Health Organization, Geneva.

Molineaux, L., G. R. Shidrawi, J. L. Clarke, J. R. Boulzaguet, and T. S. Ashkar. 1979. Assessment of insecticidal impact on the malaria mosquito's vectorial capacity, from data on the man-biting rate and age-composition. Bull. WHO 57:265-274.

Ross, R. 1911. The Prevention of Malaria. 2nd ed. Murray, London.

Ross, R. 1916. An application of the theory of probabilities to the study of *a priori* pathometry. Proc. R. Soc. A 92:204-230.

Committee Comment

In general, the interactions between hosts and parasites are examples of prey-predator associations that have some simplifying features—especially for human host-parasite systems, in which the population size of the human host is usually determined by other factors and much information on transmission and maintenance of the parasite is often available. Many public health studies can be viewed as examples of the interaction of environmental problems (broadly defined) with ecological principles.

The Garki project is notable in that it was very thoughtfully designed, was maintained in an integrated way over several years, and carefully interdigitated the data gathered with a relatively realistic mathematical model. Moreover, the mathematical model, based on hard data and tested against data gathered in the course of the study, could be thought of as a paradigm for a prey-predator model. It was successful in explaining the patterns of prevalence before, during, and after the intervention program.

Interestingly, during construction of the basic model, no thought had been given to the possible effects of genetic and behavioral heterogeneity in the vector mosquito population. Such effects turned out to be the most important factor working against control. That a large fraction of the mosquitoes were exophilic meant that simple preliminary estimates of the overall transmission rate and of the likely effect of insecticides were incorrect.

This story illustrates the need for careful design in the gathering of data (and the extent to which a thoughtful mathematical model can guide the process) and shows that, no matter how carefully these things are thought through, unexpected complications are likely. Beyond these platitudes, the story also illustrates a specific theme found in much contemporary ecological theory. It seems increasingly likely that spatial heterogeneity, with different dynamic processes going on in separate patches, will be the most important factor in the overall persistence of many natural prey-predator or host-parasite associations (Hassell and May, 1985). The Garki project is certainly one example: behavioral heterogeneity in the mosquito population was, on the one hand, the most important factor in maintaining infection in the presence of an intervention program and, on the other hand, a factor not initially reckoned with (in a preliminary analysis drawn from the conventional traditions of ecological theory, which too often treat the world as homogeneous).

Reference

Hassell, M. P., and R. M. May. 1985. From individual behaviour to population dynamics. Pp. 3-32 in R. Sibly and R. Smith, eds. Behavioural Ecology. Blackwell, Oxford, Eng.

16

Protecting Caribou During Hydroelectric Development in Newfoundland

Single species are often judged to be the valued ecosystem components likely to be adversely affected by construction and operation of a development project. When that happens, ecological information and monitoring studies are oriented specifically toward the biology of the target species. Even if such a target species is well known, however, predicting the impact of the project on it can be difficult, and the project might have to proceed on a contingency basis, taking cautionary measures and altering them if effects on the target species are different from those anticipated. The effects of the Upper Salmon Hydroelectric Development on woodland caribou constitute an example of this type of problem.

Case Study

DAVID J. KIELL and EDWARD L. HILL, Environmental Services
Department, Newfoundland & Labrador Hydro, St. John's,
Newfoundland

SHANE P. MAHONEY, Newfoundland Wildlife Division, Department
of Tourism, Recreation and Youth, St. John's, Newfoundland

INTRODUCTION

An analysis completed in 1977 of projected power requirements for insular Newfoundland, compared with installed generating capacity, suggested that a shortfall would occur by about the end of 1982. A number of alternatives were available to meet this shortfall, but preliminary information suggested that the Upper Salmon Hydroelectric Development (USD) was the most economical option.

The USD is in south-central Newfoundland, approximately 170 km southwest of Gander and 50 km northwest of St. Albans, on Bay d'Espoir (Figure 1). It exploits a portion of the available head between Meelpaeg Reservoir and Round Pond, which are elements of the existing Bay d'Espoir hydroelectric project.

Before the USD, water flowed from Meelpaeg Reservoir through Crooked Lake and Great Burnt Lake down the North Salmon River to Round Pond (Figure 1). Water from Cold Spring Pond flowed down the West Salmon River to Godaleich Pond and then into Round Pond.

A dam on the North Salmon River now diverts water from Great Burnt Lake through two diversion canals into Cold Spring Pond. A dam on the West Salmon River creates a reservoir at Cold Spring Pond. Water from this reservoir is transmitted to the intake structure via a power canal 3.6 km long. A penstock, approximately 455 m long, delivers water to the generating station on the shores of Godaleich Brook, a tributary of Godaleich Pond. The lower 1,000 m of Godaleich Brook has been excavated and serves as a tailrace to transport water to Godaleich Pond. A channel improvement was constructed to control water levels in Godaleich Pond.

The powerhouse contains one Francis turbine, which generates 84 MW of power, using a hydraulic head of about 51 m.

Access to the USD area was achieved by construction of a 50-km road from St. Albans. The 230-kV transmission line from the USD parallels the access road within a corridor 1 km wide for most of its length and terminates at the powerhouse at Bay d'Espoir.

206

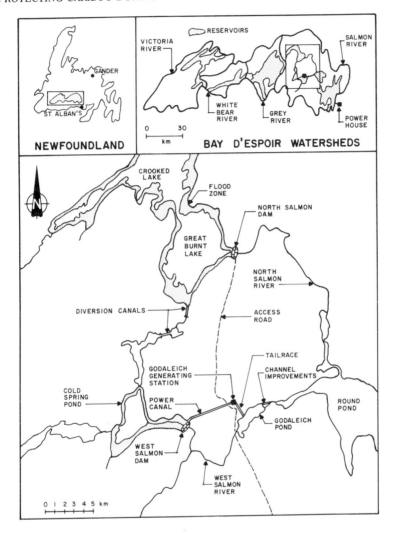

FIGURE 1   The Upper Salmon Hydroelectric Development.

A preliminary environmental analysis was undertaken in 1975 (Air-photo/Beak, 1976). Detailed engineering and environmental analyses of the USD were not begun until the spring of 1978. A study designed to evaluate the possible impacts of the USD on wildlife was commissioned with other studies on fish, forest and land resources, and socioeconomics.

In November 1978, discussion between Newfoundland and Labrador

Hydro (Hydro), the proponent of the project, and the Newfoundland Wild-life Division (NWD) revealed that the potential impact of the USD on woodland caribou *(Rangifer tarandus)*, particularly the large Grey River herd, was of great concern. It was agreed that the wildlife study in the area would not be sufficiently rigorous to permit adequate decisions related to caribou.

The USD received approval in principle from the provincial government in 1979, before the preparation of the environmental impact statement (EIS). At that time, it was generally recognized that the construction and operation of the USD would affect caribou, but the type and magnitude of impact were open to debate. Therefore, Hydro and NWD were faced with the problem of how to measure the impact of the USD on caribou.

THE ENVIRONMENTAL PROBLEMS

Caribou and the Upper Salmon Development

The USD is within the range of the Grey River caribou herd, one of the largest herds in Newfoundland (approximately 5,000 animals). The calving ranges of two smaller herds, the Sandy Lake (approximately 250 caribou) and the Pot Hill (approximately 500), are about 30 and 50 km, respectively, northeast of Godaleich Pond. Although the two smaller herds were included in the impact study, this case study deals exclusively with the Grey River herd.

The Grey River herd winters between the USD and the southern coast of the island. Its migration route to and from its summer range north of Great Burnt Lake is through and around the USD area (Figure 2). However, caribou can be found within or near the USD area throughout the year.

The Grey River caribou usually calve in the Wolf Lake-Dolland Pond area in late May to early June, but some have been known to calve near the southern end of Cold Spring Pond. Traditionally, these caribou have used the Cold Spring Pond area as part of their postcalving range during the last half of June and early July. This is where the animals strengthen the cow-calf bond and take advantage of emerging vegetation to regain the energy they have lost during the winter, pregnancy, and parturition. The cows require a highly nutritious diet at this season, because of the demands of lactation. After calving, the caribou gradually move north along both sides of Cold Spring Pond. By the middle of July, most have moved away from the upper Salmon area and onto their summer range.

The USD area is not merely a migration corridor through which caribou move as quickly as possible from calving grounds to summer range. It also contains essential grazing range and terrain used to minimize

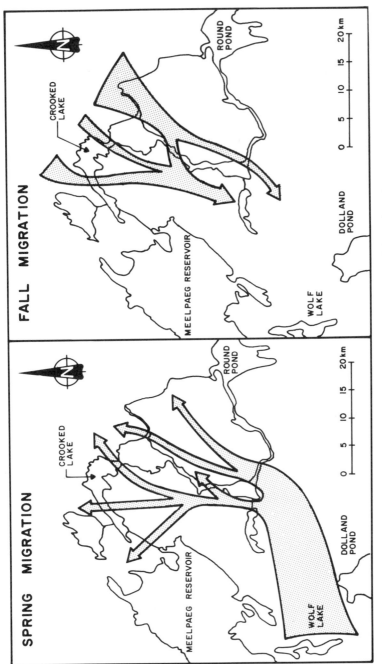

FIGURE 2   General spring and fall migration routes of the Grey River caribou herd through project area.

harassment by predators and insects. The apparently erratic movements with reversals of direction and stationary periods (other than those caused by construction activity) are normal in early summer. It is also normal for some animals to spend the summer in the USD area.

In this brief discussion of the movement patterns of the Grey River caribou, specific locations—such as point "x" where they winter and point "y" through which they migrate—have been avoided. The purpose of the avoidance was to indicate the inherent variability in caribou movements. A general range and general migration routes can be described for a herd, but its precise location at any time is very difficult to predict. An exception to this broad generalization is that caribou are usually faithful to a calving area. A number of factors can account for these relatively unpredictable movements: snow; other weather conditions; vegetation; harassment by predators, insects, or hunters; disturbance by developments, such as roads, pipelines, or hydroelectric projects; and caribou wandering.

During the last decade, much information has been accumulated about the effects of developments on wildlife. Some species, such as moose (*Alces alces*), benefit when clearing operations result in the growth of their preferred food items. Others, such as caribou, might be adversely affected. It has been shown that linear developments—such as roads, pipelines, and transmission lines—can block, delay, or deflect caribou movements (Banfield, 1974; Child, 1974; Klein, 1971). The amounts and types of traffic and human activity associated with such developments can affect the reactions of caribou to them (Horejsi, 1981; McCourt et al., 1974). Their responses also vary with group size, sex, and age. Females accompanied by young exhibit the strongest reactions (Cameron et al., 1979). The effect of blocking, delaying, or deflecting caribou movements might be to reduce or eliminate the use of part of their range. These effects might not become apparent for years.

In addition to the physical barrier effect, harassment can result in decreased rates of growth, development, and reproduction (Geist, 1975). Injuries and death from accidents could occur, as could avoidance of areas, which would lead to loss of important resources and ultimately a population decline. Increases in human access itself can also have an impact (Bergerud, 1979), for example, in the form of increased hunting pressure.

The sensitivity of caribou to development is debatable. Bergerud (1974) stated that caribou "seem to be both highly adapted and highly adaptable." He observed that caribou in Newfoundland do not have an aversion to roads or railroads themselves, but that traffic is a stimulus for flight. Banfield (1974) noted that caribou "are quick" to use cleared seismic lines for travel routes.

Ecological Questions and Issues

Ecological questions needed to be answered before the impact of the USD on caribou could be determined, including the following:

- Will the migration and distribution patterns of the herd change?
- Will calf production and mortality change?
- Will the age and sex ratios of the herd change?
- Will there be a measurable behavioral reaction to the USD?
- What will be the effects of increased human access to the previously isolated caribou range?

It was also recognized that measurement of changes was not sufficient. The reasons for changes and the long-term implications for the herd also needed to be known, particularly if corrective or preventive measures were to be undertaken. Both Hydro and NWD agreed that the key issue was whether statistically significant changes in population characteristics, such as herd size and age and sex ratios, could be attributed to the USD.

There were points of agreement and disagreement about the makeup of an adequate study. It was agreed that the measurement of changes in population characteristics (productivity and age and sex ratios) over time would be the ultimate indicator of the impact of the USD on caribou. Recognized wildlife techniques were available to obtain these measurements.

The most likely magnitude of impacts was debated early in the conceptualization of the study. The design of the study depended on such predictions. If many small impacts were expected, the study would have to be designed to measure them. However, if one or two major events were expected, a study of a different kind would be in order. The methods to be used, the area and duration of the study, and the analytical techniques to be used depended on the decisions reached.

NWD was concerned that construction of the USD would have a detrimental impact on the size of the herds. It argued that the main elements of the USD (the West Salmon dam, the power canal, the intake, and the penstock complex) were perpendicular to the general direction of movement of the Grey River herd from the postcalving area to the summer range. As a result, the USD could act as a barrier to normal migration and distribution. In the worst scenario, the postcalving migration would be blocked, and the herd would not migrate past the USD. If the blockage occurred for several consecutive years, it was feared, the caribou would lose their migratory tradition, and their range would in effect be halved.

Presumably, the population would eventually decrease to the carrying capacity of the new range. It was considered more likely, however, that the animals would move through the area and that the stress of confronting the barriers to migration would cause an energy deficit sufficient to reduce survivorship, reproductive success, and ultimately herd size. It was also hypothesized that caribou might migrate around the USD on the west side of Cold Spring Pond and so not use habitat within the USD area. Loss of this habitat might have a major impact on the annual energy budget of at least part of the herd.

Hydro's environmental staff shared Bergerud's (1974) opinion that caribou are relatively adaptable animals. Although there was reason for concern over the possible impacts of the USD on caribou, it was thought that any alterations in migration and distribution would be localized and brief. This opinion was supported by Bergerud (personal communication) when he was given the details of the proposed development. Hydro's position was that any project-related impacts would not have a significant effect on herd productivity. Hydro's biologists continued to argue that, to evaluate the impacts of a hydroelectric project on caribou, a study should emphasize individuals and groups of animals near the development. This approach would permit the identification of interactions that could be used to clarify the causes of impacts, if any, at the population level. It would also facilitate the development of effective mitigation measures for this and future projects.

OBTAINING RELEVANT ECOLOGICAL KNOWLEDGE

Population Characteristics, Migration, and Distribution

A study was developed during the winter of 1978-1979 to document the productivity and other population characteristics of the herds. It was not clear whether impacts would be severe enough to manifest themselves at this level, but the ultimate indicator of impact would be a significant decrease in productivity. Therefore, these important characteristics had to be measured.

NWD anticipated a large impact as a result of the physical and behavioral barriers presented by the USD. It suggested that this impact would be reflected in changes in herd migration and distribution patterns. Therefore, NWD proposed a study to document caribou movement and distribution on their range (8,000 km^2) with radiotelemetry. It also proposed an evaluation of the range aimed at understanding and predicting the implications of major changes in migration and distribution patterns.

Hydro was critical of some aspects of the proposed studies. There was agreement that productivity and other population characteristics had to be monitored. However, Hydro's biologists argued that impacts would probably be more subtle than major shifts in movement patterns and habitat uses. They suggested that remote sensing techniques, such as radiotelemetry, were too coarse to detect and document impacts and that field work near the USD area was more appropriate.

With some reluctance, Hydro agreed to cofund part of the studies proposed by NWD. The study was undertaken by NWD personnel and continued for 6 years (1979-1985)—through planning, construction, and 2 years of operation of the USD.

Behavior

In 1981, Hydro initiated a study to observe systematically the activities of caribou near the USD (the experimental area) and in control areas. The time devoted by caribou to distinct activities—such as feeding, lying, standing, and running—would be used to prepare activity budgets for animals in the experimental and control areas. Disturbance or harassment would alter activity budgets by increasing the incidence and duration of unproductive activities. It was hypothesized that disturbance of caribou by construction activities would be reflected in their activity budgets, even if major shifts in migration and distribution could not be detected. A similar study design was used to examine the reaction of caribou to the trans-Alaska pipeline and haul road (Roby, 1978). Observations were made through the summer of 1984, but most intensively in 1981 and 1982, when construction was going on.

Sensitivity Criteria

The environmental impact statement for the USD identified a number of mitigating measures that would facilitate caribou movement nearby. Among these measures was a restriction of blasting to times when fewer than 100 caribou were within 1.6 km of the blast site. The 1.6-km distance was based on a comment by Bergerud (1974) that one Newfoundland herd overwintered about 1 mile (1.6 km) from the trans-Canada highway. The sensitivity criterion of 100 animals was chosen arbitrarily.

As the USD proceeded, it was decided that similar protection was required for other construction activities, such as vehicular traffic on access roads and construction at various civil works areas and at borrow pits and quarries. The criteria varied with the sensitivity of caribou at the time of the activity—calving (May-June 15), postcalving aggregation (June 15-July 1), and migration (July). More protection was afforded caribou during calving than after calving and during migration, because they are more sensitive to disturbance at that time. The criteria also provided for an early warning system, so that project managers could be alerted if a large number of caribou were moving into an area and construction activities might have to be temporarily curtailed. When caribou were moving through the area, daily counts were taken from a helicopter (whenever weather permitted), to locate animals in reference to USD activities and to determine whether some elements should be shut down to minimize adverse impacts on caribou. NWD approved the criteria before implementation. Brazil (1983) and Northcott (1984) have described how the criteria were developed.

The main purpose of these mitigating measures was to facilitate decision-making with regard to caribou. They were not envisioned as integral parts of the caribou studies. However, as the caribou locations were mapped and data began to accumulate, a valuable picture of caribou movements in and around the USD emerged.

Conclusions

Although the data from the component studies have not all been completely analyzed, some conclusions have been reached.

• The productivity and the size and age structure of the herds studied have not changed significantly.

• The migration and distribution patterns of the Grey River herd appear to have been somewhat affected by construction of the USD. Calving activity was not observed near Cold Spring Pond during construction and by 1984 had not returned to the level reported in 1979. Rutting activity has not been observed in the USD area since 1979. Large numbers of migrating caribou have not been observed during the fall near Godaleich Pond since 1980. The major fall movement appears to occur west of Cold Spring Pond. The postcalving aggregation was concentrated west of the West Salmon dam during construction. Use of the area between the West Salmon dam and Godaleich Pond increased in 1983 and 1984 to form a pattern similar to that observed before construction. Dispersal from the postcalving aggregation appeared to change during construction as animals avoided crossing the North Salmon road.

• The maximal number of caribou observed in the USD area during the postcalving aggregation decreased from 2,100 in 1979 and 1980 (before construction) to 1,307 in 1981 and 821 in 1982 (during construction).

Numbers increased after completion of construction to 1,777 in 1983 and 2,039 in 1984 (Northcott, in press).

• Caribou appear to have become sensitized to construction activities in 1982. The proportion of behavior associated with energy expenditure or a state of disturbance increased significantly from 1981 to 1982, but then decreased in 1983 and 1984. This suggests that the major cause of disturbance was construction activity, rather than the structures (Hill, in press).

• The effectiveness of the sensitivity criteria as a mitigating measure has been questioned (Brazil, 1983; Northcott, 1984). Much of the criticism is related to problems associated with determining when the criteria were met and with stopping work. Although it cannot be proved, it can be argued that these measures helped some caribou to cross the USD area. These mitigation techniques require further refinement, if they are to have generic applications.

USES OF ECOLOGICAL KNOWLEDGE AND UNDERSTANDING

Valued Ecosystem Components

A preliminary environmental analysis of the USD was undertaken in 1975-1976 (Airphoto/Beak, 1976). Only two of the 16 recommendations pertaining to the predesign, design, and construction phases of the USD were related to wildlife:

• "Undertake field surveys of wildlife population, habitat, movement, and breeding characteristics. Place special emphasis on the ecology of the Godaleich delta and North Cold Spring Pond region" (Figure 1).

• "Recognize caribou migration requirements in design and construction of the access road, transmission lines, diversion canals and during reservoir filling, especially between Cold Spring Pond and Great Burnt Lake."

Caribou were not recognized as an ecosystem component of greater importance than many others, even though the study was managed by a government committee that included a representative of NWD. However, in November 1978, NWD voiced its concern that a general wildlife inventory was insufficient. It suggested that studies at the USD concentrate on caribou, for the following ecological and socioeconomic reasons:

• Caribou were a species of recreational and commercial importance.
• Caribou represented the wilderness experience to Newfoundlanders and therefore had aesthetic value.

- Caribou had historically been hunted by local Indians and were important to the Indians' heritage.
- Known range preferences and patterns of migration and distribution conflicted directly with proposed construction activities.
- Caribou had been perceived to be very sensitive to human disturbance.
- Caribou populations had been steadily recovering from serious overhunting in the early 1900s, and it was feared that this trend could be halted or reversed if the USD substantially reduced the Grey River herd.

Therefore, during the fall of 1978, woodland caribou populations came to be seen as a valued ecosystem component. Monitoring studies and mitigating measures were adopted in response to this new concern.

Significance of Impacts

The Newfoundland government approved construction of the USD in the spring of 1979, in spite of possible environmental impacts. Approval was based on the need to prevent the energy shortfall that was projected to occur in 1982. The government, however, required that all reasonable mitigation to reduce possible impacts on caribou be undertaken and that the cooperative study be undertaken. These requirements were stipulated in the Order-in-Council that authorized Hydro to proceed with the USD. Thus, the political decision regarding the relative significance of the two resources (hydroelectric and caribou) was made very early in the development of the study program.

On a technical level, Hydro and NWD biologists generally agreed that a statistically significant change in herd productivity would demonstrate that an important impact had occurred as a result of the project. This was a "primary" level of significance.

Impacts that might not have a direct effect on herd productivity (e.g., changes in distribution or behavior) were assigned "secondary" significance. They were important for two main reasons. First, such impacts are often easy to mitigate; for example, vehicular traffic could be curtailed to allow animals to cross access roads. Second, many small impacts could accumulate; therefore, if a statistical decrease in productivity were observed, an analysis of secondary impacts might help to explain it.

Establishment of Boundaries

Establishing spatial boundaries for the studies was not difficult once the scope of the studies was determined. The decision that herd productivity was the ultimate measure of impact required an investigation of animals in their whole range. Calving and postcalving areas were emphasized because of the vulnerability of animals around the project area. However, some investigations—particularly those regarding predation, distribution, and range quality—were undertaken on the summer and winter ranges.

The temporal boundary for the NWD-Hydro cooperative study was established shortly after the start of the first field season. It was thought that any substantial USD-related impacts would occur during construction, but that it was necessary to continue the study for at least 2 years into the operation phase, to determine whether the animals were becoming accustomed to the presence of the USD within their range.

The main thrust of the behavior study was to monitor changes in activity budgets of individuals and groups of caribou as they confronted the USD. This required that the study concentrate on areas near the USD, although control data were obtained away from the experimental area.

The sensitivity criteria established for the USD specifically defined spatial boundaries. The boundaries were directly related to the location of construction activities and varied with the caribou's life-history stage. Inasmuch as the sensitivity criteria were implemented to make decisions on construction activities, the temporal boundary was defined by the construction schedule of the USD.

Development and Implementation of Study Strategy

Detailed planning of the study program began after two political decisions were taken: that the development would be built and that caribou constituted a valued ecosystem component. Therefore, the program was oriented toward the monitoring of impacts, rather than assisting in making a "go, no-go" decision, which is usually the underlying goal of environmental impact assessment.

Impact prediction was one objective during the first year of the cooperative NWD-Hydro study. However, in following years, the focus was on monitoring impacts.

In this context, Hydro and NWD biologists began developing the required studies. The program's first study was to monitor the population characteristics, migration, and distribution pattern of the herd. The study had to be undertaken in a technically competent manner, within a schedule and within a reasonable budget. Various options were examined, including the use of consulting firms. However, it was decided that such a study would afford the third author (S. P. M.) an opportunity to undertake doctoral studies at the University of Calgary. This provided a valuable academic orientation to the study and reduced the cost and administrative burden to Hydro, the proponent.

It was also agreed that NWD would manage the study, provide technical assistance from its existing staff, and provide some of the aircraft time. Hydro would pay for the remainder of the flying time and the radiotelemetry equipment and drugs that were needed. Thus, the study developed as a cooperative effort involving the responsible resource agency and the proponent. Further comments on this relationship have been provided by Mahoney (1983) and Kiell (1984).

As previously discussed, Hydro initiated behavior studies near the USD. But this was not a unilateral action; NWD supported the initiative, helped to develop the terms of reference for the study, and reviewed proposals received from consultants. During the study, data were shared with NWD, and in the later stages NWD assisted in obtaining some of the necessary control data.

The main expense and responsibility of implementing the sensitivity criteria were borne by the proponent. Helicopter surveys of the USD area, in cooperation with the Newfoundland Department of Environment, provided valuable information on caribou numbers and distribution to both NWD and the consultants conducting the behavior study. Telemetry surveys by NWD assisted in the scheduling of these monitoring flights. Trained observers for the surveys were provided periodically by NWD and the consultants.

The overall strategy was one of cooperation. There were periods of disagreement, but generally this approach was satisfactory for the proponent, NWD, and the graduate student.

Development of Predictions and Hypotheses

The program involved three distinct groups, or interests: Hydro, NWD, and the academic interest, each with a staff of biologists. Therefore, hypothesis-testing, although not always specifically mentioned, was fundamental to the design of the studies. The testable null hypotheses included the following:

• The reproductive rates of the herd before and after the USD would not differ.
• The age and sex ratios of the herd would not change during the study.
• The numbers of caribou using the construction area as a postcalving site would not change during the study.
• The behavior of caribou encountering the USD area would not change

from one year to another and would not differ between control (undisturbed) situations and disturbed circumstances.

- Caribou mortality patterns would not change during the study.

These null hypotheses were tested with accepted wildlife techniques, and the data were analyzed in a statistically rigorous manner.

Monitoring

The primary objective of the program was to monitor the impact of the USD on caribou. The program treated the USD as an experiment. In 1979-1980, control (predevelopment) data were obtained on caribou population biology for comparison with similar data obtained during and immediately after construction. Control data for the behavior study were obtained in the same year as the experimental data, but were collected away from the influence of the USD. These studies were designed so that any changes due to the USD would be detected and statistically verified.

Information from these studies was often used to make USD-related decisions. For example, knowledge of the migration routes of caribou from telemetry data obtained in 1979 was used to determine areas of the reservoirs that should be cleared to facilitate future caribou migration (Kiell, 1981). Underlying the reservoir-clearing program was the assumption or prediction that caribou would use these areas in future years.

Information from the sensitivity-criteria surveys was provided directly to the decision-makers who planned the daily activities during construction. Data from this study served such a specific use that they were not expected to contribute to the general knowledge of caribou movements or behavior. Hydro in particular saw this study in a very restricted way. In retrospect, slight changes in the design of the program would have yielded more rigorous data and would have added another element to the knowledge of caribou migration, distribution, and behavior.

Cumulative Effects

The study program was designed to examine the impact of the USD on caribou at the individual and population levels. It was necessary to collect data on caribou ecology before, during, and after construction, to evaluate cumulative effects. A short-term study would have been unable to detect or assess cumulative effects.

One of the deficiencies in the program has been the lack of thought regarding the effect of increased human access caused by the USD. It now appears that this may be the most important impact of the development, because of general disturbance and increased human legal or illegal hunting. The problem has been addressed in policy, but no data from this program were collected to assist in policy development.

SOURCES OF ECOLOGICAL KNOWLEDGE

Ecological Facts

The main product of the program was ecological facts about caribou that could be compared before, during, and after construction of the USD. Well-documented wildlife research techniques were used to obtain information on herd size, sex and age ratios, calf production and survival (productivity), and habitats. Radiotelemetry allowed researchers to determine general migration routes and spatial and temporal distribution patterns. A specifically designed study documented the effect of the USD on caribou behavior.

Before the study program was implemented, information about the biology of the Grey River herd was obtained from published papers, mostly by Bergerud, in the late 1960s and early 1970s (e.g., Bergerud, 1971) and from NWD data files. Studies of barren-ground caribou (northern Canada and Alaska) and of reindeer (Scandinavia) provided some guidance for interpretation of Newfoundland data, although natural history information was not directly applicable. Experiences of local people and wildlife technicians provided a valuable historical perspective during the program.

The study program supplied most of the data needed for decision-making during construction and operation of the USD. These data were used in making decisions, such as on where to clear the reservoir and whether vehicular traffic should be allowed to operate as usual. Specific, high-quality, timely information was needed for these decisions; management data obtained by NWD in the 1960s were not adequate.

Time and budget constraints did not preclude studying some ecological elements, but were important in determining the depth of studies and precision of the data. For example, studies on the winter range were minimized by logistics.

Theory and General Principles

The theoretical framework of basic animal ecology and ethology formed the foundation of the studies. Population dynamics and bioenergetics underlay the approach to the program. It was felt that, if there were significant impacts from the USD, they would affect the energy budget of the caribou and ultimately be reflected in changed reproductive and mortality rates.

Specific Models

Airphoto/Beak (1976) reviewed available data on caribou and developed an overview of caribou migration and distribution patterns around the USD. On the basis of data from the first year of the study, Mahoney (1980) improved this overview to depict the general boundaries of calving and postcalving areas, migration corridors, and summer and winter ranges. A similar picture was produced for each later year to help organize data and to see whether there were marked differences between years. Figure 2 is an example of the types of information summarized in these descriptions.

Analog Studies

The development of oil resources in the Canadian and American North in the last decade required construction of pipelines to markets in the South or to tanker ports. Many studies have been undertaken to investigate the impact of pipelines, vehicular traffic, compressor noise, helicopter harassment, etc., on caribou migration, e.g., those of Banfield (1974), Roby (1978), Cameron *et al.* (1979), Miller and Gunn (1979), Hanson (1981), Horejsi (1981), and Smith and Cameron (1983). These studies were useful to the program at the USD, because they highlighted the issues that needed attention, provided suggestions on study techniques, assisted in formulating study design, and generated a group of experienced resource people with whom to discuss ideas.

Pilot Studies

The speed with which decisions were made and studies initiated precluded the use of pilot studies. If time had allowed, pilot studies to check the effectiveness of radiotelemetry equipment might have been useful. Development of receiving equipment, the use of radio collars, and calibration of equipment took place as the program proceeded.

The Project as an Experiment

The considerable flexibility in this program was exercised when there was a compelling reason to do so. For example, as the program developed, collars were installed almost exclusively on calves. Calves were considered to be most vulnerable to potential disturbance, and their migration and distribution patterns were considered to reflect those of cows. It was thought that more and better data would be obtained from this approach.

The evolution of the caribou sensitivity criteria also illustrates how activities were modified according to need. Originally, the criteria applied only to blasting (Newfoundland and Labrador Hydro, 1980). As construction activities multiplied, the criteria were modified to include all project activities and to be cognizant of the differing degrees of sensitivity to disturbance expected during the calving, postcalving, and migratory periods. This flexibility worked to protect the caribou around the USD.

Expert Judgment

Table 1 summarizes the expertise used and the role of the various ecologists in the program. The program was being funded and administered mainly by NWD and Hydro (both of which agencies had environmental staffs), so the need for purely academic input was identified at an early stage. This group of ecologists helped to design the studies, provided expert advice on request, and reviewed study components periodically to ensure a high degree of scientific integrity. Two scientists, D. R. Klein and E. Reimers, moved to the USD from Alaska and Norway, respectively, to gain firsthand experience with the situation. The technical input from these experts was valuable, and the confidence that they instilled in study administrators and less experienced biologists was very important.

CONTRIBUTION OF ECOLOGICAL KNOWLEDGE

This case had three attributes that contribute to its usefulness: scientific methods were applied in designing the program, sound environmental management strategies were adopted in scoping the studies and organizing their implementation, and ecological knowledge was used to address the problems at hand.

The main features of the scientific method are the framing of testable hypotheses and the design of studies in such a way that statistical confidence can be attached to the results through spatial and temporal control and bounding of observations. These elements were aspects of the study described here.

All environmental studies that contribute to decision-making are constrained. Prudent management of time, funds, and staff resources is essential for success. In this case, the decision by NWD and Hydro to undertake the studies in-house with the assistance of selected outside experts (e.g., academic advisors and consultants) was a major step toward maintaining scientific flexibility and obtaining the most from the resources available.

TABLE 1   Expertise Used During Caribou Studies at the USD

Group	Affiliation	Personnel	Experience and Role in Program
Professors	Brock University	A. Banfield	Telephone discussions early in study development; experience in analogous studies
	University of Victoria	A. Bergerud	Telephone discussions; broad experience with Newfoundland caribou
	Acadia University	D. Dodd	Review of early reports
	University of Calgary	V. Geist	Ph.D. advisor; study design advice; visited site
	University of Alaska	D. R. Klein	Technical advice on behavior study; visited site
	University of Oslo	E. Reimers	Technical advice on behavior study; visited site; familiarity with Scandinavian experience
	Memorial University of Newfoundland	D. J. Stewart	Assistance with behavior study design
Graduate student	University of Pennsylvania	D. Roby	Review of early reports; experience on analogous studies
Government	NWD	Staff (3)	Administration; wildlife techniques
	Newfoundland Department of Environment	Staff (2)	Administration; experience in environmental impact assessment
	Department of Fisheries and Oceans	J. Rice	Assistance with behavior study design and statistical analysis
Proponent	Hydro	Staff (6)	Management and administration with environmental background; wildlife biology techniques
Consultants	Northland Associates	T. Northcott	Wildlife biology; experience with Newfoundland mammals

Finally, it was recognized early in the program that an ecological approach was needed to answer the complex questions about the interaction of caribou with a hydroelectric development. Information of different types was obtained from various sources. Personal communication with scientists knowledgeable about the biology of caribou in Newfoundland and about the interaction between caribou and industrial development elsewhere provided an ecological perspective. Data in the primary and "gray" literature provided a relatively good model of caribou distribution and migration patterns in and around the USD and identified the boundaries of important habitats, but did not provide much insight into the effects of building a hydroelectric project in their range. Original research with standard wildlife management techniques (population census, aging, and classification techniques), modern methods (radiotelemetry), and behavior analysis (time and energy budgets) generated specific ecological data on the animals. These data were interpreted in the context of experiences elsewhere (analog studies) and ecological principles, such as population dynamics and energetics.

REFERENCES

Airphoto/Beak (Airphoto Analysis Associates Consultants Ltd. and Beak Consultants Ltd.). 1976. Upper Salmon/Cat Arm: Environmental Impact Assessment (Preliminary). Airphoto Analysis Associates Consultants Ltd., Toronto, Ont.

Banfield, A. W. F. 1974. The relationship of caribou migration behaviour to pipeline construction. Pp. 797-804 in V. Geist and F. Walther, eds. The Behaviour of Ungulates and Its Relation to Management. International Union for Conservation of Native and Natural Resources, Gland, Switz.

Bergerud, A. T. 1971. The Population Dynamics of Newfoundland Caribou. Wildl. Monogr. 25:1-55.

Bergerud, A. T. 1974. The role of the environment in the aggregation, movement and disturbance behaviour of caribou. Pp. 552-584 in V. Geist and F. Walther, eds. The Behaviour of Ungulates and Its Relation to Management. International Union for Conservation of Native and Natural Resources, Gland, Switz.

Bergerud, A. T. 1979. Access: Greatest threat to caribou. West. Guidelines 2:5-9.

Brazil, J. 1983. Reactions of Caribou Observed During Construction at the Upper Salmon Hydroelectric Project, October, 1979-August, 1981. Unpublished manuscript. Newfoundland Wildlife Division, St. John's, Nfld.

Cameron, R. D., K. R. Whitten, W. T. Smith, and D. D. Roby. 1979. Caribou distribution and group composition associated with construction of the trans-Alaska pipeline. Can. Field-Nat. 93:155-162.

Child, K. N. 1974. Reaction of caribou to various types of simulated pipelines at Prudhoe Bay, Alaska. Pp. 805-812 in V. Geist and F. Walther, eds. The Behaviour of Ungulates and Its Relation to Management. International Union for Conservation of Native and Natural Resources, Gland, Switz.

Geist, V. 1975. Harassment of large mammals and birds. Unpublished report to the Berger Comm., University of Calgary, Calgary, Alta.

Hanson, W. C. 1981. Caribou (*Rangifer tarandus*) encounters with pipelines in northern Alaska. Can. Field-Nat. 95:57-62.

Hill, E. L. In press. A preliminary examination of the behavioural reaction of caribou to the Upper Salmon Hydroelectric Development in Newfoundland. In McGill Subarctic Research Papers. Vol. 40. McGill University Centre for Northern Studies, Montreal.

Horejsi, B. L. 1981. Behavioural responses of barren ground caribou to a moving vehicle. Arctic 34:180-185.

Kiell, D. J. 1981. Development of a reservoir preparation strategy. Can. Water Res. J. 7:112-131.

Kiell, D. J. 1984. Environmental decision-making during planning, construction, and early operation of the Upper Salmon Hydroelectric Development in Newfoundland, Canada: A case study. Pp. 352-374 in Proceedings of the Facility Siting and Routing '84 Symposium, Banff, Alberta. Environmental Protection Service, Environment Canada, Ottawa, Ont.

Klein, D. R. 1971. Reaction of reindeer to obstructions and disturbances. Science 173:393-398.

Mahoney, S. P. 1980. The Grey River Caribou Study. Newfoundland Wildlife Division and Newfoundland and Labrador Hydro, St. John's, Nfld.

Mahoney, S. P. 1983. The trend toward bio-politics. Caribou News 3(3):12-13.

McCourt, K. H., J. D. Feist, D. Doll, and J. J. Russell. 1974. Disturbance studies of caribou and other mammals in the Yukon and Alaska, 1972. Arctic Gas Biological Report Series. Vol. 5. Canadian Arctic Gas Study Ltd. and Alaskan Arctic Gas Study Co., Calgary, Alta.

Miller, F. L., and A. Gunn. 1979. Responses of Peary caribou and musk oxen to helicopter harassment. Occasional Paper 40. Canadian Wildlife Service, Minister of Supply and Services Canada, Ottawa, Ont.

Newfoundland and Labrador Hydro. 1980. Upper Salmon Hydroelectric Development: Environmental Impact Statement. Newfoundland and Labrador Hydro, St. John's, Nfld.

Northcott, P. L. Impact of the Upper Salmon Hydroelectric Development on the Distribution and Movement of the Grey River Caribou Herd in 1982. Unpublished manuscript. Newfoundland Department of Environment.

Northcott, P. L. In press. Impact of the Upper Salmon Hydroelectric Development on the Grey River caribou herd. In McGill Subarctic Research Papers. Vol. 40. McGill University Centre for Northern Studies, Montreal.

Roby, D. D. 1978. Behavioral Patterns of Barren-Ground Caribou of the Central Arctic Herd Adjacent to the Trans-Alaska Oil Pipeline. Unpublished master's thesis, University of Alaska, Fairbanks.

Smith, W. T., and R. D. Cameron. 1983. Responses of caribou to industrial development of Alaska's Arctic slope. Acta. Zool. Fennica 175:43-45.

Committee Comment

Development projects, such as hydroelectric dams, have impacts on ecosystems both during construction and during their operation. Measures designed to reduce undesirable impacts of these two phases of a project are often different, as they were in the case of the Upper Salmon Hydroelectric Development. Construction of the dam was accompanied by much vehicular traffic and blasting. The latter ceased on project completion, but traffic, made possible in part by access roads built for construction, was clearly a continuing and underestimated problem whose nature changed as people began to use the area for purposes other than those occurring during construction.

During the planning phase, caribou were perceived as the ecosystem component of greatest concern and value. There were reasons to believe that the large Grey River herd might be affected during migration and calving. Planners, however, were confronted with conflicting information that suggested that caribou either might be very sensitive to the project or might be adaptable and become accustomed readily to the kinds of disturbance caused by construction and operation of the dam. As a result, although a great deal was known about caribou biology and this information was used extensively during the planning and construction of the project, it was necessary to develop a plan for monitoring the behavior of animals and taking cautionary measures when animals were close to the construction sites, even if their responses to construction activity could not be predicted. In addition, roads were designed to be readily crossed by caribou and revegetation was encouraged, in the hope that the roads would pose minimal barriers to caribou movements once the project was completed.

The monitoring during construction revealed that use of the general project area by caribou was greatly reduced, but that use increased after construction. The long-term effects of the project remain to be determined. Indeed, one of the disappointing aspects of the project is that monitoring was carried out for only 2 years after project completion. This is insufficient to determine whether caribou are moving through the project area as they did before construction, whether the efforts to make roads readily crossable have been successful, whether caribou are crossing the new lakes where old trees were cleared to facilitate their crossing, and whether there are long-term changes in the site and behavior of the herd.

A major problem, and one that is likely to occur regularly in similar projects, is that public access to previously roadless areas is greatly facilitated by roads built for project construction. The general disturbance and illegal hunting in the area might be much more serious for the long-term welfare of the Grey River caribou herd than was the disturbance during construction itself. The problem of increased use of areas after construction of projects needs to be addressed formally and incorporated into policies governing such projects, perhaps including methods of limiting access of people to the areas at especially critical times.

17

Conserving a Regional
Spotted Owl Population

Human influence on the environment has reduced the populations of many plants and animals. Small populations are vulnerable to extinction, because of the difficulty of finding mates and because of random fluctuations, disease, unfavorable weather, and other catastrophic events. When populations of a species are small and fragmented (a common situation when habitat destruction is the cause of the reduction in population size), long-term survival can also be threatened by genetic deterioration—the deleterious effects of inbreeding and loss of genetic variability.

The survival of spotted owls is of increasing concern, because their required old-growth forest habitat is being reduced and fragmented by logging. The planning for spotted owl management described here is based on theoretical population genetics and ecology, as well as on quantitative natural history—observations on reproductive ecology, dispersal, and foraging behavior. Although the underlying theory is not yet fully developed, it is applicable to many attempts to conserve populations of vertebrates with low reproductive potential.

Case Study

HAL SALWASSER, USDA Forest Service, Wildlife and Fisheries
Staff, Washington, D.C.

INTRODUCTION

Maintaining the full diversity of native vertebrates is a legal mandate
of the federal government (Endangered Species Act of 1973, as amended;
National Forest Management Act of 1976), as well as a policy of many
state resource management agencies. In the late 1970s, the United States
Department of Agriculture (USDA) Forest Service adopted regulations
(36 CFR 219) requiring habitats to be managed to maintain viable populations of all native vertebrates in the national forest system—over 190
million acres of land and water.

Since the establishment of the national forests in the early 1900s, state
and federal policies and lack of attention have resulted in the loss of
species, such as wolves (*Canis lupus*) and grizzly bears (*Ursus arctos*),
from some forests. Land and resource management planning under the
National Forest Management Act of 1976 is intended to prevent further
loss of species.

The spotted owl (*Strix occidentalis*) constitutes a major test of the policy
of maintaining species in national forests. The northern subspecies (*S. o.
caurina* Merriam 1898) inhabits mature and old-growth coniferous forests
in the Cascade, Klamath, and Coast Range mountains of Washington,
Oregon, and California. It appears to need stands of trees more than 24
in. in diameter, a multilayered canopy more than 70% closed, and large
standing and fallen dead trees (Forsman *et al.*, 1984). Such forests have
extremely high commercial timber value (Heinrichs, 1983), and therein
lies the spotted owl dilemma: the kind and amount of habitat required for
survival of each pair of owls is a resource highly valued by an industry
that is the economic backbone of the Pacific Northwest.

Further fragmentation of old-growth forest might impede dispersal of
owls and isolate populations that are too small to survive for long. Thirty
to forty years ago, half the original 15 million acres of old-growth forest
in the Pacific Northwest remained (Franklin, 1984), and the spotted owl
was not an issue in forest management. Timber harvests in the last few
decades have removed nearly all the easily accessible lowland old-growth
forest, and the much-reduced spotted owl population now exists primarily
in rugged, mountainous terrain.

A major purpose of the national forests is to sustain yields of different kinds of resources. Timber, wildlife, and wilderness are specifically mentioned in the law (Multiple-Use Sustained Yield Act of 1960; Wilderness Act of 1964). What is the most efficient way to manage forests to maintain both a continuous flow of timber products and viable populations of spotted owls and their coinhabitants of old-growth forests?

By the mid-1970s, the biology of the spotted owl was under study, and inventories had been initiated. In 1974, biologists began to develop management guidelines based on the new information, to prevent the owl from declining so much that formal listing as a threatened species would be necessary. Research and inventories have continued, and the management guidelines have evolved to incorporate new findings. In 1981, the guidelines were revised, largely under the influence of the work of Soulé (1980), to include new theories and observations concerning the genetic basis of population viability.

This chapter presents an overview of an evolving population management plan for the fragmented spotted owl population over a large area of the Pacific Northwest. The plan attempts to integrate management of individual national forests over the whole Pacific Northwest region. Rather than attempting to review the considerable research on spotted owl biology in detail, this discussion incorporates demographic estimates based on the research.

THE BASIC PROBLEM AND APPROACH

One objective of the Forest Service has been to develop plans for the national forests in the Pacific Northwest that protect resident populations of the spotted owl while allowing the multiple uses of the forest required by law. The primary threat to the owl is further reduction and fragmentation of its old-growth forest habitat through logging. Thus, the plan must deal with the effects of logging on the suitability of habitat for long-term maintenance of individual pairs, populations in specific national forests, and the whole regional population. Hence, the problem is to manage logging and related activities so that remaining old growth will support long-term survival of spotted owls in all national forests in the Pacific Northwest. Four general ecological issues are most important:

• It is necessary to determine the habitat characteristics required for the survival and successful nesting of individual pairs of owls. This ultimately involves detailed studies of owls to determine patterns of habitat use and factors that influence habitat quality, the relationship between habitat quality and home range size, and specific requirements for nesting, foraging, roosting, and dispersal. Several studies are being conducted.

Owls have been censused to establish which habitats they occur in and with what frequency and to determine the spacing between pairs. Radio-telemetry has helped to determine patterns of habitat use and the size of home ranges of individual birds, and cast pellets have been examined to determine what prey are taken. These studies have provided the input for habitat-suitability index models that can evaluate habitat quality on the basis of specific features (Laymon *et al.*, in press).

• The distribution of habitats is critical to the successful dispersal of juveniles and movement of adults into suitable vacant habitat. Radio-telemetry has been used to determine patterns of dispersal of juveniles and to determine what kinds of habitats they will not cross. Adults with radio transmitters have also been followed to determine their movements between areas and throughout the year. The information can be used to develop criteria for minimal distances between patches of habitat during dispersal and for resident birds and to determine the best configuration of habitat that connects populations in adjacent forests.

• The population in each forest must be large enough to withstand normal environmental fluctuations and random demographic changes without becoming locally extinct. In addition, the regional population must be sufficiently large and well distributed to withstand severe environmental fluctuations and reductions and even extinction of local populations. These requirements are being studied with a general model of regional population dynamics and with demographic values determined for the spotted owl from field studies and from the literature on better-studied species. The model is evolving to incorporate estimates of adult and juvenile mortality, range of dispersal, reproductive success, and other variables.

• An effective population large enough to minimize the deleterious genetic effects associated with small and isolated populations must be maintained. The northwest spotted owl population is a *metapopulation* (Levins, 1970), i.e., a large regional population made up of many smaller populations of varied sizes, densities, and degrees of isolation from one another. Metapopulations are subject to two genetic problems: a loss of average individual fitness through an increase in the frequency of breeding of close relatives (inbreeding depression) and a reduction in potential for evolutionary adaptation because of a loss of genetic variability. Both these problems can be reduced by maintaining a number of relatively large local populations with substantial gene flow among them. A model similar to that used for population dynamics is being used with values for the spotted owl from field studies and from the literature on more intensively studied species.

Planning for protection of the spotted owl began in the early 1970s and continues today. An interagency task force of state and federal scientists and managers developed interim spotted owl management guidelines in the middle and late 1970s; later research led to a revision in 1981. The focus was initially on the kinds and amounts of habitat needed by individual pairs. The late 1970s saw increased academic interest in long-term population viability as the threat of mass extinction of species became widely recognized. The result has been more research on factors underlying extinction (e.g., Frankel and Soulé, 1981; Schonewald-Cox, 1983; Soulé and Wilcox, 1980). A 1982 workshop that included management biologists and academic ecologists studying population viability used the spotted owl as a basis for developing a general risk-assessment planning process for long-term population management (Salwasser *et al.*, 1984). The approach has been revised to incorporate new information on other species and the results of a second workshop that was held in the fall of 1984.

Debate concerning a plan for managing spotted owls has focused on several issues. First, what are the specific habitat requirements of individual pairs of owls? The management unit for individual pairs or small groups of pairs with contiguous home ranges is called a spotted owl management area (SOMA). The type, amount, and distribution of habitat that should constitute an adequately managed SOMA have been controversial. Underestimation of minimal home range size or quality of sites needed or too much fragmentation of habitat within a SOMA could reduce the probability of survival and reproduction of a pair. Second, are the demographic estimates used in the management models sufficiently accurate? At issue are estimates of adult and juvenile mortality, reproductive rate, dispersal distance, habitat occupancy, and related population characteristics. Third, are the models themselves adequate and appropriate? Are demographic or genetic problems more important to long-term viability? The current guidelines are based on the assumptions that genetic problems are more critical at the scale of the regional population and that preventing genetic deterioration will prevent demographic collapse. Demographics and biogeography are assumed to be most critical at the scale of a forest population.

The overriding issue concerns the minimal regional population size necessary for long-term survival and consequently the amount and distribution of old-growth forest to be provided in the future. Points of view range from the position of some environmental interest groups, that too much old growth has already been cut and that habitat management criteria (particularly minimal SOMA characteristics) are inadequate, to the view of some representatives of the timber industry, that current management guidelines are too stringent and that the current estimate of minimal acceptable population size is too large. The current guidelines were under administrative appeal in March 1985, and legal action might be taken by parties at both extremes.

ECOLOGICAL KNOWLEDGE USED IN DEVELOPING THE MANAGEMENT GUIDELINES

The management guidelines are based on data, principles, and theory. In addition to field research on the biology of spotted owls themselves, the guidelines rely heavily on the findings of recent theoretical and empirical research on the long-term viability of populations of other species, both in captivity and in the wild. Studies on spotted owls are used to estimate values for general models of population viability based on this recent literature.

Studies of the Ecology of Spotted Owls

Research on spotted owls has focused on breeding biology, foraging ecology, habitat use, and general distribution. Because spotted owls are nocturnal and now inhabit rugged, mountainous terrain almost exclusively, field research is very difficult, and much of their biology is still poorly known.

Individual pairs of owls do not reproduce every year, and clutch size averages only about two eggs (Forsman, 1980). In addition, the survival of juveniles appears to be extremely low; no juveniles with radio transmitter harnesses have survived to breed. Hence, spotted owls appear to have a very low reproductive potential and thus poor ability to recover from reductions in population size.

Spotted owls in the Northwest appear to be nonmigratory, and radio-telemetry has shown that adults move over extremely large home ranges in the course of a year. A home range can be as large as 8,300 acres for an individual owl and 10,400 acres for a pair (Forsman et al., 1984). The old-growth coniferous forests occupied by spotted owls in the Pacific Northwest are over 200 years old and have several layers, with many standing dead trees and new trees coming up in gaps where old ones have fallen. The multilayered structure of these forests is believed to affect foraging success, partly through an effect on the abundance of the owl's prey—primarily arboreal rodents, such as flying squirrels, voles, and wood rats. Multilayered forests also provide cool microsites that allow owls to avoid heat stress, to which they are apparently sensitive. The broken-off tops of mature trees characteristic of old-growth forest also provide nest sites for the owls (Forsman, 1980). Spotted owls are not known to breed in young second-growth forests and are rarely found there.

Although it is generally assumed that home ranges must be larger in areas with lower habitat quality, and home ranges are indeed larger in heavily logged areas (Forsman, 1980), there is no good evidence of a general correlation between habitat quality and home range size within a study area. Territories might be abandoned as a result of timber harvest (Forsman *et al.*, 1984). In any year, not all territories are occupied; the average occupancy rate can be as low as 50% and does not exceed 75% of potentially suitable sites. The minimal home range size that would adequately support a breeding pair is still being debated and probably varies with the individual pair, terrain, forest type, and forest distribution. Sizes estimated from censuses and radiotelemetry studies range from 740 acres to 8,300 acres in heavily logged areas (Forsman *et al.*, 1984; Marcot, 1978). Logging in a pair's home range reduces habitat quality, not only by reducing the amount of high-quality habitat, but also by fragmenting the forest. Fragmentation leads to greater travel time for foraging and greater exposure to predators, such as great horned owls. The effects of habitat removal can be minimized by retaining travel corridors of old-growth forest to connect the larger stands.

Some juvenile spotted owls have been reported to disperse as far as 100 miles from their areas of birth, although most do not travel that far (Gutierrez, personal communication). It is not known why no radiotagged juveniles have survived to breed. They might be too inefficient in finding suitable habitat or suffer high mortality when occupying low-quality habitats during dispersal. It is also possible that the radio transmitters themselves increase the chance of death in young and inexperienced owls. Adults apparently do not like to cross large open areas (Forsman *et al.*, 1984), and they use corridors of old-growth forest, when available, to travel between old-growth stands, even if much longer distances must be covered.

Long-Term Population Viability

Studies of population viability have focused on how and why the risk of extinction increases as populations become smaller and more isolated from other populations. Interest has been not only in the fate of local populations, but also in how long-term viability of a regional population or an entire species can be affected by changes in the size, makeup, and distribution of its constituent local populations.

Reviews of the factors that have led to recent extinctions indicate that natural agents—such as predation, competition, parasitism, and disease—have rarely been the cause of extinction and that the reduction, alteration, and isolation of habitats through human activity are ultimately more important (Frankel and Soulé, 1981; Hester, 1967; Soulé, 1983; Terborgh and Winter, 1980; Ziswiler, 1967).

Two categories of factors increase the risk of extinction of small and isolated populations (Shaffer, 1981; Soulé, 1983; Terborgh, 1974): internal changes and external stresses. Population productivity can decrease as a result of random changes in fertility rate, litter size, sex ratio, death rate, immigration rate, and so on. Behavioral dysfunction occurs in some species below a threshold population size, often disrupting breeding. Genetic drift can lead to loss of genetic variation, with a loss of potential for adapting to environmental change. Studies of captive populations have also shown that inbreeding depression occurs when fewer individuals—all closely related—are available to mate; this situation often decreases reproduction and survival (Soulé, 1980). External factors also have more severe effects as populations become smaller and more isolated. Habitat change, intense predation or competition, fire, drought, and floods can leave too few individuals for recovery to occur. In short, when populations become very small, the probability increases dramatically that random or even regular periodic events will reduce population numbers so much that recovery can occur only if individuals immigrate from other populations. As populations become more fragmented and isolated, the supply of potential immigrants diminishes.

For a population to survive, it must have demographic resilience, i.e., resistance to extinction due to random demographic changes or environmental fluctuations. Theoretical studies suggest that as few as 10 adults can sustain a population for decades; but, for any given species, this minimum depends on social system, reproductive potential, generation length, the nature of random events, and other factors (May, 1973). Some empirical studies suggest that 50 adults might be necessary for a reasonable chance of surviving for several decades (Shaffer, 1981). For a population to survive, it must also maintain adequate average individual fitness in the face of inbreeding depression (Chambers, 1983) and must have sufficient genetic variability to allow adaptation to environmental change (Soulé, 1980). Populations lose genetic variation through random changes in gene frequency (Kimura, 1983). In small populations, inbreeding is more likely and genetic variation is more likely to be lost through genetic drift. Genetic variation arises through mutation and immigration, and rough estimates indicate that, to avoid extinction through genetic deterioration, a population must be about 10 times larger than that necessary to maintain demographic resilience (Soulé, 1980).

The initial guidelines for management described below were based on the assumption that adequate demographic resilience will be maintained if a population is managed to minimize genetic deterioration. Current planning is assessing the roles of demographic variation and population fragmentation as well.

In a genetically "ideal" population, individuals have an equal probability of mating with each other, fecundity is constant, population size is constant, and generations are discrete. Few real populations meet these criteria (Hartl, 1980; Kimura, 1983; Kimura and Crow, 1963; Wright, 1938). To predict random genetic change in a population of average size N, when the population characteristics depart from the ideal, an *effective* population size, N_e, is often used in lieu of the actual census size. The calculation of N_e typically incorporates adjustments for deviation from the ideal in sex ratio, variation in litter size or survival of offspring to reproductive age, overlapping of generations, random mating, and fluctuation of population number (Kimura, 1983). N_e is used to calculate inbreeding coefficients, and it is usually less than the census N—sometimes less than 20% of it. Formulas for estimating the effects of population structure on N_e have been given by Franklin (1980), Hartl (1980), Frankel and Soulé (1981), Kimura (1983), and Thomas and Ballou (1983).

Corrections for three factors are often incorporated into the calculation of N_e: sex ratio different from 1:1, variation in offspring survival, and population fluctuations.

If not all adults breed, some genes are less likely than others to be passed on. A biased sex ratio can create such an effect, and N_e should be adjusted as follows:

$$N_e = 1/[1/(4N_m) + 1/(4N_f)], \qquad (1)$$

where N_m and N_f are the numbers of adult males and females.

The reproduction of genes is biased when the production of offspring varies between parents. The effect of this variation on N_e is given by:

$$N_e = 4N/(2 + V), \qquad (2)$$

where V is the variance in survival of offspring per parent and N is the actual population size.

When a fluctuating population is well below the average size, genes can be reproduced in a biased fashion as a result of genetic drift. The longer a population remains small (i.e., fails to recover from decrease), the greater the effect. It will be more pronounced, therefore, in species with low powers of increase. The effect on N_e is as follows:

$$N_e = t/(1/N_{e1} + 1/N_{e2} + \dots + 1/N_{et}). \qquad (3)$$

where t is the number of generations stipulated and N_{et} is the effective number (N adjusted for sex ratio and offspring variance) in generation t.

The total effect of these three factors can be approximated by:

$$N_e = NSOP, \qquad (4)$$

where S, O, and P are the ratios of N_e to N calculated for sex ratio, offspring variance, and population fluctuation, respectively. This approximation should be sufficiently precise for general wildlife habitat management, in which case other information is likely to be much less precise than the estimate of N_e.

The inbreeding coefficient, F, is calculated with the overall value of N_e as derived above. Empirical data on the actual effects of inbreeding in small populations of normally free-ranging and outbred species of mammals have shown that even a small amount of inbreeding is correlated with reduced fecundity and reduced survival of offspring (Ralls and Ballou, 1982, 1983; Ralls *et al.*, 1979). Free-ranging populations normally have behavioral and ecological patterns that keep natural inbreeding low. The formula for F is as follows:

$$F_t = 1 - (1 - [1/(2N_e) + 0.5]t, \qquad (5)$$

where F_t is the inbreeding coefficient in generation t, t is the number of generations from time zero, and N_e is the effective population number during the period of interest.

Some of the variables in these formulas cannot be measured in wild populations, and Soulé (1980) cautioned that basing management on estimates of N_e can yield only rules of thumb, rather than reliable quantitative results. As a rule of thumb, Franklin (1980) and Soulé (1980) suggested that an N_e of 500 or more might approach the balance point between random loss and addition of genetic variation for many species, and Lande (1980) suggested that populations of several hundred are at little more risk than very large ones. The life history and population structure of a species must be considered before such a rule of thumb is applied, but there is now little basis for determining a more specific threshold value for any species.

Soulé (1980) has proposed that wild populations are at risk of extinction because of genetic factors when F_t reaches 0.5 (Figure 1). Because Formula 5 does not consider the mitigating effects of migration between populations or mutations, it overestimates the inbreeding coefficient, sometimes substantially (Hartl, 1980; Kimura, 1983; Figure 2). Figure 1 shows that, assuming a generation time of 2 years for the spotted owl, a regional population of 500 or more will be sufficient to provide protection against genetic deterioration for many centuries. It is assumed that the inbreeding coefficient calculated for the regional population also applies

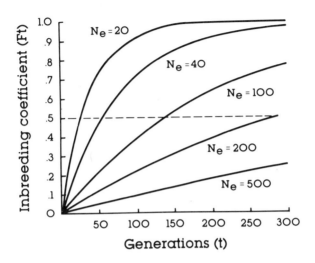

FIGURE 1 Inbreeding coefficient, F, increases as function of effective population number, N_e, and number of generations. At low N_e, F approaches dangerous extent of inbreeding in fewer generations.

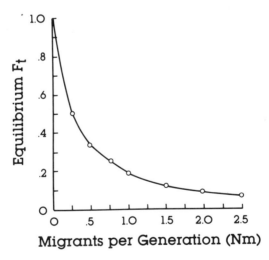

FIGURE 2 Higher migration rates of reproductively successful individuals into small population from larger population offset effects of inbreeding.

to all local forest populations, because they are not totally isolated from each other.

THE PLANNING PROCESS FOR SPOTTED OWLS

The planning process for spotted owls is derived from a general protocol (Salwasser *et al.*, 1984) and now consists of eight steps (Figure 3).

• *Step 1*: The northern spotted owl in the Pacific Northwest was identified as a species of concern, for several reasons. The owl is obligately dependent on a habitat that is now rare (old-growth coniferous forest) and that is being reduced rapidly, and individuals require large amounts of the habitat. The population is becoming increasingly fragmented, as a result of continued timber operations, and some individual forest populations might be close to the lower limit of adequate short-term demographic resilience—about 50-100 individuals. Planning is required now, so that the regional population does not decrease below the size necessary for long-term survival. Because the spotted owl has the most stringent requirements for old-growth habitat, managing old growth to protect spotted

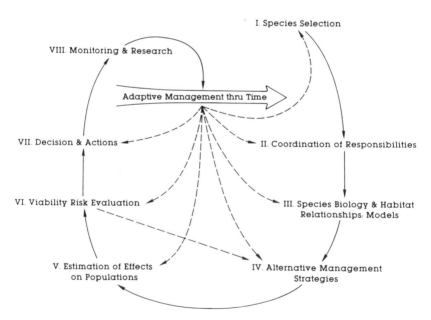

FIGURE 3   Eight-step process for planning viable population and analyzing risk.

owls will probably also protect most, if not all, other species that depend on old-growth forest (Raphael and Barrett, 1984).

• *Step 2*: Planning responsibility is assigned on the basis of range and distribution of the species. Censuses have been performed on all national forest land and evaluated in conjunction with information from adjacent lands to identify areas that might become effective biological reserves for the species, regardless of ownership and prevailing land use.

• *Step 3*: Habitat requirements for spotted owls and the best distribution of habitat within and between individual forest populations are determined, and the results lead to development of habitat capability models that describe the full range of habitats over which the species occurs (Nelson and Salwasser, 1982). The models can be used to guide habitat planning and to determine the effects of different land-use patterns on the species (Laymon *et al.*, in press). On the basis primarily of radiotelemetry studies of spotted owl habitat use in Oregon, a minimally suitable year-round home range for a pair is presumed to include 300 contiguous acres of mature to old-growth forest and 700 acres within 1.5 miles of the nest site (Forsman *et al.*, 1984).

To facilitate interchange among local populations and to make the occupancy of suitable habitat more likely, each managed habitat area must be within the normal dispersal distance of the species—6-12 miles, according to the results of radiotelemetry studies. The key concern is that loss of a piece of habitat not lead to permanent isolation of a local population. The best pattern of habitat distribution would entail several connections among suitable pieces of habitat, so that the loss of one connection would not isolate any piece. Linear patterns should be avoided.

• *Step 4*: Population and habitat requirements are translated into land-use planning variables. This is accomplished by assigning each individual forest (or other planning unit) a quota for the number of pairs to be maintained (as determined from the overall risk analysis, discussed below) and by specifying how habitat in each forest is to be managed. In the national forests, this plan must be flexible to accommodate multiple land uses. There will inevitably be alternative plans for meeting these multiple objectives, each of which will involve trade-offs among individual goals, such as protecting spotted owls.

• *Step 5*: The alternative management strategies are projected to estimate their effects on key population dimensions, such as N, N_e, and the structure of local demes. The projections use habitat and population simulation models with explicit assumptions about rates of systematic changes and the importance of random variation.

For each planning alternative, two substeps are performed: estimation of the population size in each forest that the planned habitat distribution can support (habitat capability) and estimation of the effective population size (N_e).

Habitat capability for each forest can be estimated with the habitat management criteria for owl pairs described above and with censuses of owls and maps of projected habitat distribution. The formulas presented earlier can be used with population data on spotted owls to estimate N_e both for individual forests and for the regional population (Formulas 4 and 5). Spotted owls breed monogamously, and, because there is no indication of a biased sex ratio, the coefficient S is assumed to be 1.0. Data on variance in offspring production are few, but studies have shown that complete reproductive failure occurs in some years (Gutierrez, personal communication). Assuming that variance in offspring production would be at the high end of the range for vertebrates, the coefficient O is assigned a value of 0.66 (inferred from Crow and Morton, 1955).

Accurate counting of the nocturnal spotted owl is difficult, because the failure of individuals to respond to imitated or recorded calls might indicate only unresponsiveness, rather than absence. Population fluctuation in spotted owls is probably small, because populations of other species with low reproductive rate and high adult survivorship generally fluctuate little. It is assumed that populations do not decrease to less than 50% of the average more frequently than 1 year in 5 and that it takes about 3 years to recover from a decrease to 50% of average N. Use of these assumptions in Wright's (1938) harmonic-mean formula gives a value of 0.76 for the coefficient P. Combining these coefficients (Formula 4) yields an estimated N_e of half the census N (1.0 × 0.66 × 0.76). Thus, N_e for the northwest regional spotted owl population, currently estimated at more than 2,000 adults, would be about 1,000, or 500 pairs. The N_e for any isolated population can be calculated similarly by multiplying the census number by 0.5.

- *Step 6*: An effective degree of protection (Table 1) is determined for a species through a risk analysis of the estimated demographic, generic, and geographic results of the management alternatives. The eventual management goals will depend explicitly on the degree of protection desired for the species and will be in essence a value judgment involving a cost-benefit analysis of possible protective measures. Obviously, the greatest protection would be obtained if no more habitat were altered or removed, but that is not always practical.

Assessing the degree of protection with each alternative involves evaluating the expected size of the whole regional population and the size and degree of isolation of each forest unit. If isolation is not expected, the potential demographic resilience of a forest population can be evaluated

TABLE 1   Degrees of Protection for Species, Expected Population
Viability, and Habitat Required for Use in National Forest Planning

Degree of Protection	Viability	Population Pattern Supported
1	*Individual*: Survival likely for only a few years to a few decades	Several individuals, isolated in forest; no interchange with species out of forest
2	*Individual*: Survival likely up to several decades, depending on N and distribution	Family, social group, or small population isolated in forest; deme of 10-30 adults
3	*Short-term local population resilience*: Survival likely for 1 to a few decades	Several reproductive or social groups isolated in forest; deme of 30-60 adults
4	*Mid-term local population resilience*: Survival likely for several decades	Well-distributed forest population, isolated from rest of species; deme of 60-100 adults
5	*Long-term local population resilience*: Survival highly likely for several decades to a century	Well-distributed forest population with at least degree 4 protection; part of population with N_e in mid-100s
6	*Short-term adaptability*: Survival of populations likely for a century	Well-distributed forest population, with at least degree 4 protection; part of population with N_e in mid-100s
7	*Mid-term adaptability*: Survival of populations highly likely for a century	Well-distributed forest population(s), with at least degree 4 protection; part of population with N_e of 500-1,000
8	*Long-term adaptability*: Survival of populations likely for millennia	Distinct, well-distributed forest populations each with at least degree 4 protection; part of population with N_e greater than 1,000, whose demes could diverge genetically
9	*Evolutionary fitness*: Survival of populations highly likely for millennia	Distinct, well-distributed forest populations, each with at least degree 4 protection; part of population whose demes with N_e greater than 1,000 could diverge genetically

NOTE: Adapted from Schonewald-Cox (1983).

with the habitat capability model, because some immigration can be expected to offset the risk of temporary population declines. Local populations that are not isolated from others are assumed to experience the
same inbreeding coefficient as the regional population as a whole (Figure

1). When a proposed alternative is expected to result in isolation of a local population, Steps 5 and 6 should be performed for that population. Land-use planning should evaluate the likelihood that planned or unplanned events will eliminate adjacent habitats at the weakest points in the habitat distribution.

• *Step 7*:   This step, the decision process, involves consideration of other resource concerns and values for the areas of land in question. Biological assessments are blended with social, political, and economic issues. The decisions often entail a social preference among competing uses of the land.

• *Step 8*:   A monitoring and research program is developed. It should reflect the degree of protection and the potential environmental costs of management and should allow for evaluation of critical assumptions used in the risk assessment protocol (Salwasser *et al.*, 1983). It should also stipulate the variables to be measured and the frequency of measurement and should address the issue of measurement reliability.

A monitoring program has three major goals. First, compliance with the plan must be monitored. For example, if 100 SOMAs are allocated to one national forest, it must be determined whether all SOMAs are being maintained properly and whether the actual spatial relationships of the various SOMAs are acceptable. Second, monitoring must show whether management is achieving its resource goals. For example, are the SOMAs supporting as many pairs of owls as expected, is reproductive rate adequate for long-term population stability, and is genetic variation being maintained? Third, information must continually be gathered and used to update and revise the plan. Some of this information must come from experimental research.

Key assumptions in current planning involve dispersal behavior of juveniles, the nature of the owls' dependence on stands of old-growth forest for reproduction and survival, and the adequacy of population sizes for maintaining demographic resilience. These assumptions are being studied by a number of agencies (Ruggiero and Carey, 1984). There are no plans to conduct intensive research on genetic factors, as long as at least 1,000 adults are maintained in the regional population. A demographic model for the spotted owl is being modified to incorporate genetic considerations and the effects that demographic variation might have on viability.

CONTRIBUTION OF ECOLOGICAL KNOWLEDGE TO THE CASE STUDY RESULTS

Until recently, efforts to protect populations were based primarily on site-specific habitat management, to ensure the survival and reproductive success of individuals and breeding pairs. Although such efforts are still critical to species management, it is now recognized that local populations cannot be protected for long without studies of interactions among populations that might exchange individuals—studies that have been central to developing a regional management plan for the spotted owl.

But metapopulation and other genetic models are only as good as the values that go into them. In spite of recent research on spotted owls, there is little assurance that the values of variables used in the models are accurate. Many must be derived from studies on unrelated species. Information on distribution and abundance is generally accurate, but knowledge of survivorship, reproductive rate, dispersal, and other demographic characteristics is sparse. Thus, it is essential to view current management plans and their assumptions as hypotheses to be tested. Both monitoring and research must be involved in the tests. The current plan is flexible and adaptive, and it takes the uncertainty of the effects of our proposed management actions into account.

ACKNOWLEDGMENTS

Contributions to this chapter have come from many people. Michael Soulé's lucid writing on conservation biology prompted the planning process described. Eric Forsman, Rod Canutt, and Dean Carrier worked out the basic method for specifying suitable habitats and distribution (Step 3). Jack Ward Thomas showed how to use the species habitat niche to represent resource needs in a practical way (Steps 3 and 4). Karl Siderits and Bob Radtke pioneered the use of diversity standards to provide for the habitat needs of all wildlife (Step 4). John Lehkuhl, Ed Harshman, and Daniell Jerry tested the application of population theories to determining the number of individuals needed (Step 5). Michael Soulé, Daniel Goodman, Michael Gilpin, James Brown, Linda Joyce, Tom Hoekstra, Mark Shaffer, Curt Flather, Dick Holthausen, Bill Burbridge, Brad Gilbert, Charlie Phillips, Maureen Beckstead, Tom Burke, and Paul Brouha assisted in workshops, brainstorming, and review of manuscripts to develop the analytical strategy and process.

REFERENCES

Chambers, S. M. 1983. Genetic principles for managers. Pp. 15-46 in C. M. Schonewald-Cox, S. M. Chambers, B. MacBryde, and W. L. Thomas, eds. Genetics and Conservation: A Reference for Managing Wild Animal and Plant Populations. Benjamin/Cummings, Menlo Park, Calif.

Crow, J. F., and N. E. Morton. 1955. Measurement of gene frequency drift in small populations. Evolution 9:202-214.

Forsman, E. D. 1980. Habitat Utilization by Spotted Owls in the West-Central Cascades of Oregon. Ph.D. thesis, The Oregon State University, Corvallis.

Forsman, E. D., E. C. Meslow, and H. M. Wight. 1984. Distribution and biology of the spotted owl in Oregon. Wildl. Monogr. 87:5-68.

Frankel, D. H., and M. E. Soulé. 1981. Conservation and Evolution. Cambridge University Press, New York.

Franklin, I. R. 1980. Evolutionary change in small populations. Pp. 135-149 in M. E. Soulé and B. A. Wilcox, eds. Conservation Biology: An Evolutionary-Ecological Perspective. Sinauer, Sunderland, Mass.

Franklin, J. F. 1984. Characteristics of old-growth Douglas-fir forest. Pp. 328-334 in Society of American Foresters. New Forests for a Changing World. Society of American Foresters, Bethesda, Md.

Hartl, D. L. 1980. Principles of Population Genetics. Sinauer, Sunderland, Mass.

Heinrichs, J. 1983. The winged snail darter. J. For. 81:212-262.

Hester, J. 1967. The agency of man in animal extinctions. Pp. 169-192 in P. S. Martin and H. E. Wright, eds. Pleistocene Extinctions: The Search for a Cause. Yale University Press, New Haven, Conn.

Kimura, M. 1983. The Molecular Theory of Evolution. Cambridge University Press, New York.

Kimura, M., and J. F. Crow. 1963. The measurement of effective population number. Evolution 17:279-288.

Lande, R. 1980. Genetic variation and phenotypic evolution during allopatric speciation. Am. Nat. 116:463-479.

Laymon, S. A., H. Salwasser, and R. H. Barrett. In press. Habitat suitability index models: Spotted owl. U.S. Fish Wildl. Serv. Biol. Rep.

Levins, R. 1970. Extinction. Pp. 77-107 in M. Gerstenhaber, ed. Some Mathematical Questions in Biology. Vol. II. American Mathematical Society, Providence, R.I.

Marcot, B. G. 1978. Prolegomena of the Spotted Owl (*Strix occidentalis*) in Six Rivers National Forest. Tech. Rept. USDA For. Serv. Six Rivers National Forest, Eureka, Calif.

May, R. M. 1973. Stability and Complexity in Model Ecosystems. Princeton University Press, Princeton, N.J.

Nelson, R. D., and H. Salwasser. 1982. The Forest Service wildlife and fish habitat relationship program. Trans. N. Am. Wildl. Nat. Resour. Conf. 47:174-183.

Ralls, K., and J. Ballou. 1982. Effect of inbreeding on juvenile mortality in some small mammal species. Lab. Anim. 16:159-166.

Ralls, K., and J. Ballou. 1983. Extinction: Lessons from zoos. Pp. 164-184 in C. M. Schonewald-Cox, S. M. Chambers, B. MacBryde, and W. L. Thomas, eds. Genetics and Conservation: A Reference for Managing Wild Animal and Plant Populations. Benjamin/Cummings, Menlo Park, Calif.

Ralls, K., K. Brugger, and J. Ballou. 1979. Inbreeding and juvenile mortality in small populations of ungulates. Science 206:1101-1103.

Raphael, M. G., and R. H. Barrett. 1984. Diversity and abundance of wildlife in late successional Douglas-fir forest. Pp. 352-360 in Society of American Foresters. New Forests for a Changing World. Society of American Foresters, Bethesda, Md.

Ruggiero, L. F., and A. B. Carey. 1984. A programmatic approach to the study of old-growth forest-wildlife habitat relationships. Pp. 328-334 in Society of American Foresters. New Forests for a Changing World. Society of American Foresters, Bethesda, Md.

Salwasser, H., C. K. Hamilton, W. B. Krohn, J. F. Lipscomb, and C. H. Thomas. 1983.

Monitoring wildlife and fish: Mandates and their implications. Trans. N. Am. Wildl. Nat. Resour. Conf. 48:297-307

Salwasser, H., S. P. Mealey, and K. Johnson. 1984. Wildlife population viability—A question of risk. Trans. N. Am. Wildl. Nat. Resour. Conf. 49:421-439.

Schonewald-Cox, C. M. 1983. Conclusions: Guidelines for management: A beginning attempt. Pp. 414-446 in C. M. Schonewald-Cox, S. M. Chambers, B. MacBryde, and W. L. Thomas, eds. Genetics and Conservation: A Reference for Managing Wild Animal and Plant Populations. Benjamin/Cummings, Menlo Park, Calif.

Shaffer, M. 1981. Minimum population sizes for species conservation. BioScience 31:131-134.

Soulé, M. E. 1980. Thresholds for survival: Maintaining fitness and evolutionary potential. Pp. 151-169 in M. E. Soulé and B. A. Wilcox, eds. Conservation Biology: An Evolutionary-Ecological Perspective. Sinauer, Sunderland, Mass.

Soulé, M. E. 1983. What do we really know about extinction? Pp. 414-446 in C. M. Schonewald-Cox, S. M. Chambers, B. MacBryde, and W. L. Thomas, eds. Genetics and Conservation: A Reference for Managing Wild Animal and Plant Populations. Benjamin/Cummings, Menlo Park, Calif.

Soulé, M. E., and B. A. Wilcox. 1980. Conservation Biology: An Evolutionary-Ecological Perspective. Sinauer, Sunderland, Mass.

Terborgh, J. 1974. Preservation of natural diversity: The problem of extinction-prone species. BioScience 24:715-722.

Terborgh, J., and B. Winter. 1980. Some causes of extinction. Pp. 119-134 in M. E. Soulé and B. A. Wilcox, eds. Conservation Biology: An Evolutionary-Ecological Perspective. Sinauer, Sunderland, Mass.

Thomas, W. L., and J. Ballou. 1983. Equations and population management. Pp. 414-446 in C. M. Schonewald-Cox, S. M. Chambers, B. MacBryde, and W. L. Thomas, eds. Genetics and Conservation: A Reference for Managing Wild Animal and Plant Populations. Benjamin/Cummings, Menlo Park, Calif.

Wright, S. 1938. Size of population and breeding structure in relation to evolution. Science 87:430-431.

Ziswiler, Z. 1967. Extinct and Vanishing Animals. Springer, New York.

Committee Comment

As we continue to alter and fragment the habitat of more and more species of plants and animals, we add to the list of species that are composed of metapopulations at an alarming rate. Only recently has it been widely recognized that such species cannot be protected against the threat of extinction by managing habitat for their local constituent populations. This recognition is the crucial first step toward maintaining the long-term viability of threatened species, but the scientific issues raised will not be easy to deal with, and the social and political steps required will not be easy to accomplish. Habitats of many species are distributed without regard for national boundaries and are often in countries where pressure for modifying those habitats is intense.

Even if we had control over habitat destruction, we would be far from understanding how to protect these species. Theoretical research in meta-population management is in its infancy, and the data needed for emerging models are difficult to acquire and are lacking for most species. The spotted owl management plan constitutes one of the first attempts to incorporate metapopulation modeling into usable management guidelines. But, as pointed out by Salwasser, even the basic issue of whether demographic or genetic constraints are more critical is being hotly debated, and only very general rules of thumb for determining minimal population sizes are available. In Salwasser's spotted owl case study, genetic constraints are assumed to be more stringent, and the model used is based primarily on limiting genetic deterioration. Until more research is done on demographic models, we cannot be confident that this assumption is valid.

In a recent study, Lande (1985) suggested that demographic factors might be more critical than genetic factors to the survival of spotted owls. Lande used two independent analytical methods—a basic life-table analysis and an analysis of habitat occupancy—and concluded that current management plans for the spotted owl will eventually result in demographic collapse. Lande's study demonstrates the importance of exploring every available analytical approach before making irreversible habitat management decisions.

In addition to the deterministic approaches used by Lande, Shaffer (1981) and others have been developing a Monte Carlo simulation approach for determining minimal viable population size. Various demographic values can be used in an iterative stochastic simulation model that projects population size and makeup for many generations. By simulating both environmental and demographic variability, the models clearly demonstrate that population viability is a matter of probability. How long do we wish the population to avoid extinction? What risk (probability of extinction) are we willing to accept? Management decisions ultimately depend on what are essentially value judgments.

On a metapopulation level, we must also address the issue of whether near-term (centuries) protection against demographic collapse is more important than providing the potential for long-term evolutionary change. Overall management plans aimed at these two different goals will often be very different. Is one very large population—less likely to become extinct—more desirable than a system of smaller, connected populations that are more likely to facilitate evolutionary change in response to locally changing environments? Clearly, protecting a species against demographic collapse is a first priority, for, unless we maintain a viable population, we will not even have the chance to tackle the longer-term problems of loss of genetic variation.

The management plan presented in this chapter will undergo many changes with a better understanding of metapopulation dynamics and genetics and with more demographic and ecological data on spotted owls. The clear and accessible presentation of the plan, however, is of great value, because it focuses the debate and research needed for the achievement of its goals.

References

Lande, R. 1985. Report on the Demography and Survival of the Spotted Owl. Paper prepared for the National Wildlife Federation, Portland, Oreg.

Shaffer, M. 1981. Minimum population sizes for species conservation. BioScience 31:131-134.

18

Restoring Derelict Lands
in Great Britain

Reclamation of derelict lands is of increasing importance for industrialized societies, both because quantities of lands so affected are increasing and because the value attached to restoring natural vegetation on these lands is increasing. Early reclamation attempts tried to produce vegetation that looked pleasing and that reduced soil erosion. More recently, society has demanded the creation or restoration of more complex natural communities of plants and animals, even though this goal is more difficult and expensive. The task requires knowledge of the tolerances of plants for the environmental conditions; of the ways in which plants compete with one another for access to water, light, and mineral resources; and of the course of succession of plant species after establishment of pioneering vegetation. Less attention has been paid to restoration of animal communities, in part because animals often invade these sites naturally once suitable vegetation has been established, and in part because less is known about members of animal communities and how to manipulate them.

The Longstone Edge reclamation project in Derbyshire, England, represents the current state of the art of land reclamation. The project drew on existing ecological knowledge about plant competition and succession, but also included experiments to help develop new theories more applicable to the specific conditions of reclamation with which the investigators had to work. A particular part of the challenge—one that is generally a component of such projects—was to produce vegetation that would be self-regenerating once established, so that costly maintenance would not need to be continued for long periods. In Britain, most derelict sites are on very poor soils; the specifics of the Longstone Edge results are, accordingly, applicable to many other sites in that country. In North America, however, many sites being restored are on richer soils, and techniques adapted to those conditions, as well as to the more continental climates prevailing in most of North America, will need to be developed to meet challenges there.

Case Study

PETER WATHERN, Department of Botany and Microbiology, University College of Wales, Aberystwyth

INTRODUCTION

Bradshaw and Chadwick (1980) estimated the amount of derelict land in the United Kingdom in April 1971 at over 55,000 hectares (ha), and the consensus is that the amount has increased steadily in recent years. Much of this dereliction is the result of mining activities, and pressures on resources in the United Kingdom are such that mineral deposits are likely to be worked with little or no concern for their location. Thus, mining operations often occur in some of the most scenically attractive and environmentally sensitive areas, such as national parks. Examples include Dartmoor (kaolin), Yorkshire Dales (heavy metals and fluorspar), and the Peak District (heavy metals, fluorspar, and limestone). Many areas, including western and northeastern Wales and Cornwall, carry a legacy of dereliction from mines long defunct.

In the United Kingdom, reclamation of despoiled land has been a fertile area for research for more than 2 decades, since the pioneering work in the Lower Swansea Valley of Wales reported by Hilton (1967). Reclamation research has been geared to developing techniques for achieving three main objectives: to restore land so that it can be used productively, to remove local sources of environmental pollution associated with toxic materials in discarded wastes, and to make a site visually attractive again. Reclamation has generally been achieved by establishing a vegetation cover that prevents further weathering and allows such productive uses as agriculture, forestry, housing, manufacturing, and recreation. Thus, reinstatement can be regarded as an essential component of the recycling of land through a succession of uses.

In some instances, derelict land detracts markedly from an otherwise pleasing landscape and must be reclaimed in the interests of visual amenity, irrespective of a readily apparent productive after-use. The major concern in the reinstatement of such sites is the creation of an area that, at worst, blends with the surrounding landscape or, preferably, is pleasing in its own right. This case study describes experimental work undertaken to develop techniques for dealing with such situations.

Longstone Edge, Derbyshire, is a limestone scarp situated in a prominent position in the heart of the Peak District National Park. It visually dominates an area intensively used for recreation. Fluorspar deposits are found in a vertical vein (locally known as a rake) running for about 1.5 km along its crest. The deposits are worked by a combination of opencast and deep mining techniques. The area has a long history of lead mining, and the steep scarp face of Longstone Edge is covered with waste heaps tipped to a depth of some 5 m by medieval miners. These waste heaps of gangue minerals, such as fluorspar and barytes, are being reworked by the mining company. Much of the site, including the medieval waste heaps, is covered with species-rich calcareous grassland, and heath dominated by heather (*Calluna vulgaris*) has developed on superficial loessic drift (Balme, 1953; Pigott, 1962). The limestone grassland contains several rare species, including *Epipactis atrorubens*, *Minuartia verna*, *Orchis apifera*, and the terricolous lichen *Cetraria islandica*. As a result of mining activities, seminatural vegetation is being destroyed over much of the site.

The general objective of the experimental work described here was to learn whether it would be possible to recreate the seminatural grasslands being destroyed by mining operations. The experimental areas were flat plots beside the worked-out rake. In addition, plots were established on the steep south-facing slopes of the Edge itself. The mining company had attempted reinstatement of some areas that provided a valuable contrast with the experimental plots. The experiments were set up in February-September 1972 and covered about 2 ha. The experiments are continuing, in an attempt to establish the long-term fate of the sown vegetation—the only real criterion of success in reinstatement.

ENVIRONMENTAL PROBLEMS IN REINSTATEMENT

Derelict land reclamation has been the subject of intensive research in the United Kingdom so long that few waste materials or mine sites present insurmountable problems for reinstatement. The existence of large amounts of derelict land reflects the limited resources that have been made available for such work, rather than a lack of appropriate techniques. As a result of the work mainly of A. D. Bradshaw and his co-workers at Liverpool University since the early 1960s, even seemingly intractable wastes contaminated by toxic substances can be revegetated with ecotypes able to tolerate the presence of these materials. Toxic metals present such a special problem that there is probably no alternative to the use of this approach. However, the treatment of nontoxic wastes has been much debated. The primary consideration has been to develop a recipe that can be used to produce a quick green cover, irrespective of its appropriateness as a long-term solution. This search for universal recipes that eliminate, rather than accommodate, constraints on vegetation development has colored much of the research on derelict land reclamation. The major constraints on plant growth on derelict land, apart from possible toxic effects, are low nutrient status, low soil organic content, drought stress, and, if the material has been mechanically sorted, uniform particle size.

Abandoned mineral workings and waste materials—such as alkali wastes (Lee and Greenwood, 1976), quarries (Davies, 1976; Hodgson, 1982), and wastes contaminated with heavy metals (Holliday and Johnson, 1979; Johnson, 1978; Johnson et al., 1978)—often develop into important preserves for seminatural vegetation. The evidence suggests that there is no need to reclaim derelict or abandoned mineral workings, because an interesting vegetation will develop naturally, given sufficient time. But that stance neglects important changes that have rendered old sites unreliable as models. The changes affect all three main phases of the diversification of sown vegetation (Wathern, 1977):

• *Transfer*. Propagules of the species must be dispersed from an established stand and arrive at the reinstated site.
• *Establishment*. Propagules must become established and develop into mature plants.
• *Periodic recruitment*. From time to time, new individuals derived from the original source or from the new population must become established, and the rate of recruitment should at least equal the death rate, if the population is to be sustained.

The abandoned mineral workings that are now biologically so interesting began their development in the early decades of this century. The landscape over much of the United Kingdom was different from that seen today. Seminatural vegetation was the basis of much of the land use. Wood products were derived largely from broad-leaved woodlands, rather than softwood plantations. Pastoral agriculture relied on the use of seminatural grassland for grazing and hay. Thus, the countryside had a basic matrix of seminatural vegetation that could serve as a source of propagules for the diversification of abandoned mine workings. Transfer distances were short and within the capabilities of most species. The preponderance of wind-dispersed species on the alkali wastes highlighted by Lee and Greenwood (1976) indicates the importance of isolation, in that these waste areas are many miles from potential seed sources. Species with other types of dispersal mechanism usually move only a short distance with each generation and can move long distances only incrementally through a chain of suitable habitats. At present, however, species are becoming concentrated in fewer and fewer refuges. It is likely, therefore, that a quarry site abandoned today will be many miles from potential sources of propagules and that transfer distances will be too great for many species.

In addition, because mining and quarry operations have changed over the years, it is unlikely that modern sites will acquire the same ecological interest as the old sites. Large-scale workings are preferred, and mineral extraction produces uniform conditions over extensive areas. A modern limestone quarry lacks the physical diversity of an old quarry with its uneven face, rubble-strewn floor, and scattered waste heaps, which account for much of the biological interest. Such diversity can be created only by a program of treatment after quarrying operations have ceased. Many of the old skills have been lost, but Humphries (1977) describes how, for example, a quarry face could be treated.

The final objection to a laissez-faire attitude to site rehabilitation is not biological. Society will no longer tolerate the visual intrusion of a derelict mineral working or quarry over a long period in the hope that interesting habitats will eventually develop. Industry must be seen to be rehabilitating environments degraded by its activities. The challenge in derelict land reclamation is to reinstate a site so that its visual impact is reduced, while maintaining its potential for development into a biologically interesting area.

The traditional approach to reclamation of derelict land for visual amenity has been based on agricultural techniques. In the period 1967-1972, the mining company made considerable efforts to revegetate areas on Longstone Edge where extraction was complete or where there was a temporary halt in operations. The approach adopted was to sow agricultural seed mixtures (*Lolium perenne*, *Dactylis glomerata*, and *Phleum bertolonii*) on raw subsoils with heavy fertilizer applications. It soon became apparent that this agricultural approach would provide no long-term solution. Whenever such grass-seed mixtures were sown, a green cover developed briefly. After a short period of vigorous growth, swards became moribund, the grasses showed signs of nutrient deficiency, and standing crops declined. On these free-draining soils lacking organic matter, nutrients, particularly nitrogen, were leached out. Periodic applications of fertilizers became the main form of management used by the mining company. Clearly, there would be considerable advantages in developing some form of single-application treatment that would lead to stable vegetation. It would be more visually acceptable to the public and to the controlling authority, the Peak Park Planning Board, and it would have obvious economic benefit.

Either of two approaches could be adopted: constraints on the productivity of high-yield species could be eliminated, or constraints on the system could be accommodated by creating some form of low-productivity system whose demands for nutrients would be better attuned to the ability of the system to supply them. The first approach has been used successfully in the revegetation of kaolin wastes in Cornwall. Kaolin wastes are nutrient-deficient, drought-stressed environments (Bradshaw *et al.*, 1978), and fertilizers must be added regularly to maintain vegetation cover (Bradshaw *et al.*, 1975). To overcome this problem, a biological source of nitrogen was sought. Rhizosphere fixation was inadequate to achieve the target of about 700 kg/ha estimated by Bradshaw *et al.* (1975) to be necessary to produce a self-sustaining system. The potential of nitrogen fixation by nodule bacteria in legumes was investigated in a series of trials. The results showed the use of legumes to be a means of overcoming nutrient limitations (Jefferies *et al.*, 1981); *Lupinus arboreus*, the tree lupine, was the most effective species (Palaniappan *et al.*, 1979), but is not native to the United Kingdom. Although sowing legumes is useful for reinstating some nutrient-deficient substrates, it was rejected for Longstone Edge, because the seminatural vegetation of Longstone Edge comprises calcareous grassland and *Calluna* heath, both of which are naturally nutrient-deficient systems. A productive agricultural sward created on the mineral working would be out of context with the surrounding vegetation.

Consequently, it was decided to attempt to produce a low-productivity vegetation in harmony with the existing nutrient-deficient system. The potential of these swards appeared great. Grime (1973a) has argued that environments moderately stressed by drought, nutrient deficiency, or grazing support diverse communities. In contrast, high-productivity systems are often species-poor. These low-productivity grasslands could be expected to develop into diverse swards akin to the existing seminatural vegetation.

Local populations of plants could not be used as the sole source of propagules in this research, because of the difficulty of interpreting the experimental results. Because the natural arrival of propagules, mainly seeds, in the experimental plots would be a chance occurrence, it would be impossible to determine whether the failure of a particular species to invade an area resulted from a lack of propagules or from an inability to establish under the prevailing conditions. The ability of a sward to accommodate a new species is termed its receptivity (Wathern, 1976). Natural seed rain provides a flow of only a small irregular number of propagules into experimental areas; that makes it difficult to measure the receptivity of contrasted swards. Therefore, we diversified the swards artificially by introducing known quantities of seeds of a range of native species, hoping to create a visually pleasing community rapidly.

The decision to introduce native species deliberately raised an important issue, which has since achieved greater prominence. Deliberate introductions are discouraged by scientists, mainly because they disrupt natural distribution patterns, particularly of local ecotypes. However, because plant distribution in the United Kingdom has been modified drastically by human activity over a long period, that objection has little validity. In these experiments, only commonly occurring species were used, and they were collected from local seed sources whenever possible. This is an ethical consideration that requires more thorough discussion, because failure to transfer species artificially might doom them to local extinction. The ethics of introductions in the construction of natural ecosystems are discussed by van der Hoek (1982).

APPROACHES TO RECREATING SEMINATURAL GRASSLAND

Experiments were established to investigate whether the yield of species on the Longstone Edge subsoils was related to relative growth rate (RGR). Monocultures of *Lolium perenne* (RGR_{max}, 1.30), *Festuca rubra* (1.18), and *F. ovina* (1.00) showed an inverse relationship between yield and RGR_{max} after 3 years. Thus, the species with the greatest potential for growth, *L. perenne*, performed least well. Other trial areas were sown with two contrasting mixtures—one consisting of *F. ovina* and *F. rubra* and the other consisting of *L. perenne*, *Dactylis glomerata* (1.31), and *Poa trivialis* (1.40)—and monitored for 5 years.

After 1 year, the *Lolium* sward had almost twice the standing crop of the *Festuca* sward. After 2 years, both grasslands had increased in yield, but they did not differ significantly from one another. After 5 years, the *Lolium* sward had the same standing crop as it had after 2 years; in the intervening 3 years, it had become moribund. In contrast, the *Festuca* sward continued to increase its standing crop throughout the experiment. The results of this experiment, described more fully by Wathern and Gilbert (1979), confirmed that nutrient-deficient subsoils could be reinstated successfully with low-RGR grasses.

Grasses, however, provide only the general matrix of the community in seminatural grasslands. In the United Kingdom, many of the distinctive elements, and certainly the visually dominant species, are dicotyledons. Indeed, the seminatural grasslands of the carboniferous limestone are renowned for their floral richness. Consequently, in parallel with the trials on low-productivity swards, the feasibility of recreating seminatural grass-lands was investigated. Experiments were established as long-term trials in September 1972 and still cannot be considered complete. For the first 5 years, the plots were monitored fairly intensively, but they have since been assessed only infrequently.

One area of Longstone Edge provided an opportunity to investigate the effect of a grass seed mixture on the rate of invasion by native species under natural conditions. Three years before experiments on Longstone Edge began, the company had regraded a large area. Half this area had been sown with a grass seed mixture, and the remainder had been inex-plicably left fallow. The vegetation of the two sections was surveyed, but no important differences in the number of adventive species could be found. Initial seeding with a mixture producing an open sward did not appear to affect the rate of diversification. This suggested that the two-part objective of reinstatement—reducing visual impact without signifi-cantly affecting sward receptivity—could be achieved. The second major conclusion from this work was that, although diversification occurred naturally, most of the adventives were species that produced large numbers of highly dispersable seeds. It appeared that, without deliberate introduc-tion, the more characteristic species of climax limestone grasslands with larger, less readily dispersed seeds might not become established in the short term.

In the light of these observations, experiments were set up to establish the most appropriate means of diversifying swards and to determine the receptivity of contrasted swards to native species. The *Festuca-* and *Lol-ium*-based mixtures used in the low-productivity sward experiment de-scribed above were used. Various sources of propagules of native species were investigated. Hand-collected seed was broadcast at known rates in the swards and fallow areas. After 2 years, the *Festuca* sward contained 22 sown species; six species showed an establishment rate of over 10%. The *Lolium* and fallow areas contained 15 and 17 species, respectively, at low frequency. After 5 years, although the frequency of native species had increased in all areas as a result of additional recruitment, self-set seed, and vegetative spread, the differences among the *Festuca*, *Lolium*, and fallow plots were even more marked than they had been after 2 years. Thus, not only was the *Festuca* sward initially more receptive to native dicotyledons, but this increased receptivity was sustained over a long period. Adventive species also showed higher frequencies in *Festuca* areas and after 12 years included important elements of the limestone grassland flora.

In a separate experiment, sowings along a soil depth gradient showed that sward performance determined establishment rates for native species.

Establishment was greatest in swards of intermediate density. In such swards, competition is not severe, but there are enough plants to ameliorate the microclimate of developing seedlings. Dense swards eliminate native species by competitive exclusion, whereas failure rates of seedlings in open situations are high because of harsh environmental conditions (Wathern and Gilbert, 1978). These experiments suggest that low-productivity swards increase the rate of diversification by native species and that sowing seed of native species by hand can create visually pleasing grasslands rapidly.

Other sources of native species were investigated. From casual observations on the site, it was noted that fragments of old turf had become established, particularly in compacted bulldozer tracks, providing small islands of seminatural grassland. Therefore, turf stripped from an area about to be reworked by the mining company was used in the diversification trials with an oversowing of the *Festuca* mixture (Gilbert and Wathern, 1980). The use of seed-rich topsoil on this site was not successful, but the approach should not be dismissed, inasmuch as it has been successful in other experimental areas. Seed-rich *Calluna* litter (Gimingham, 1960) has been used to produce a *Calluna* heath community on an acidic subsoil road verge on highway A57 west of Sheffield (Gilbert and Wathern, 1976). Similarly, Farmer *et al.* (1982) and Tacey and Glossop (1980) have since used seed-rich topsoils for the reinstatement of mine areas.

To obtain additional information on the fate of man-made swards, a chronosequence of grasslands up to 150 years old was analyzed. A sample of 69 quadrats covering the complete range of grassland age was recorded. Soils were taken for laboratory analysis to determine whether edaphic factors were controlling the composition of the vegetation. For grasslands less than 25 years old, specifications for the original seed mixtures were known. The detailed procedures, particularly the seed mixtures, used to create the older grasslands could not be discovered. The major finding of this study was that soil nutrient status was the most important factor in controlling succession in urban grasslands (Wathern, 1976, 1980).

USES OF KNOWLEDGE

Of the ecosystem modifications listed by Wathern (1984), the most important changes resulting from development that influence reinstatement appear to be simplification, the opening of previously closed nutrient cycles, and the loss of vegetation types. The objective of revegetation should be to redress that progression. Many of the objections to development are based on amenity, so it seems appropriate for reinstatement to aim at creating vegetation that has or can rapidly acquire amenity value. The application of agricultural technology to this problem, however, has often failed to achieve this objective. In the case reported here, ecological knowledge was used to develop a more subtle approach to reinstatement.

The revegetation of mineral workings on Longstone Edge was designed to satisfy five criteria: reinstated swards should be inexpensive to produce, result from a single simple treatment, require no intensive management, be visually pleasing, and resemble surrounding seminatural vegetation. The research involved field observations and theoretical considerations aimed at understanding the relationships between reinstated grasslands and the environment and then application of this information in devising a reclamation scheme. The application of ecological considerations pervaded the study. However, ecological knowledge figured most prominently in four ways: production of low-productivity swards, sward diversification, assessment of the potential of soil seed banks, and urban grassland succession.

Low-Productivity Swards

The most important concept used in this phase of the work involved the ecological importance of relative growth rate. Hunt (1970) and Grime and Hunt (1975) measured the RGR of 115 species in a uniform high-nutrient environment. The RGR_{max} of each species was determined. This study showed that low-productivity environments are characterized by species with low RGR_{max}, whereas productive systems contain species with high RGR_{max}.

The agricultural species used previously in landscape reinstatement have high RGR_{max}. When sown on Longstone Edge, they initially produced a green cover, because artificial fertilizers provided the high concentrations of nutrients required for their growth. These skeletal soils, however, have a low organic content. Under these conditions, nutrients are not held within the system, but rapidly leach out. For the highly productive species to flourish, fertilizers must be added regularly. From Hunt's observations, it can be predicted that such soils would naturally carry species with low RGR. If those species were sown during reinstatement, the results should be more satisfactory than those obtained with agricultural species. Thus, the work on RGR provides the theoretical basis of the reinstatement of low-productivity swards. In turn, the experiments provide a field trial of the importance of growth rate in explaining plant distribution.

Sward Diversification

High RGR, which reflects the ability of a species to use resources that are freely available, is only one of the characteristics by which one plant can gain a competitive advantage over another. Structure and height affect competition for light, and the ability to produce litter that persists from one growing season to the next is an aid to occupying space when a plant is not actively growing.

Grime (1973a) argued that these four attributes can be amalgamated into an index of competitive ability. Species with high competitive index ("competitors" in Grime's terminology) grow in productive environments where resources, such as nutrients and water, are freely available. Such environments are species-poor, because competitors eliminate other species that are less able to exploit the resources. Environments where resources or physical and chemical constraints severely limit plant production, described as "stressed," are also species-poor. Only a few species are adapted to tolerate extreme conditions of, for example, drought or heavy-metal contamination. Such species are characterized by very low RGR_{max}.

Composition of a particular community, however, is dictated not only by the growth rates of the constituent species and by the physical and chemical characteristics that can limit the full realization of this potential. The degree of disturbance of the environment, such as grazing, is a third factor. Poverty of species is associated with a high incidence of disturbance.

Grime advocated a triangular ordination as the means of analyzing community composition, the axes being the relative importance of competition, stress, and disturbance. At the extremes, species-poor communities are present. Diverse communities are found where competition, stress, and disturbance are all moderate, and they are characterized by species that exhibit the "C-S-R" strategy in Grime's terminology. C, S, and R stand for competitive, stress-resistant, and ruderal (weedy). Such plants are "confined to habitats in which competition is restricted to moderate intensities by the combined effects of stress and disturbance . . ." (Grime, 1977). The Longstone Edge reinstatement areas have this combination of features, so they should ultimately develop into rich communities of C-S-R forbs. The major representatives of C-S-R strategies in the U.K. flora are small tussock grasses, small deep-rooted forbs, small stoloniferous species, forbs with short rhizomes, legumes, and small sedges and rushes (Grime, 1979).

The major grasses used successfully to form the low-productivity sward on Longstone Edge were two small tussock-formers, *Festuca ovina* and a nonstoloniferous variety of *F. rubra* (var. *fallax*). The grasslands were diversified by using the seeds of wildflower species that could be collected in the area. Many achieved high rates of seedling establishment and conversion to mature plants. Several of the species that Grime (1979) lists among a short illustrative group of U.K. C-S-R strategists established successfully from seed in the Longstone Edge trials. In contrast, several tall herb species included for comparative purposes showed initial high rates of seedling establishment, but then failed to survive.

Thus, Grime's concepts of community ecology are directly relevant to derelict land reclamation. Although the concepts were known in general terms at the start of this research, the detailed design of the experiments was dictated by pragmatic considerations, such as seed availability. Before the trials, it was known that diverse natural grasslands were characterized not only by species with low growth rates, but also by environmental constraints on plant productivity. Such considerations influenced the basic suppositions underlying the diversification trials. The substantial theoretical developments that took place during the project permitted a more thorough and systematic analysis of the results and set the conclusions in a broader context.

Soil Seed Banks

Established ecological theory was used to appraise the potential of soil as a source of propagules for diversification. Initially, only the general characteristics of soil seed banks had been determined. Seed banks from beneath particular types of seminatural vegetation have now been described (Brown and Oosterhuis, 1981; Donelan and Thompson, 1980; Thompson and Grime, 1979), but when our experiments were being formulated, most of the literature was related to the weed populations of arable systems (see Harper, 1977).

Spray (1970) and Grime (1978) characterized the soil seed bank of tall herb communities. On the basis of high numbers of some native species, Spray suggested that seed-rich topsoil could be used to diversify reinstated vegetation. Experiments were undertaken to establish whether his techniques could be used on a large scale. Soil from beneath limestone turf was taken from a part of the site about to be worked for minerals. In laboratory germination tests, this soil was found to contain seeds of various native species. The field trials were not successful, but the reasons for the failure have never been adequately explained.

Succession in Urban Grassland

Sown vegetation is not static. Over the course of time, it changes as a result of a complex array of interacting factors. Succession is a phenomenon that must be accommodated by any vegetational scientist involved in derelict land reinstatement. In the United Kingdom, few data on grassland successions are available. Wells (1965) monitored the changing composition of a sown agricultural grassland. Of particular interest, however, is the description of a complex of grassland types that have developed on formerly arable land on Salisbury Plain after various periods of abandonment (Wells et al., 1976). Changing composition was associated with nutrient accumulation within the ecosystem.

Analysis of a chronosequence of sown urban grasslands provided some insight into the fate of vegetation created for visual amenity. In Sheffield, extensive areas of grassland have been produced for public open space around municipal housing in recent years. The age of the grasslands, up to 25 years, and the seed mixtures originally sown were known from the records of the local authority. Older grasslands were discovered around large country houses in the vicinity. In the case of the latter grasslands, the composition of the sowing mixtures could not be determined, but the chronosequence could be extended over a 150-year period with estate records. The results of this survey are discussed by Wathern (1976, 1980).

Diversification is the most important successional phenomenon for revegetation studies. The process in natural successions is accompanied by the accumulation of materials, particularly nutrients, within the system (Wells et al., 1976). In the urban grassland chronosequence, no increased diversity with age could be detected; older swards did not contain more species. However, the soils beneath these grasslands are subject to leaching. Grime (1973a) has shown that the number of species in seminatural grassland in the Sheffield region is inversely related to the acidity of the soil. Inasmuch as few species tolerate extremely acidic soils, a chronosequence in which there is a decrease in pH is likely to be one in which the number of potential recruits also decreases. Grime's data make it possible to separate the effects of diversification from pH changes. For each quadrat in the urban grassland survey, the number of species present was divided by the number of species anticipated in a soil with the pH of that quadrat. Values of a relative diversity index (Wathern, 1976), when plotted against age, showed that as a grassland gets older it acquires more of the species that are capable of growing at the appropriate pH. The results confirm the conclusion that succession is important in molding the composition of sown grasslands and indicate that many species invade only after long periods. Invasion of *Calluna* into these grasslands under conditions of natural seedfall, for example, takes about a century.

The second major conclusion with important implications for the diversification trials concerned the significance of soil nutrient status in determining composition. This was assessed with the *Rumex acetosa* seedling bioassay developed by Rorison (1967), a standard technique used in Sheffield. Increasing relative diversity was associated with decreasing nutrient status. These results are at variance with the observations of Wells *et al.* (1976), but it should be noted that soils in the urban grassland chronosequence initially had high nutrient concentrations as a result of fertilizer applications to topsoils. The implications with respect to the triangular ordination of community characteristics are discussed below. The observations suggested that nutrient deficiency in the limestone subsoil of the mineral workings on Longstone Edge should not be regarded as a major constraint on vegetation development, but potentially as an asset that should lead ultimately to high diversity if an appropriate method can be developed.

SOURCES OF ECOLOGICAL INFORMATION

The overall objective of the Longstone Edge trials was to create seminatural vegetation resembling the surrounding diverse grasslands on the nutrient-deficient subsoils of the reinstated mineral workings. The research had two basic assumptions: sward production depends on the relation of soil nutrient status to the growth rate of constituent species, and diversity depends on sward productivity. Each experiment was established to address these factors, with the intention of combining the results in the development of an approach for the reinstatement of this and similar sites. The important ecological sources in this case, therefore, were those which either aided identification of soil nutrient status and relative growth rate as key issues in the research or influenced experimental design.

As an illustration of the application of ecological theory to the resolution of ecological problems, the Longstone Edge revegetation trials are unusual. The development of the experimental work preceded the publication of much of the ecological theory to which it is most closely allied. This apparent anomaly can be explained only with a thorough understanding of the situation in which the work was conducted.

The evolution of an experimental research project is a complex process and one in which it is exceedingly difficult to unravel the detailed history of an idea. In a research environment like a university, many important innovations begin in a few chance remarks, in discussions with colleagues working in related fields, or in questions raised in research seminars. It is often difficult later to apportion progress among these causal factors.

Chance occurrences often change the course of research or provide valuable information that permits a more comprehensive analysis of results. For example, the soil-gradient experiment described earlier provided much information on the effect of sward performance on the establishment of native species. The experiment was not part of the original scheme; in some ways, its establishment was accidental. The company requested the use of the low-productivity sward mixture on this part of the site and, because surplus seeds of native species were available, this was added almost as an afterthought. Therefore, although planning is an essential component of good research, no program should be so intensely planned that there is no room for exploiting chance occurrences.

The factors that influenced the evolution of approaches to the reinstatement of disturbed land can be considered in four categories: ecological facts, ecological theory, analog studies, and expert judgment.

Ecological Facts

The major impetus in experimental design came from the field observations of research workers on site, who analyzed unsuccessful attempts at reclamation undertaken by the company, and from a detailed appraisal of the likely causes of failure in the light of ecological knowledge. A number of published and unpublished sources aided the study of low-productivity swards and grassland diversification.

The experimental grass sowings on Longstone Edge were designed to assess the performance of different species on limestone subsoil and to investigate the feasibility of creating low-productivity swards. Initial species selection therefore had to take account of the commercial availability of seed, as well as of potential productivity. Hunt (1970) compiled data on the relative growth rate of a range of species found in different habitats in the Sheffield region. These data confirmed the differences in growth rates of the species selected for the grass monoculture and the low-productivity sward experiments and allowed these differences to be quantified. The urban grassland survey described above was also important in species selection. The results showed that the younger grasslands on soils with high nutrient status were dominated by *Lolium perenne* swards, whereas older nutrient-poor soils carried grasslands dominated by fine-leaved *Festuca* species.

Although the older urban grasslands were more interesting, there was no evidence from an initial analysis of the data that they were more diverse than younger grasslands. Not until after the publication of the Grime

(1973a) paper, after the Longstone Edge trials were established, were the data reinterpreted to take account of the pH effects discussed above. The effect of modifications of soil nutrients on species composition in the United Kingdom, however, has been known for many years, with respect to upland grasslands (Milton, 1940), lowland meadows (Brenchley and Warington, 1958), dunes (Willis, 1963), and limestone grasslands (Jeffrey, 1971; Jeffrey and Pigott, 1973).

The major conclusion from these experiments is that grassland composition can be manipulated by changing nutrient status with fertilizer applications. The park grass experiment described by Brenchley and Warington (1958) resulted in the production of grassland types totally different from a common sward over the course of a century of contrasted fertilizer applications to trial plots. Most important, these investigations showed the high diversity of swards on soils of low nutrient status. This information motivated our expectation that rich grassland would develop on the nutrient-poor subsoils of the reinstated areas.

Data on the relative growth rate and competitive ability of different species proved to be the most valuable unpublished materials available to the study. The interrelationships among environmental characteristics, competitive ability, and plant strategies were already being investigated by members of the Unit of Comparative Ecology at Sheffield University. Grime proposed that nutrient-rich communities were species-poor because of the competitive dominance of the constituent species. A competitive index, quantifying this dominance, had been developed and calculated for 441 species present in the Sheffield region. The index was based on maximal height, structure, and maximal accumulation of persistent litter. The data collected by Hunt (1970) on relative growth rates were also included. Only incomplete data related to 115 species were available, so growth rate was not included formally in the index. The index confirmed the high competitive ability of agricultural grasses compared with that of the low-productivity sward mixture. In more recent discussions of competition (Grime, 1973a), growth rate has been incorporated into the index as a result of the availability of additional data.

The Unit of Comparative Ecology at Sheffield has been collecting systematic data on the characteristics of grasslands in the Sheffield region for many years. These data were used to produce a computerized data base on grasslands and their constituent species, but the first analysis of grasslands in the area (Lloyd et al., 1971) was too general to be valuable during this study.

The data were also used to produce an ecological atlas of common grassland plants (Grime and Lloyd, 1973), including detailed information on the environmental characteristics of the habitats in which each of 94 species occurs. This atlas has been described as a practical aid for land-scape managers involved in, for example, revegetation, but that appears to be a post hoc rationalization of its function. Selection of species for landscape reinstatement is determined by pragmatic considerations, such as seed availability, ease of establishment, and likelihood of success. These considerations are not addressed in the atlas. For example, productivity of the systems in which a species occurs, acknowledged elsewhere by Grime as a major factor in determining distribution, is not considered. The atlas therefore has the wrong emphasis to be valuable in revegetation.

Consequently, the information on distribution was not used in the re-vegetation trials. Species from a wide range of habitats were selected for the diversification experiments. It was felt that this approach would be better, because it would provide information on the conditions necessary not only for success, but also for failure.

Ecological Theory

The results of the seral study on urban grasslands and the subsequent diversification trials illustrate well the implications of the theory of plant strategies and community ecology for reinstatement studies. There were important changes in the status of this theoretical background over the course of the work described here. At the outset, the theory was still being developed by Grime in Sheffield. It was not published until most of the experiments had been set up, although the general tenor of the ideas was known. Thus, during formulation of the reinstatement work, the urban-grassland study and papers on the relationship between diversity and nu-trient status were more important.

Preliminary analysis of the experimental work began in 1974, after publication of the basic theoretical concepts (Grime, 1973a,b, 1974). The theory has been further developed by Grime (1977, 1979), but not all aspects have been accepted without dissent (see, for example, Newman, 1973). With this more comprehensive theoretical base, the results of the experimental work became more important and could be set within a broader context. The results demonstrate that diverse vegetation can be created artificially under conditions of moderate stress, as the theory pre-dicts, and that the techniques used constitute a worthy strategy for rein-statement of disturbed land for visual amenity.

Analog Studies

No reports of previous attempts to reinstate mineral workings to semi-natural vegetation were discovered during a systematic search. Work on diversification of road verges, however, is relevant to the study. Additional diversification trials were established on road verges in the Sheffield area, because they have the potential to make a major contribution to the reserves of seminatural habitats in the United Kingdom (Mellanby, 1974). The results were reported by Wathern and Gilbert (1978). A number of previous attempts at establishing native species in verges have been reported in the conservation press, and these were investigated in detail.

Most cases were in the form of "rescue" operations, in which populations of rare species were threatened by highway construction. Generally, mature plants were returned to the verge after a period in a nursery to bridge the construction phase. The most spectacular success has been the mass planting of native *Narcissus pseudonarcissus* (daffodil) on the M40. There were a few attempts at more general diversification. *Calluna* turf was used successfully for reinstatement at the Devil's Punch Bowl, Surrey, after roadwork and at Cumbernaud New Town. Commercially available herbs have been used on road verges in southeastern England. The results of this work have never been published. Significantly, some members of voluntary organizations had suggested more general diversification of highway verges in their areas, but such proposals were invariably abandoned after objections on the grounds that distribution patterns would be disrupted. The main value of the analog studies was in confirming that artificial diversification could be achieved. In general, the labor-intensive approaches were rejected as being inappropriate for large-scale projects.

Experimental work in the Netherlands on landscape reinstatement was of particular interest, but it was not seen until the Longstone Edge trials had been established. The Dutch experience was useful in interpretation of the experimental results, because a similar approach had been adopted independently in many instances. For example, the problem of low-productivity swards on sandy soils was investigated early (Hoogerkamp, 1971), and trials on the establishment of *Calluna* on highway verges had shown the importance of a nurse crop in seedling establishment, as was found in the United Kingdom (Gilbert and Wathern, 1976).

Expert Judgment

It should be clear from this discussion that a university provides a unique atmosphere for research. The interplay of personnel leads to a rapid evolution of ideas. The Longstone Edge study was undertaken in the Department of Landscape Architecture of Sheffield University as a doctoral research project. The supervisor was an experienced ecologist, and the research student a botany graduate. There were also important links with the Botany Department, in particular its Unit of Comparative Ecology.

These contacts occurred during a formative and productive period for the Unit, when ideas concerning plant strategies and community ecology were being developed. The work on the comparative ecology of individual species and the composition of plant communities might seem esoteric, but the revegetation work on Longstone Edge has shown its relevance to reinstatement. Although not published, information was freely available to the study, and informal discussions around this work were useful in the development of ideas concerning reinstatement.

Consultations with other organizations were not generally important in the evolution of the work. However, discussions with experts outside the university system, particularly government scientists in the Institute of Terrestrial Ecology (ITE) and the Nature Conservancy Council (NCC), took place. In addition, a British Council travel fellowship enabled the work undertaken in the Netherlands to be discussed with government scientists at IBS Wageningen in 1974.

The responses of research scientists differed. Most research workers were concerned primarily with investigating the effects of management on grassland composition. A few research scientists at ITE, however, were also interested in creating seminatural grassland; indeed, they set up experimental work in 1973 and 1978 (Wells, 1983). A guidance manual on diversification has also been prepared (Wells *et al.*, 1981). The discussions in 1971 generally concerned the need to develop techniques, rather than how this might be done.

Sowings of wildflower species became quite common in the Netherlands at one time (Cole and Keen, 1976). During the discussions with Dutch scientists, however, the ethics of the approach were called into question. This hinged not on theoretical considerations of disrupted distributions of different ecotypes, but on a more fundamental point: whether people should interfere in the development of an ecosystem.

The attitude of conservation officers in the United Kingdom was defensive and antagonistic. Many saw within the techniques a potential diminution in their ability to object to development proposals on conservation grounds, because they believed that any suggestion that plant communities could be recreated after development would undermine their position. This attitude reflected a failure to understand the concepts underlying the techniques. There is no suggestion of constructing communities that contain all manner of rare species. The approach seeks merely to redress somewhat the losses of seminatural vegetation that are occurring. The alternative is that yet more highly productive grazing land will be created in reinstatement, irrespective of location. This misconception concerning the overall approach, unfortunately, continues.

CONTRIBUTION OF ECOLOGICAL KNOWLEDGE

A Peak Park Planning Board official has described the Longstone Edge trials discussed here as "the most exciting piece of reinstatement work which has been undertaken within the area." The excitement had two sources. First, the resulting sward is diverse and contains many of the elements of seminatural calcareous grassland. The presence of these characteristic species is the result of both deliberate introductions of propagules and chance invasions. Even 12 years after establishment, the grasslands are still acquiring additional species. Second, development of these techniques marks an important departure from previous approaches to reinstatement.

Formerly, reclamation was based on agricultural technology with remedial measures, so the environment was modified primarily by fertilization to support the species sown. The scheme adopted on Longstone Edge was based on a more objective ecological approach. In the simplest terms, it can be described as an attempt to sow species capable of surviving under the environmental regime of the limestone subsoils. The change in emphasis was possible because the objectives of reinstatement were reconsidered and ecological theories concerning the control of community composition were reappraised. Development of theory was particularly useful in two respects: it could be used to explain the composition of seminatural communities in the area and the grassland chronosequence studied early in the research (it was on the basis of these field observations and the literature on the effects of fertilizers on grassland diversity that the more intuitive experiments were established), and, during interpretation of the experimental results, it enabled success and failure to be explained, rather than merely noted.

There is one major criticism of the work. Covering no more than 2 ha, the plots are at best large trials. With one exception, they have yet to be scaled up. Although increased scale should not affect the outcome of reinstatement, it would establish whether the approach is usable on an industrial scale. The possibility of developing a "wildflower" mixture for reinstatement schemes based on commercial sources of native species has been considered and rejected. Commercial seed is prohibitively expensive, but there are more serious ecological objections to such a mixture. The research was developed in an attempt to avoid the previous approach—namely, producing a recipe to be applied in all situations—even if it could guarantee a visually attractive grassland. The species that are available in commercial quantities constitute inappropriate ecotypes. For example, native populations of *Lotus corniculatus* figured prominently and successfully in the experiments, and this species seems to be useful in reinstatement, whereas commercial material is a robust, erect ecotype from central Europe that is selected for its high yield and ease of harvesting and is quite unlike the calcareous grassland ecotype found on Longstone Edge. Hand collection of local ecotypes is far more satisfactory, but its feasibility and, more important, its impact on local populations are not known. Some form of bulking up might be required, because large quantities of seed could be involved. Wells (1983), for example, recommends up to 2 kg/ha for some species. These practical considerations must be resolved.

Work on *Calluna* has been scaled up by other workers, and the creation of *Calluna* heathland has generated considerable interest. Large areas of moorland and heath over much of the United Kingdom are degraded and in need of reinstatement. In addition, serious visual intrusion has resulted from the use of traditional methods of reinstatement in the revegetation of such developments as pipeline rights of way. Reinstatement of *Calluna* heathland is now feasible on a large scale as a result of further field trials (Meaden, 1984; Moorland Restoration Project, 1983; Putwain *et al.*, 1982).

The Longstone Edge research has stressed the necessity for a long-term perspective in analyzing the results of reinstatement schemes. Results over short periods can have misleading implications for small-scale pilot schemes. Differences that were encountered in these trials over time could be explained by applying theories of community ecology.

The main value of the case described here concerns the interrelationship between theoretical ecology and applied ecology. The study has provided experimental support for theories of community ecology, and theoretical considerations, particularly of plant strategy and community ecology, help to weld the experimental work into a coherent whole. This unity converts what might have been regarded merely as an interesting set of empirical observations on a set of reinstatement trials on a small site in Derbyshire into a radical alternative strategy for the revegetation of derelict and despoiled land.

ACKNOWLEDGMENTS

The work described here was undertaken in collaboration with my research supervisor, Oliver Gilbert, and formed the basis of a doctoral thesis submitted to the University of Sheffield in 1976. The work was funded by a research studentship from the Natural Environment Research Council. The assistance of the mining company, Laporte Industries Ltd., and Jack Harwood in providing facilities for this work is gratefully ac-knowledged. The research should be regarded as the joint work of Oliver Gilbert and me, but the observations and interpretation of events included in this report are my responsibility.

REFERENCES

Balme, O. E. 1953. Edaphic and vegetational zoning on the carboniferous limestone of the Derbyshire Dales. J. Ecol. 41:331-344.

Bradshaw, A. D., and M. J. Chadwick. 1980. The Restoration of Land. Blackwell, Oxford, Eng.

Bradshaw, A. D., W. S. Dancer, J. F. Handley, and J. C. Sheldon. 1975. The biology of land revegetation and the reclamation of the china clay wastes in Cornwall. Pp. 363-384 in M. J. Chadwick and G. T. Goodman, eds. The Ecology of Resource Degradation and Renewal. Br. Ecol. Soc. Symp. 15. Blackwell, Oxford, Eng.

Bradshaw, A. D., R. N. Humphries, M. S. Johnson, and R. D. Roberts. 1978. The restoration of vegetation on derelict land produced by industrial activity. Pp. 249-274 in M. W. Holdgate and M. J. Woodmand, eds. The Breakdown and Restoration of Ecosystems. Plenum, New York.

Brenchley, W. E., and K. Warington. 1958. The Park Grass Plots at Rothamsted, 1856-1949. Rothamsted Experimental Station, Harpenden, Eng.

Brown, A., and L. Oosterhuis. 1981. The role of buried seeds in coppiced woods. Biol. Conserv. 21:19-38.

Cole, L., and C. Keen. 1976. Dutch techniques for the establishment of natural plant communities in urban areas. Landscape Design 116:31-34.

Davies, B. N. K. 1976. Wildlife, urbanisation and industry. Biol. Conserv. 10:249-291.

Donelan, M., and K. Thompson. 1980. The distribution of buried and viable seeds along a successional series. Biol. Conserv. 17:297-312.

Farmer, R. E., M. Cunningham, and M. A. Barnhill. 1982. First-year development of plant communities originating from forest topsoils placed on southern Appalachian minesoils. J. Appl. Ecol. 19:283-294.

Gilbert, O. L., and P. Wathern. 1976. Towards the production of extensive *Calluna* swards. Landscape Design 114:35.

Gilbert, O. L., and P. Wathern. 1980. The creation of flower-rich swards on mineral workings. Reclam. Rev. 3:217-221.

Gimingham, C. H. 1960. Biological flora of the British Isles: *Calluna vulgaris* (L.) Hull. J. Ecol. 48:455-483.

Grime, J. P. 1973a. Control of species density in herbaceous vegetation. J. Environ. Manage. 1:151-167.

Grime, J. P. 1973b. Competitive exclusion in herbaceous vegetation. Nature 242:344-347.

Grime, J. P. 1974. Vegetation classification by reference to strategies. Nature 250:26-31.

Grime, J. P. 1977. Evidence for the existence of three primary strategies in plants and its relevance to ecological and evolutionary theory. Am. Na. 111:1169-1194.

Grime, J. P. 1978. Interpretation of small-scale patterns in the distribution of plant species in space and time. Pp. 101-124 in A. H. J. Freyson and J. W. Woldendorp, eds. Structure and Functioning of Plant Populations. Elsevier-North Holland, Amsterdam.

Grime, J. P. 1979. Plant Strategies and Vegetation Processes. John Wiley & Sons, Chich-ester, Eng.

Grime, J. P., and R. Hunt. 1975. Relative growth rate, its range and adaptive significance in a local flora. J. Ecol. 63:521-534.

Grime, J. P., and P. S. Lloyd. 1973. An Ecological Atlas of Grassland Plants. Edward Arnold, London.

Harper, J. L. 1977. Population Biology of Plants. Academic Press, London.

Hilton, K. J. 1967. The Lower Swansea Valley Project. Longmans Green, London.

Hodgson, J. G. 1982. The botanical interest and value of quarries. Pp. 3-11 in B. N. K. Davies, ed. Ecology of Quarries. ITE Symposium 11. Institute of Terrestrial Ecology, Cambridge, Eng.

Holliday, R. J., and M. S. Johnson. 1979. The contribution of derelict mineral and industrial sites to the conservation of rare plants in the United Kingdom. Min. Environ. 1:1-7.

Hoogerkamp, M. 1971. Probleme bei Ansaaten an Strassenrändern. Rasen 2:85-86.

Humphries, R. N. 1977. An ecological approach to the revegetation of limestone quarries in the United Kingdom. Pp. 2-39 in E. J. Perry and N. A. Richards, eds. Limestone Quarries. Allied Chemical Corporation, Jamesville, N.Y.

Hunt, R. 1970. Relative Growth Rate: Its Range and Adaptive Significance in a Local Flora. Ph.D. thesis, University of Sheffield.

Jefferies, R. A., A. D. Bradshaw, and P. D. Putwain. 1981. Growth, nitrogen accumulation and nitrogen transfer by legume species established on mine spoils. J. Appl. Ecol. 18:945-956.

Jeffrey, D. W. 1971. The experimental alteration of a *Kobresia*-rich sward in Upper Teesdale. Pp. 79-89 in E. Duffey and A. S. Watt, eds. The Scientific Management of Animal and Plant Communities for Conservation. Br. Ecol. Soc. Symp. 11. Blackwell, Oxford, Eng.

Jeffrey, D. W., and C. D. Pigott. 1973. The response of grasslands on sugar-limestone in Teesdale to application of phosphorus and nitrogen. J. Ecol. 61:85-92.

Johnson, M. S. 1978. The botanical significance of derelict industrial sites in Britain. Environ. Conserv. 5:223-238.

Johnson, M. S., P. D. Putwain, and R. J. Holliday. 1978. Wildlife conservation value of derelict metalliferous mine workings in Wales. Biol. Conserv. 14:131-148.

Lee, J. A., and B. Greenwood. 1976. The colonisation by plants of calcareous wastes from the salt and alkali industry in Cheshire, England. Biol. Conserv. 10:131-149.

Lloyd, P. S., J. P. Grime, and I. H. Rorison. 1971. The grassland vegetation of the Sheffield region. J. Ecol. 59:863-886.

Meaden, D. P. 1984. The Restoration and Creation of Heather Moorland Vegetation. Ph.D. thesis, University of Liverpool.

Mellanby, K. 1974. The changing environment. Pp. 1-6 in D. L. Hawksworth, ed. The Changing Flora and Fauna of Britain. Academic Press, London.

Milton, W. E. J. 1940. The effects of manuring, grazing and cutting on the yield, botanical and chemical composition of natural hill pastures. I. Yield and botanical composition section. J. Ecol. 28:326-356.

Moorland Restoration Project. 1983. Phase 2 Report: Re-vegetation Trials. Moorland Restoration Project, Bakewell, Eng.

Newman, E. I. 1973. Competition and diversity in herbaceous vegetation. Nature 244:310.

Palaniappan, V. M., R. H. Marrs, and A. D. Bradshaw. 1979. The effect of *Lupinus arboreus* on the nitrogen status of china clay wastes. J. Appl. Ecol. 16:825-831.

Pigott, C. D. 1962. Soil formation and development on the carboniferous limestone of Derbyshire. I. Parent materials. J. Ecol. 50:145-156.

Putwain, P. D., D. A. Gillham, and R. J. Holliday. 1982. Restoration of heather moorland

and lowland heathland with special reference to heathlands. Environ. Conserv. 9:225-235.

Rorison, I. H. 1967. A seedling bioassay on some soils in the Sheffield area. J. Ecol. 55:725-741.

Spray, M. 1970. Management of roadsides, hedgerows and grasslands for nature conservation and amenity. Pp. 28-33 in P. J. Grime, ed. People and Plants. Derbyshire Naturalists' Trust, Matlock, Eng.

Tacey, W. H., and B. L. Glossop. 1980. Assessment of topsoil handling techniques for rehabilitation of sites mined for bauxite within the Jarrah Forest of Western Australia. J. Appl. Ecol. 17:195-201.

Thompson, K., and J. P. Grime. 1979. Seasonal variation in herbaceous seed banks. J. Ecol. 67:893-922.

Van der Hoek, D. 1982. New man-made nature. Pp. 287-288 in S. P. Tjallingii and A. A. van der Veer, eds. Perspectives in Landscape Ecology. Centre for Agricultural Publishing and Documentation, Wageningen, Neth.

Wathern, P. 1976. The Ecology of Development Sites. Ph.D. thesis, University of Sheffield.

Wathern, P. 1977. The ecology of herb establishment and survival in swards and its relevance to grassland reinstatement. Rasen 8:102-108.

Wathern, P. 1980. The creation of natural vegetation after development. Pp. 5-20 in Proceedings of the Second International Conference on Environment Protection. Vol. II. Scientific Society for Building, Budapest.

Wathern, P. 1984. Ecological impact assessment. Pp. 273-292 in B. D. Clark, A. Gilad, R. Bisset, and P. Tomlinson, eds. Perspectives on Environmental Impact Assessment. Reidel, Dordrecht, Neth.

Wathern, P., and O. L. Gilbert. 1978. Artificial diversification of grassland with native herbs. J. Environ. Manage. 7:29-42.

Wathern, P., and O. L. Gilbert. 1979. The production of grassland on subsoil. J. Environ. Manage. 8:269-275.

Wells, T. C. E. 1965. Changes in the botanical composition of a sown pasture on the chalk in Kent, 1956-65. J. Br. Grassl. Soc. 22:277-281.

Wells, T. C. E. 1983. The Creation of Species Rich Grasslands. Pp. 215-232 in A. Warren and F. B. Goldsmith, eds. Conservation in Perspective. John Wiley & Sons, Chichester, Eng.

Wells, T. C. E., J. Sheail, D. F. Ball, and L. K. Ward. 1976. Ecological studies on the Porton Ranges. Relationships between vegetation, soils and land-use history. J. Ecol. 64:589-626.

Wells, T. C. E., S. A. Bell, and A. Frost. 1981. Creating Attractive Grasslands Using Native Plant Species. Nature Conservancy Council, Shrewsbury, Eng.

Willis, A. J. 1963. Braunton Burrows: The effects on the vegetation of addition of mineral nutrients to the dune soils. J. Ecol. 51:353-374.

Committee Comment

An increasingly important activity of environmental scientists is the reinstatement and reclamation of derelict land that has been modified for various reasons with various outcomes. Reclamation practices are most highly developed in the densely populated countries of western Europe, where human modifications affect virtually every piece of terrain. Much of this experience is summarized in a valuable book by A. D. Bradshaw and M. J. Chadwick, *The Restoration of Land*, published in 1980. Increasing interest in this subject in the United States is reflected in the launching of a new journal, *Restoration and Management Notes*, published by the University of Wisconsin Arboretum. As in most fields of applied ecology, early attempts at restoration relied almost entirely on an empirical approach—try a variety of approaches and see which ones work best. They were also highly constrained by cost, availability of materials for revegetation purposes, and the objectives expressed by society for reinstatement projects.

Happily, the combination of several factors has resulted in important changes in the practice of restoration. Society is placing increasing value on land restoration, plant ecology has seen substantial recent development of relevant theory and data, and the goals of restoration have become much more sophisticated than merely getting something pretty and green. The project at Longstone Edge is an admirable example of applied ecology. It showed the importance of drawing on a variety of sources—such as ecological facts, ecological theories, analog studies, and the reasoned judgment of persons familiar with the local areas—if sound solutions are to be achieved.

The Longstone Edge project took its direction from the recognition that most derelict land in Britain occurs on nutrient-poor soils and that attempts to make such areas fertile in the long run require repeated fertilization, which, because of its cost, is seldom feasible. But it was observed that diverse natural grasslands often occurred on nutrient-poor sites; apparently, therefore, the site could be ecologically interesting without being productive. These important observations played a role in the development of the powerful experimental design that was used in this project.

Species-rich swards can be established on nutrient-poor soils when constituent species are poor competitors characterized by low maximal growth rates and when little competitive exclusion of species occurs. Under rich soil conditions, a few of the strongest competitors often come to dominate—that is, interestingly, an empirical observation for which there is no satisfactory theoretical explanation. We do not understand competition among plants in enough detail to be able to identify the mechanisms of interactions. Restoration projects might turn out to be especially valuable for gaining further knowledge about these fundamental processes. They are carried out on a scale larger than is usually possible in purely academic research and, given current interest in them, are likely to be followed long enough to reveal aspects of the processes whose outcomes require decades and longer to be seen.

Better understanding of competition among plants might offer new insights into the practical problems of reinstatement. For example, the Longstone Edge project used primarily information on relative growth rates to assess long-term success of species on the site. Other factors that influence interactions among plants, such as growth form and the relative allocation of resources to different tissues and organs, might also be used in the selection of species for seeding of experimental plots.

Similarly, since the initiation of the Longstone Edge project, considerable progress has been made in understanding the role of particular combinations of nutrients in determining the outcome of competition among plants, particularly algae. The experimental results and a theoretical interpretation of them have been provided by Tilman (1982). Because derelict lands are likely to be diverse in their nutrient status, even if they are mostly on poor soils, better understanding of the importance of particular combinations of soil nutrients for the success of competing plant species should enable both better selection of species for colonization and the cultivation of specific genotypes for use in particular circumstances.

Restoration ecology might also contribute substantially to our knowledge and understanding of seed banks in the soils and their roles in the development of vegetation after disturbance. In the Longstone Edge project, important use was made of topsoil with seeds; this technology appears to have great potential. New advances in the understanding of the causes of seed longevity and the specific conditions that result in termination of dormancy in various species might reveal which species have seeds that persist well in different circumstances.

An important lesson of the Longstone Edge project is that in restoration ecology there are unlikely to be formulas of broad applicability that can be used in restoring most derelict lands. Solving problems for a particular site will inevitably require knowledge of its environmental conditions and of the genotypes of local plants that are potential colonizers of the site. Nonetheless, general concepts—such as the relationships among soil nutrients, growth rates, and competition—are likely to be broadly applicable. In addition, society is likely to have multiple goals for reclaimed lands. Many sites in North America, for example, unlike those in Britain, are on potentially valuable agricultural land or are in highly productive sites that are to be restored to some seminatural state. Thus, apart from highly specific knowledge of local conditions, continued developments in the theory of competition, plant succession, seed ecology, and plant nutrition are likely to remain vital to restoration ecology far into the future.

References

Bradshaw, A. D., and M. J. Chadwick. 1980. The Restoration of Land. Blackwell, Oxford, Eng.

Tilman, D. 1982. Resource Competition and Community Structure. Princeton University Press, Princeton, N.J.

19

Optimizing Timber Yields in New Brunswick Forests

Management of commercially important species to increase yields is usually a local matter, independent of conditions and management actions undertaken elsewhere. That is especially true in forestry, in which individual trees are stationary, long-lived, and easily countable. Nevertheless, plot-by-plot management can create long-term supply problems on a larger scale. These larger-scale effects are the most important ones that influence the vitality of economic activity based on exploitation of the resource. This case study illustrates an attempt to link local decisions with regional ones where the valued ecosystem component (wood) is readily identified and the processes producing wood are well known. The major uncertainties are associated with changes in scale, and they require a balancing of scientific and social issues.

Case Study

THOM A. ERDLE, Forest Management Branch, New Brunswick
Department of Natural Resources, Fredericton, Canada
GORDON L. BASKERVILLE, Faculty of Forestry, University of New
Brunswick, Fredericton, Canada

INTRODUCTION

Silviculture and wood technology are highly developed, and the factors
that influence wood production rates are well known, despite the very
long cycle of the crop. Like many other areas, New Brunswick depends
heavily on industrial forestry. However, the forest base probably cannot
continue indefinitely to supply the quantity and quality of wood required
by industry without a large and expensive effort to develop the forest
resources. Recognition of this need by government and industry forestry
decision-makers has resulted in a strong commitment to carry out the
necessary development, which entails legislative changes in forest administration, changes in the allocation of raw material to industrial wood
consumers, and intensified action aimed at making the forest biologically
more productive (Bird, 1980).

Increasing the productivity of the forest requires special consideration,
because it includes direct intervention in the development of a natural
system. Tools for direct biological intervention in forest development
include harvesting, protection, and silviculture; and the use of those tools
requires explicit decisions on how much, where, and when to implement
them. Their implementation initiates a biological response that alters forest
development and production of wood, and the response must be forecast
to permit the design of useful intervention strategies. This is all made
difficult by disparities of scale, in both time and space, between the stand
level at which the actions are taken and the forest level at which the
response must be assessed.

Stands, considered here as small homogeneous communities of trees
(10-50 hectares), collectively constitute a forest. A forest can be enormously complex, encompassing many thousands of stands and many
hundreds of thousands of hectares. Forest management attempts to deal
with complexity of this scale by orchestrating the implementation of harvesting, silviculture, and protection at the stand level.

The direct results of local interventions are immediate and visible, as
stands are harvested, established, and protected. But the interventions

276

alter local dynamics that govern the development of stands and that contribute to the unfolding of forest dynamics. Because of the complexity of the forest and the long periods associated with stand growth, forest-level responses triggered by local interventions are not readily apparent and can accumulate into unforeseen, undesirable, and essentially unalterable proportions.

How can the time and scale differences between the stand level at which actions are taken and the forest level at which management is planned and evaluated be bridged to permit the formulation of local intervention tactics that address the forest management problem in the most appropriate, biologically realistic manner? The case described here, involving a 300,000-ha forest management license in New Brunswick, is an attempt to construct a part of that local-global bridge. The results of similar analyses have led to management decisions that affect a large fraction of the forests of New Brunswick.

THE ENVIRONMENTAL PROBLEMS

The issue in this case is impact assessment, i.e., designing local interventions that change the development of a biological system and forecasting the nature and extent of the change (impact) to permit evaluation of the desirability of proposed interventions.

Three entities must be considered with respect to wood supply: the quantity of wood, the quality of available wood, and the timing of availability. Analyses have revealed that development of New Brunswick's forest under management restricted solely to harvesting and protection, as has been the practice, yields wood whose volume is insufficient to meet industrial demand, whose quality is below minimal standards, and whose availability is discontinuous and erratic (Baskerville, 1982). Studies have shown that incorporation of silviculture, particularly tree planting and tending in the present case, can mitigate these supply problems (Baskerville, 1983). This case study addresses the specific questions: What are the probable forest-level responses to an array of local, stand-level decisions regarding the amount of planting and density at which plantations are established, and what are the immediate and long-term impacts of the decisions on wood supply?

The restriction of the present analysis to planting is solely for illustrative purposes. Stand spacing, thinning, and fertilization are other powerful silvicultural tools. The procedure described here could well be extended to those other tactics.

In planting, two major points are at issue. First, at what rate (hectares per year) should plantations be established? Tree planting, an expensive undertaking initially, forces a continuing commitment to tend planted stands as they develop. Second, at what tree density (trees per hectare) should plantations be established? High volume per unit area and large trees are both desirable in a stand. However, the density dependence of individual tree growth in a stand is such that volume per stand and volume per tree cannot be maximized simultaneously. The former tends to be directly related to stand density, the latter inversely related to it. Thus, any chosen planting density represents a trade-off between volume per unit area and volume per tree, and the impact of this trade-off must be forecast for the proposed strategy and expressed in a form suitable for review by a decision-maker.

Sophisticated routines have been developed that help to optimize stand performance (Brodie and Kao, 1979; Hann *et al.*, 1983). But they concentrate on stand-level responses and ignore the interplay of stand dynamics, governed by the interactions among trees in a stand, and forest dynamics, governed by the collective development of stands in the forest. Failure to place stand development in the context of forest development has separated the elegant silvicultural solutions from the forest management problem and has impeded evaluation and selection of efficient means to address overall wood-supply concerns.

THE APPROACH

The approach presented here is an analytical framework that presents probable outcomes of a wide array of decisions and yet retains the linkages between trees, stands, and the forest. The framework is built around six steps, as follows:

Step 1. Select a specific forest holding and determine the future wood supply and the precise nature of the biological limitations on wood supply produced by restricting management actions to harvesting and protection. This is the biological definition of the economic problem of wood supply.

Step 2. Identify an array of remedial measures that can be applied to the strategic problem of wood supply. These are tactical approaches and, in this case study, are limited to the rate and density at which plantations are established.

Step 3. Forecast the stand-level outcomes of each alternative in Step 2.

Step 4. Link the stand-level responses into the forest mosaic and forecast the new forest dynamics that would result from implementing each of the planting interventions of Step 2.

Step 5. Translate forest-level outcomes into performance indicators relevant to the wood-supply problem (identified in Step 1).

Step 6. Assemble the performance indicators of all the alternatives in Step 2 to create a response surface (impact statement) that displays relationships between stand-level decision variables and forest-level performance indicators.

USES OF KNOWLEDGE AND UNDERSTANDING

Valued Ecosystem Component

Within the six-step framework described above, the valued ecosystem component is the wood supply, as described by the performance indicators at the stand and forest levels. These indicators can be divided into two sets. The first set, at the forest level, includes short-term indicators of maximal sustainable harvest and associated unit cost and long-term indicators of potential harvest expansion and associated unit cost. These are the most important measures, because they are related directly to the wood supply. The second set, at the stand level, includes accumulated salable volume and average tree size at any point in the life of the stand. Although these are not the direct basis of decision-making, they collectively contribute to the forest-level indicators.

Significance of Impacts

The significance attached to changes in forest-level indicators brought about by the exercising of a set of stand-level control options is largely a function of the goals of the decision-making agency. Interpretations of significance can vary considerably between and within public and private agencies. Therefore, interpretations of significance are deliberately omitted here; "significance" has meaning peculiar to each agency, given its current perspective on the problem. The magnitude of reforestation effects on wood supply must be shown in a form that enables decision-makers to impose their own values and to draw their own conclusions regarding the significance of the impacts of a wide variety of strategies.

Bounding the Problem

The scale differences between forest-management control of stands and silvicultural control of trees necessitate different spatial bounding. At the forest level, spatial bounds can be logically established around the wood-supply base for a mill or group of mills controlled by one management agency. In this case, that supply base comprises 300,000 ha in northwestern New Brunswick. At the stand level, the spatial bounds must be expressed with a resolution consistent with silvicultural intervention (plantations), usually about 20 ha. The density dependence of tree growth operates at a very local level, so there is little interaction among non-neighbors. For a given stand structure, this leads to a perfectly linear relationship between stand production and size of stand (or group of stands of a given size).

Two considerations were involved in imposing temporal bounds at the stand and forest management levels. Because of the long period required for stand development and the societal importance of the forest as a long-term renewable resource, the appropriate period for examining wood flow is necessarily long. Furthermore, the forest age structure has a biological memory: effects of disturbances, such as harvesting and planting, will surface and persist as the forest develops. A time horizon of 80 years was therefore deemed appropriate. In wood-supply problems, it is not sufficient merely to know the wood yield of a plantation or the volume of wood available from a forest. The timing of wood availability, and therefore the timing of supply problems, is of utmost importance. Detailed time resolution within the long-term horizon is necessary, to address timing in design of strategies. Consequently, the 80-year horizon was divided into 2-year steps for forecasting forest development and for tracking performance indicators.

For purposes of this chapter, it was decided not to consider the impact of spruce budworm on the forest and not to consider the numerous other issues that pervade real forest management. The budworm has played a major role in the development of the New Brunswick forest, as reflected in the present age-class structure of the forest. Yields from forest lands also are influenced by the sizes of budworm populations. Nonetheless, attempting to capture the complexities of the budworm-forest interaction would so dominate the analysis as to obscure the central issue considered here—assessing the cumulative effects of harvesting practices. (For an overview of the budworm problem, see Baskerville, 1976.)

Study Strategy Development

A clear understanding of population dynamics at two levels is required for strategy formulation. At the stand level, dynamics are governed by the density-dependent forces that influence growth of individual trees. At the forest level, dynamics are governed by the structure and development patterns of the various types of stands that make up a forest. Density-dependent stand development and age-dependent forest development must be functionally integrated to provide a path for local intervention to be projected to the forest level.

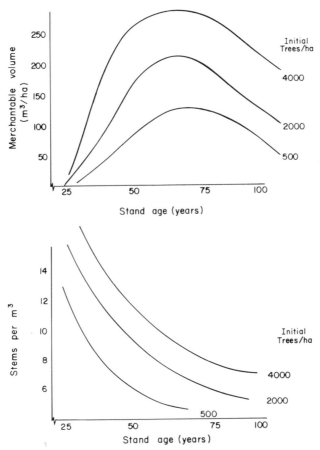

FIGURE 1   Stand development over time characterized by merchantable volume (above) and stems/m³ (below) for various initial stand densities.

The results of density-dependent tree competition in plantation development can be captured in two stand variables and their change over time: salable volume per hectare (m³/ha) and average tree size (trees/m³). Typical patterns for stands of three different initial densities are shown in Figure 1. Within limits, the density dependence of tree growth shifts the patterns up in higher-density stands and down in lower-density stands. For each stand type or plantation in the forest, these relationships must be quantitative.

For a stand to be considered economically available for harvest, minimal thresholds in salable volume per hectare and average tree size must be satisfied. These thresholds are the criteria that stands must meet to be

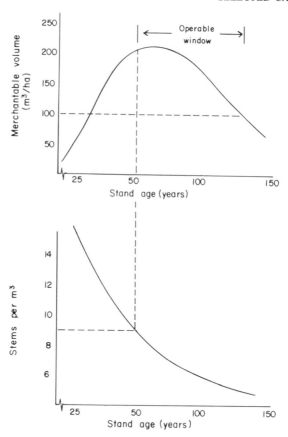

FIGURE 2 Determination of operable window for stand that defines its time of availability for harvest. Broken horizontal lines represent operability constraints of 100 m³/ha (above) and 9 stems/m³ (below). Broken vertical lines indicate earliest age at which both constraints are satisfied.

recruited into the operable (or available) volume inventory. The age range over which these thresholds are satisfied defines the timing of a stand's availability for harvest (Figure 2). For a given set of volume and tree-size thresholds, initial stand density can be silviculturally controlled to influence the timing of the stand's availability for harvest (Figure 3).

The potential power of stand-level control of availability becomes readily apparent in the context of forest-level dynamics. Because of forces of origin, species mixture, site fertility, and stocking, each of the many stands in a forest has its own pattern of development. A description of a forest shows the stage (age class) of each stand in its development pattern. Forest

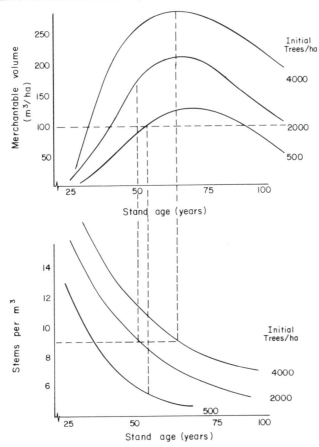

FIGURE 3   Effect of initial stand density on timing of stand availability for harvest. Operability thresholds, indicated by broken horizontal lines, are 100 m³/ha (above) and 9 stems/m³ (below). Broken vertical lines represent earliest time of availability for stands at each initial density.

dynamics are generated by the collective development of the constituent stands as they progress (age) along their own patterns of development. Stands do not influence one another's growth, but they are linked at other levels. First, there is an analytical relationship between stands, in that the abundance of each developmental stage collectively constitutes an age-class structure for the forest. Second, interventions, such as harvesting and silviculture, performed in one place can force the withholding of such interventions in some other place.

Managing the forest for wood supply is primarily a matter of regulating the availability of stands. That is achieved through a balancing of the liquidation of mature stands with the recruitment of immature ones across the lower threshold of operability. It requires a harvest schedule that defines the rate and sequence at which the harvest will proceed through the age structure of the forest. The rate of harvest is constrained by the timing of the availability of replacement stands as mature, operable ones are harvested. Availability of wood over time is a function of the initial forest age-class structure, particularly the relative abundance of stands in each age class, and the rate of development of these stands to operability. Consequently, any actions that hasten or retard availability of young stands for harvest have obvious implications for the rate at which mature ones can be harvested. That is the path by which density control at the stand level influences wood supply at the forest level.

The forest age-class structure both determines availability and constrains the manner in which implemented stand-level actions translate into forest-level responses. As a result of the natural and man-made forces behind their development histories, New Brunswick forests have irregular and different age structures. A fixed set of stand actions, carried out on similar sites, will yield identical stand-level results that are wholly independent of the forest in which they are applied. However, the same set of stand actions might generate dramatically different responses, depending on the age structure of the forest. This case study deals with one such initial structure.

Hypotheses

The use of these conceptual tools to manage forest-level responses requires testing and evaluation of three hypotheses.

• A quantitatively specific hypothesis must be made about how stand density affects tree growth and, ultimately, stand growth. It must be comprehensive enough to address such a question as: What will be the pattern of volume per hectare and average tree size over the life of a black spruce (*Picea mariana* (Mill.) B.S.P.) plantation established in northern New Brunswick at an initial density of 500, 1,000, 2,500, or 4,000 stems/ha? The variables expressed in the forecast must be, at least, average tree size and volume per hectare, because of their important role in defining stand availability for harvest.

• A hypothesis is necessary with respect to change over time in the minimal thresholds of tree size and stand volume required for stand operability (utilization standards). Rapid changes in use indicate that operability constraints change as a result of better technology in harvesting and product development. Instead of attempts to forecast how utilization standards would change, a range of reasonable future values was established, and the impact of these values on the forest-level indicators were analyzed. All the outcomes were incorporated into the impact statement in the form of a response surface, so that decision-makers could locate their own expectations with respect to changing operability limits and assess their significance.

• A third hypothesis must relate development of the whole forest quantitatively to that of the component stands, particularly those for which the plantation effort is contemplated. This hypothesis is also the basis for a forecast. The following are relevant and representative groups of questions: (1) How will the forest structure and resulting total growing stock change over time if no harvest is performed? If harvesting is carried out in the oldest stands first at a rate of, say, 400,000 m^3/year? 500,000 m^3/year? (2) How will forest structure and resulting growing stock change, if the same harvest schedules are attempted, but 4,000 stems are planted per hectare at a rate of 1,500 ha/year? 3,000 stems/ha at 3,000 ha/year? This second type of question is essential in designing a strategy of plantation use, and it highlights the necessary linkage between the local decision variables (planting rate and density) and the forest performance indicators (wood supply from the forest).

Cumulative Effects

To evaluate the cumulative effects of stand-level tactics on forest-level performance, the hypotheses with respect to stand development, operability limits, and forest development were systematically knitted together. Plantation performance was forecast for each density alternative between 500 and 4,000 stems/ha. Various sets of operability constraints were imposed on each of these, to establish the pattern of stand availability for harvest. Forest dynamics were then forecast iteratively as plantation tactics were systematically varied over all combinations of 0-4,000 hectares planted per year and 500-4,000 trees planted per hectare. For this purpose, the planting rate and density variables were changed at intervals of 500 ha/year and 500 trees/ha, respectively, to generate 72 unique planting strategies, each with a pattern of future wood availability. Performance indicators were tracked, through a model of forest development, to describe the cumulative wood-supply effects resulting from each strategy.

SOURCES OF KNOWLEDGE AND UNDERSTANDING

Generally Accepted Ecological Facts

Many empirical studies have described the effects of density on tree and stand growth for several species and locations in North America (e.g., Baskerville, 1965; Ker, 1981; Lundgren, 1981; Stiell and Berry, 1973). The findings, qualitatively displayed in Figure 1, are generally accepted by foresters, although quantification has proved more difficult for particular species and site conditions.

Major planting programs were initiated only in the 1960s, so few comprehensive data sources are available to help in quantifying the developmental relationships in plantations in New Brunswick. For efficient use of the limited data, a plantation growth model, responsive to density control, was constructed for the species and sites relevant to this case study.

The forest structure and natural stand growth are more adequately described by data sets available in the New Brunswick forest inventory and the records of the industrial land owner in this case. The forest-level characterization is presented in simplified form in Figure 4, which shows the present forest to comprise three types of stands: softwood stands dominated by fir (*Abies balsamea* (L.) Mill.), softwood stands dominated

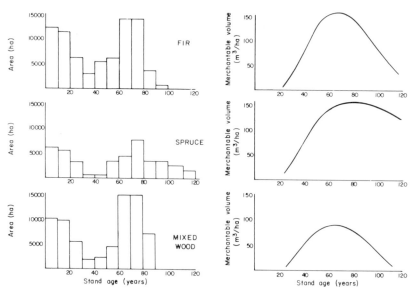

FIGURE 4   Age-class structures and yields for three stand types—fir (top), spruce (middle), and mixed wood (bottom)—that constitute forest used in case study.

by spruce (*Picea* sp.), and mixed-wood stands with fir and spruce components. The development pattern of each type, in terms of wood yield, reflects the biological characteristics of the species that make it up.

The present forest structure is a product of history and is obviously nonuniform, with a preponderance of mature stands, a moderate presence of newly regenerating stands, and a dearth of stands at stages of development between the extremes. This age structure bears the ecological imprint of the two main historical forces behind the forest's development—harvesting by man and harvesting by spruce budworm (*Choristoneura fumiferana* (Clem.)). Intensive harvesting by man over the last 30 years has created many stands in the regenerating category. The forest suffered severe "harvesting" as a result of defoliation by the spruce budworm between 1910 and 1920; large-scale destruction of the mature forest at that time resulted in a large number of regenerating stands that appear today as the sizable area in the 60- to 80-year age classes.

Most of the current stands are at or very near their peak volume. The futures of these stands, particularly fir, will be characterized by substantial decrease in volume, owing to natural decadence and breakup. Thus, even in the absence of harvesting, there will be a decrease in growing stock for the whole forest as the individual stands age. The paucity of stands in the 30- to 60-year age classes means that little in the way of replacement stands will become available for harvest as the mature stands are eliminated through harvest or by natural decline. The wood-availability problem hinges on the latter point. To maintain a constant wood flow, harvest of mature stands must be paced so that they last until sufficient young stands are available for harvest.

Together, these factors reveal that measures that hasten the entry of regenerating stands into the available inventory might have a powerful effect in overcoming the problem of continuity of available stands posed by the broken forest age structure. If cut-over sites are immediately restocked with desirable species, spatially arranged to use as much of each site as possible and to control intraspecies competition, individual stem growth will be more rapid than in untreated stands. The more rapid development of plantations increases the availability of replacement stands, which in turn permits a greater rate of liquidation of mature stands and therefore an increase in the annual wood supply. That is the principle that underlies the use of planting tactics to address the wood-supply problem for this particular forest.

The model used to forecast plantation development is based on two general principles: in the absence of any competition, individual tree height and diameter growth are bounded by inherent upper limits that are functions of species and site conditions, and realized growth (less than these limits) is regulated by the amount of growing space afforded the individual tree and its competitive status within the canopy of the stand.

Specific Models

Two models were used to make the required forecasts of stand and forest dynamics. At the stand level, a simulation model was constructed by giving quantitative form to the principles stated above and using available field data to define the maximal potential growth rates for black spruce in the region.

The model views the stand as a population of trees described by a diameter and height distribution at any time. Growing space and competitive status for each size class are determined from this distribution and applied to potential dimension increments to obtain forecast increments for the current competitive situation. The growth-estimation process is iterative, each iteration taking into account the accumulating changes in stand structure in the computation of growing space and competitive status, the two growth-regulation factors. The model is therefore dynamic enough to grow stands in a structurally realistic manner and to allow incorporation of planting-density alternatives.

Lack of long-term plantation growth data on New Brunswick made creation of the model and assessment of its predictive behavior particularly difficult. Comparisons were made against relevant data sets; in all cases, the model forecasts of volume per hectare and tree growth were qualitatively in accord with the empirical studies. Model forecasts for four initial stand densities are shown in Figure 5. The model provides a specific focus for carrying out a plantation-growth monitoring program, and the explicit assumptions behind the model are in a form amenable to specific field-testing and revision.

Forest development was forecast for various harvest and planting interventions with slight modification of an existing model. The original model was built by Erik Wang (Fraser Inc.) and parallels the concept behind the Wood Supply and Forest Productivity (WOSFOP) model of Hall (1978). This type of simulation model has seen extensive use in New Brunswick and other provinces (Cuff and Baskerville, 1982) and was used here because it was appropriate for the forest-level questions asked and because forest managers in the province are generally comfortable with it.

The forest-development model accepts as input an array of stand growth types that constitute the initial description of the selected forest, by giving stand areas, age-class structures, and yield functions (as in Figure 4); operability limits (as in Figure 3); and a rate of harvest specifying the

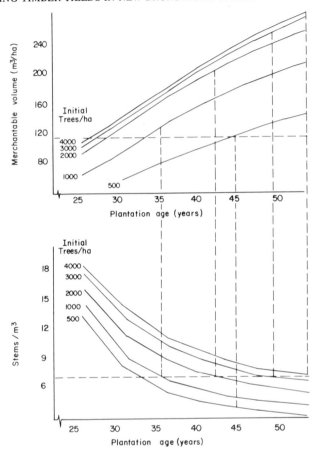

FIGURE 5   Model forecasts of merchantable volume (above) and tree size development (below) for black spruce plantations at five initial densities. Broken vertical lines represent minimal time required to satisfy current operability thresholds of 115 m³/ha and 7 stems/m³ simultaneously.

amount to be removed annually and the priority rules for allocating that harvest to the various stand types and to stages of development within a stand type. At 2-year intervals, the model advances the age-class structures for each stand type along their yield curves, performs the harvest and planting activities, and calculates forest-level indicators, such as total available growing stock, volume loss in mortality, and the evolving forest age structure.

The two models are linked, in that the stand model is an input source for the forest model. Through this linkage, stand-level tactics are carried through to the forest-level performance indicators.

CONTRIBUTION OF RESULTS TO ECOLOGICAL KNOWLEDGE

The study shows that limited preliminary data, combined with basic ecological principles, can be systematically used to display the impacts of a range of management strategies. Several hundred simulations were performed by linking the plantation-development model with the forest-development model to evaluate the forest-level outcomes of various combinations of planting rate and density. The results are presented as a set of response surfaces, or nomograms, in Figure 6. This format is particularly useful when two decision variables are to be assessed via their control over the valued ecosystem components through several indicator variables (Peterman, 1975). Decision-makers can easily review outcomes of a wide range of strategies before setting a decision process (or optimization) in motion. In this case, the decision variables are local (stand-level) implementation options of plantation density and planting rate, shown on the Y and X axes, respectively. The surface evaluation, or Z variable, represents a forest-level performance indicator associated with each stand-level decision combination. This surface is a type of prediction containing not only absolute outcomes of a host of alternative combinations of stand tactics, but also the sensitivity of the forest indicators to those tactics.

Figure 6 contains a number of noteworthy relationships between local stand actions and overall forest outcomes that have important implications for selection of a strategy. For example, Figure 6A shows that increasing the area planted annually increases the sustainable harvest immediately available from the forest, regardless of planting density. The increase in forest response surface is nonlinear, however, as evidenced by the widening intercontour gaps associated with increasing the planting rate. At the onset of horizontal contours (e.g., at 2,000 ha/year and 2,500 stems/ha), the positive immediate impact of planting on wood supply disappears altogether. Obviously, at a given density, each additional hectare planted is growing identically with the rest, but the effect of the marginal hectare on total forest performance changes dramatically with the degree of activity. Each additional hectare is progressively less "productive" in promoting immediate increased harvest. Thus, increases in planting are not accompanied by proportional harvest increases, and, beyond particular amounts of planting, there is no immediate harvest increase at all. At this point, harvesting has so drastically altered the forest structure that the resulting low abundance of mature stands constrains further immediate

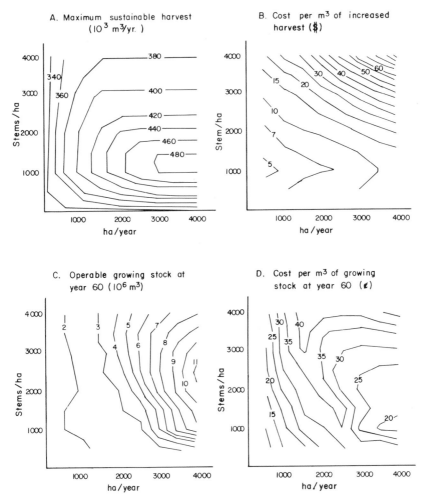

FIGURE 6 Impact of planting rate (*X* axis) and plantation density (*Y* axis) on four indicators of wood supply. Indicators A and B are related to immediate impacts; indicators C and D are related to impacts 60 years later.

harvest increases. Essentially, the biological limitation on wood supply switches from scarcity of young stands at low planting rates to scarcity of mature stands at high planting rates. The time to stand operability associated with each plantation density regulates the planting rate at which this switch occurs. This has two important implications. First, the powerful influence of forest structure on the effectiveness of planting is not evident if planting is evaluated solely on the basis of local, stand-level performance, as is common in investment approaches. Stand-level assessment would indicate that, if some planting is good, then more is better, and it is better in direct proportion to the extent of effort. Clearly, that is not the case: the linearity breaks down when plantations are viewed in the context of the whole forest. Second, the forest-structure constraint on planting effectiveness necessitates plantation tactic design specific to the forest in question. The surface in Figure 6A would be different if the same stand response were applied to a forest of different initial age structure from the one used here. Thus, management strategies embodying design of stand tactics are not freely transportable between forests.

The maximal sustainable harvest for this forest is 480,000 m^3/year (Figure 6A) and can be attained by planting 1,500 stems/ha at a rate of 3,000 ha/year. This combination, of course, pertains to a specific set of operability constraints and an 80-year horizon. The maximum resides in a very sensitive area in Figure 6A. The tightly packed contours in the lower third of the figure form a ''cliff'' that represents a high-risk zone for the decision-maker. If strategies at the edge of the cliff are pursued to maximize the harvest, then substandard plantation performance, poor competition control, ineffective protection, or bad model forecasts could have disastrous consequences. Such occurrences would mean that the sustainable harvest was over the brink and either a severely reduced sustainable harvest or an unexpected disruption in wood flow would result from implementing a harvest rate set in accordance with plantation yields that never materialize. Strategies in this high-sensitivity zone might be chosen by the risk-taking decision-maker, whereas one with a more conservative approach to risk might opt for a higher plantation density (2,000-3,000 stems/ha) and accept a slightly lower sustainable harvest to ensure a less sensitive response.

At the stand level, the highest-yield-per-hectare option is planting at a density of 4,000/ha (Figure 5). This does indeed keep stands in the most productive state, which is seen as a desirable goal by some foresters, but, as shown in Figure 6A, devastates the productivity of the forest as a whole, because it severely delays availability of plantations by reducing growth rates of individual trees. Again, this exposes the important difference between local stand response and overall forest response to selected sets of tactics.

Figure 6C contains information relevant to possible future increases in harvest, showing the growing stock available 60 years from now. Appreciable gains in growing stock occur at planting rates higher than those needed to achieve the maximal immediate harvest. Furthermore, with respect to future increases, the greatest gains are realized at higher densities than those which generate the maximal immediate harvest. Thus, high present harvests and high future harvests cannot be ensured simultaneously through the implementation of one plantation tactic. The decision-maker faces a trade-off between immediate gains and future gains. It can be resolved by striking a compromise between the two or by using a strategy with two simultaneous tactics: a low-density one aimed at immediate harvest gains and a high-density one aimed at future harvest gains. Regardless of the choice, failure to link stand and forest performance would not even reveal the problem to the decision-maker, let alone suggest potential solutions.

To illustrate the impact of operability limits on the decision-maker, several stand- and forest-level simulations were performed in which the minimal acceptable operability standards were systematically altered. The annual planting rates were fixed at the number of hectares required to maximize the current harvest for each planting density (e.g., 1,500 ha/ year at 4,000 stems/ha and 3,000 ha/year at 1,500 stems/ha). The results are presented in nomogram form in Figure 7. The X, Y, and Z axes show minimal average tree size, plantation density, and maximal sustainable harvest, respectively.

Each of the four surfaces in Figure 7 represents a different minimal-volume-per-hectare requirement. It is clear that the desirability of alternative densities, with respect to wood supply, varies considerably with the tree-size constraints that will be in effect. In Figure 7B, for example, under the stringent requirement of 6 stems/m^3, plantation densities of 1,000-1,500 stems/ha provide the maximal harvest. However, as the tree-size constraint is relaxed, both the maximal sustainable harvest and the plantation density at which the maximal harvest is realized increase. This is evidenced by the ridge that runs through the surface in the figure. That the surface elevation increases with relaxation of tree-size constraint is a reflection of the decreased time to stand operability and the consequent increased availability of stands when harvest of smaller trees is acceptable. That the ridge has a positive slope in the X-Y plane is indicative of the earlier achievement of these lower operability thresholds by higher-density (and consequently higher-volume) plantations when the tree-size constraint is reduced.

Examination of all four surfaces reveals that the increase in harvest associated with relaxed tree-size constraints holds for the minimal-volume-per-hectare constraint as well. The maximal possible harvest increases as the minimal volume per hectare is decreased from 205 m^3/ha (Figure 7D) to 70 m^3/ha (Figure 7A). Similarly, the planting density that yields the greatest harvest increases as the volume-per-hectare constraint increases. That is consistent with Figure 5, which shows the advantages of low and

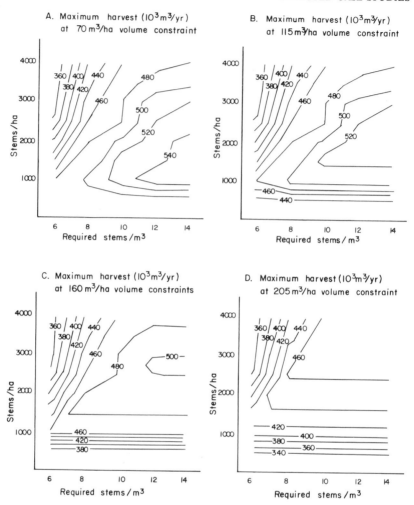

FIGURE 7 Impact on wood supply of plantation density (*Y* axis) and minimal acceptable tree size (*X* axis) at four constraints on minimal volume per hectare.

high densities to be rapid tree-size development and rapid volume-per-hectare development, respectively. Of importance to decision-making in Figure 7 is the powerful influence that utilization constraints can have over forest-level outcomes. The actual biological response underlying each surface in Figure 7 is identical. What generates the marked differences in forest outcome is a logistical harvesting constraint that the decision-maker imposes on the biological system, which regulates the availability of stands for harvest—this is no mere detail, inasmuch as utilization constraints are largely under the decision-maker's control. Figure 7 reveals likely outcomes that the decision-maker might achieve by exercising that control.

APPLICATION OF THE TOOL

The modeling process captured in Figures 6 and 7 constitutes a useful tool in policy design. Nomograms like those shown are powerful aids in structuring policy questions so that detailed modeling and analysis can be directed incisively. The diagrams are not used to choose a "best" planting policy. Rather, they are displayed to senior decision-makers, first to discover how they weight various indicators and second to discover where the decision-makers would like to be within the possible policy domain. Detailed analyses can then proceed with the indicators of choice within limits in terms of the policy variables. Thus, the nomograms are not used to answer operational questions, but rather to structure management questions. In this context, they make the scientific advisor more efficient both in the use of his own time and in establishing an understanding of the problems between advisor and decision-maker.

CONCLUSION

The case study presented here shows how a wide range of stand-level actions would influence forest-level outcomes. The purpose is to make the decision-maker aware of the importance of linking the two levels of consideration. No recommendations of optimal planting densities or planting rates are presented, because such decisions are strongly influenced by specific industrial strategies and objectives and by the degree of risk aversion of decision-makers. Furthermore, the decision-making picture is incomplete. Analysis becomes more complex as wood value, harvesting cost, and additional silvicultural tactics (such as spacing and thinning) are considered. Once the range of impacts is understood, more sophisticated analytical tools, like mathematical programing, might be effectively brought to bear on strategy design.

Different planting alternatives, applied at the stand level, generate different forest-level outcomes, because of the interaction of stand and forest dynamics. These differences highlight risks, sensitivities, and trade-offs that, although of prime importance in decision-making, would not be readily evident from stand-level or forest-level analyses alone. As a package, the analytical framework and the results it provides form a rich body of information relevant to the forest management problem of evaluating the impact of interventions on wood supply.

The techniques described in this study have been applied in the design of New Brunswick's silvicultural program. Effective control over planting is possible, because about 50% of forest land in the province is Crown land and the large timber companies need access to Crown land to maintain economical operations. Of the remaining land, half is owned by large companies and half has varied ownership. Thus, about three-fourths of the forest land in the province has been readily incorporated into a harvesting program suggested by the analysis as necessary to even the flow of timber and counteract the unfavorable age distribution of the stands.

There is not yet a firm link between stand dynamics and budworm population dynamics. Therefore, it is not possible to predict the effects of the harvesting and planting program now being implemented on population dynamics of the insect. Whether the program will have to be modified to accommodate problems generated by the budworm remains to be seen.

REFERENCES

Baskerville, G. L. 1965. Dry matter production in immature balsam fir stands. For. Sci. Monogr. 9:1-42.

Baskerville, G. L. 1976. Report of the Task Force for Evaluation of Spruce Budworm Control Alternatives. New Brunswick Cabinet Committee on Economic Development, Fredericton, N.B.

Baskerville, G. L. 1982. The Spruce/Fir Wood Supply in New Brunswick. New Brunswick Department of Natural Resources, Fredericton, N.B.

Baskerville, G. L. 1983. Good Forest Management—A Commitment to Action. New Brunswick Department of Natural Resources, Fredericton, N.B.

Bird, J. W. 1980. Forest management—A provincial perspective. Pp. 33-37 in The Forest Imperative. Proc. Can. For. Congr., Toronto, Ont., September 22-23, 1979. Canadian Pulp and Paper Association, Montreal.

Brodie, J. D., and C. Kao. 1979. Optimizing thinning in Douglas-fir with three descriptor dynamic programming to account for accelerated diameter growth. For. Soc. 25:665-674.

Cuff, W., and G. L. Baskerville. 1982. Ecological Modelling and Management of Spruce Budworm Infested Fir-Spruce Forest of New Brunswick, Canada. Paper presented at 3rd Int. Conf. on State-of-the Art in Ecological Modelling, Colo. State Univ., May 24-28, 1982.

Hall, T. H. 1978. Toward a Framework for Forest Management Decision-Making in New Brunswick. Report TRI-78. New Brunswick Department of Natural Resources, Fredericton, N.B.

Hann, D. W., J. D. Brodie, and K. H. Riitters. 1983. Optimum stand prescriptions for ponderosa pine. J. For. 81:595-598.

Ker, M. F. 1981. Early Response of Balsam Fir to Spacing in Northwestern New Brunswick. Maritime Forest Research Center Information Report M-X-129. Canadian Forest Service, Fredericton, N.B.

Lundgren, A. L. 1981. The Effects of Initial Number of Trees Per Acre and Thinning

Densities on Timber Yields from Red Pine Plantations in the Lake States. Forest Service Research Paper NS-193. U.S. Department of Agriculture, Washington, D.C.

Peterman, R. M. 1975. New techniques for policy evaluation in ecological systems: Methodology for a case study of Pacific salmon fisheries. J. Fish. Res. Bd. Can. 32:2179-2188.

Stiell, W., and A. B. Berry. 1973. Development of Unthinned White Spruce Plantations to Age 50 at Petawawa Forest Experiment Station. Publication 1317. Canadian Forest Service, Ottawa, Ont.

Committee Comment

The work reported in this case study provides a concrete example of how to assess the long-term, regional consequences of immediate, local forest-management actions in a manner useful for the design of sustainable strategies for resource development. The story is interesting as a particularly simple and clear case of effective assessment of cumulative effects, and it has proved useful in actual practice—by the Department of Natural Resources (DNR) of New Brunswick. Indeed, Baskerville initiated the analysis described here to deal with the practical difficulties he encountered as assistant deputy minister in DNR; Erdle is now responsible for applying the results of the analysis in DNR's Forest Management Branch.

The problem addressed here is widespread in forest management. A forest is a large-scale mosaic of individual stands of trees. Different stands can generally be characterized by age distributions, species mixtures, and habitat. The forest therefore is also a mosaic of stand-development trajectories, different stands reaching maturity at different times and rates. Management actions can alter those trajectories.

Although wood is considered the only valued ecosystem component in this model, additional components, such as control of spruce budworm and maintenance of an aesthetically pleasing mosaic of forest patches, have been considered in other analyses of the system (Baskerville, 1976; Clark *et al.*, 1979). The focus developed in this study was the result of a consensus that emerged among government, industry, and academic participants in the forest-management debate that, if the wood-supply problem were solved, most other concerns would be met automatically. Interestingly, the supply of wood is from the outset defined in cumulative terms over the regional scale and long periods relevant to the economics of the provincial forest industry. What matters ultimately is not the production from an individual stand of trees, but rather how the production from all the stands of the forest, taken together, can best be managed to meet society's needs.

Erdle and Baskerville studied the cumulative effects of alternative management actions by integrating local models of stand growth, regional models of forest age-structure dynamics, and a detailed inventory of the forest's existing age structure. They used available models of the relations among site quality, tree density, and growth rates of individual trees as the major sources of ecological knowledge. In essence, a commonly used wood-supply and forest-productivity model was taken "off the shelf" and provided with data relevant to the New Brunswick situation. This procedure made their task simpler and made their results more acceptable to managers already familiar with the models.

The forest model used by Erdle and Baskerville is essentially a bookkeeping one that tracks the aging of forest stands as they respond to additions to wood (through planting and natural regeneration) and removals of wood (through harvest, insect damage, and natural death). The underlying ecological theories are simple. Growth rates are assumed to be density-dependent, and the yield curve has a maximum. These results have been empirically determined, but they can also be derived theoretically from basic principles of plant competition. The simplicity of these demographic accounting models is also their main strength. They are robust, and, if solid biological data are available, they can be readily used for a wide variety of situations (see also Chapter 12). More complex, general, and rigorously tested stand models (Shugart, 1984) could have been used in the analysis and probably would have afforded improved credibility in academic circles. But there is no reason to believe that these more elegant models would have appreciably improved the results obtained by Erdle and Baskerville.

An additional aspect of ecological knowledge central to the success of the Erdle and Baskerville analysis was the existence of an accurate description of the present age distribution of provincial forests. Such baseline data on the heterogeneity of tree stands with respect to age distribution, species mixture, and growth potential are necessary to coordinate local, short-term management actions in a way that achieves desired regional cumulative consequences. Unfortunately, such data are extremely rare, especially in the case of age distribution, in inventories of forests and other renewable resources.

New Brunswick has a forest inventory data base that is one of the most accurate and useful for forecasting purposes in North America. This is in part because of the data-base shortcomings uncovered in Baskerville's earlier modelings and analyses and his later tenure in DNR. Generally speaking, however, useful background data are difficult and unglamorous to obtain, monitor, and update, although a characteristic feature of the few success stories in resource management is the existence and intelligent use of such data bases (see, for example, Chapter 12).

One final aspect of the Erdle and Baskerville analysis, the method of presenting results, is relevant not only to the assessment of cumulative effects, but also to a wide variety of efforts to provide usable ecological knowledge. They use nomograms to illustrate the trade-offs in valued ecosystem components that are likely to result from alternative management actions or other development interventions. In this forest-management case, as in many other environmental problems, interpretations of what is important vary considerably within and among government agencies, private interest groups, and the ecological profession. Moreover, as Erdle and Baskerville point out, these perceptions of significance change with time. Rather than adopting a single definition that would render their analysis usable from only a single perspective, Erdle and Baskerville have devised a framework in which users can specify (and indeed often discover) their own definitions of significance and explore the implications of their definitions for a wide range of possible decisions.

An added benefit of the nomogram approach, used to good effect in this case study, is that the spacing of the nomograms' contours of valued ecosystem components indicates the sensitivity of a predicted outcome to errors or incompleteness in implementing the decisions. Peterman (1981) has shown how minor modifications of the nomogram technique can be used to show the significance of uncertainties in the ecological models for the expected effects on valued ecosystem components. An analysis of the uncertainties inherent in the regional demographic and local competition models underlying this case study would have increased its usefulness even more.

Nomograms of this sort have been applied to environmental problem-solving in cases of renewable-resource management (e.g., Holling, 1978; Peterman, 1975, 1977; Regier, 1976), river-basin planning (e.g., Rabinovich, 1978), and regional development (e.g., Miller, 1982). They have a long history of application in business and industrial management. Nomograms are not a cure-all, especially given their restriction to two or at most three simultaneous trade-offs. But they are useful in appropriate circumstances, and they deserve wider application.

Not all the approaches and techniques for cumulative-effects assessment described by Erdle and Baskerville are relevant to the more complicated situations of, say, river-basin planning, or even to the more closely analogous situations of regional fisheries management. As difficult as forest age structure is to deal with, it is nonetheless measurable, and it provides a firm handle on the assessment of future forest development—a handle that most cumulative-effects studies will not have. In addition, the mere existence of a forest-level ''planning authority'' (the province's Department of Natural Resources) makes the case reviewed here much easier than the more typical one in which no agency is either solely responsible for or has sole power over the long-term, regional impacts of a class of related development decisions. More generally, it is worth emphasizing with the authors that a great proportion of their analysis involved the careful ''tuning'' to local conditions of a few relatively simple ecological concepts and models. No general precriptions for cumulative-effects assessment or management emerged from this study, nor are they likely to emerge from others.

References

Baskerville, G. L. 1976. Report of the Task Force for Evaluation of Spruce Budworm Control Alternatives. New Brunswick Cabinet Committee on Economic Development, Fredericton, N.B.

Clark, W. C., D. D. Jones, and C. S. Holling. 1979. Lessons for ecological policy design: A case study of ecosystem management. Ecol. Model. 7:1-53.

Holling, C. S., ed. 1978. Adaptive Environmental Assessment and Management. John Wiley & Sons, Chichester, Eng.

Miller, P. C. 1982. Simulation of socio-ecological impacts. Environ. Manage. 6:123-144.

Peterman, R. M. 1975. New techniques for policy evaluation in ecological systems: Methodology for a case study of Pacific salmon fisheries. J. Fish. Res. Bd. Can. 32:2179-2188.

Peterman, R. M. 1977. Graphical evaluation of environmental management options: Examples from a forest-insect pest system. Ecol. Model. 3:133-148.

Peterman, R. M. 1981. Form of random variation in salmon smolt-to-adult relations and its influence on production estimates. Can. J. Fish. Aquat. Sci. 38:1113-1119.

Rabinovich, J. E. 1978. An analysis of regional development in Venezuela. Pp. 243-278 in C. S. Holling, ed. Adaptive Environmental Assessment and Management. John Wiley & Sons, Chichester, Eng.

Regier, H. A. 1976. Science for scattered fisheries of the Canadian interior. J. Fish. Res. Bd. Can. 33:1213-1232.

Shugart, H. H. 1984. A Theory of Forest Dynamics. Springer, New York.

20

Control of Eutrophication
in Lake Washington

Most large cities in the world are situated on coastlines or the shores of rivers or lakes. Freshwaters and estuaries are the initial or eventual recipients of much of the waste products of technological societies. Consequently, water pollution is one of the first environmental problems to arise, and it continues to be pervasive even when discharges of wastes are reduced and waste materials are treated in more sophisticated ways. Because limnology is one of the more advanced fields of ecology, the key factors influencing the responses of lakes and rivers are rather well known, and the reasons for those response patterns are often understood. The Lake Washington case study is an example of creative interaction between the scientific community and the political arena in the development and execution of a plan that resulted in striking and rapid improvement of the quality of the waters of this lake, which was being increasingly influenced by growth of the metropolitan Seattle area.

Case Study

JOHN T. LEHMAN, Division of Biological Sciences, University of
Michigan, Ann Arbor

Lake Washington at Seattle (47°37' N, 122°14' W) is a moderately
deep (65 m), warm, monomictic basin in the drainage of the Cedar River
and Sammamish River. The lake discharges into Puget Sound via a system
of locks and canals built in 1916. Situated in an expanding metropolitan
area, Lake Washington has for years experienced varied and intense demands for transportation, recreation, and waste disposal. The condition
of the lake has changed greatly over the years in response to changes in
nutrient income brought about by the sewerage arrangements. That its
water quality today is as good as or better than at any other time in its
history is due to a unique blend of scientific judgment and public action.
During the expansion of suburban Seattle in the years after World War
II, the lake deteriorated in proportion to the pressures applied by a growing
populace. Reversing that trend by design, the citizens of the region responded voluntarily to the environmental problem. The solution was costly,
but public decision was guided by a firm statement of the problem and a
plain alternative. Scientific knowledge helped to define possible future
conditions of Lake Washington, and voting citizens selected the course.

The story of deterioration and recovery of water quality in Lake Washington on the one hand reflects changes in demographics and politics of
a city and its suburbs and on the other hand shows a development and
application of scientific thought on a problem that required special qualities
of scientific leadership and communication for public education. Action
and expenditure of public funds were linked to scientific arguments and
to quantitative predictions about conditions of the lake. From a scientific
perspective, the actions constituted an experiment and an opportunity to
refine hypotheses about how lakes function. From a civil perspective, the
case exemplifies transition from parochial, local concerns to a regional
outlook in environmental matters.

Seattle began discharging raw sewage into Lake Washington at the start
of the twentieth century. The large (86.5 km²) and deep basin became the
repository of street and septic discharges as the city expanded eastward
from Puget Sound. In 1926, however, Seattle created a bond issue for a
series of intercepting and trunk sewers to divert sewage from Lake Washington to a treatment plant on the Duwamish River that discharged directly
into Puget Sound. By 1941, the last sewer outfall into Lake Washington

302

from Seattle had been removed. Thereafter, water quality in the lake reflected development not of Seattle, but of its suburbs.

Between 1941 and 1953, 10 sewage treatment plants began operating at points around the lake, with a combined daily effluent of 80 million liters. Alternative discharge options were not as readily available to the small municipalities as they had been to Seattle. By 1953, James Ellis, a lawyer whose clients included some of the sewer districts around Lake Washington, sought diversion of sewage from Bellevue, but could not get the cooperation of neighboring districts for the necessary routing. He therefore spoke to the Seattle-King County Municipal League, proposing a system of metropolitan organization that could oversee such regional issues.

While these first steps were being discussed in the political arena, scientific investigations of Lake Washington attracted public notice with release of Technical Bulletin 18 of the Washington Pollution Control Commission, *An Investigation of Pollution Effects in Lake Washington (1952-1953)* (Peterson, 1955). This was the first substantial report of nutrient enrichment of the lake. It cited the work and data of Anderson (1954) and Comita (1953), who had conducted their doctoral studies under the guidance of W. T. Edmondson of the University of Washington. The *Seattle Times* trumpeted the report with its July 11, 1955, article, "Lake's Play Use Periled by Pollution." The *Times* returned to the issue a month later, when it reported the complaints of lakeshore residents and spectators at the Gold Cup yacht races (August 11, "Algae Increase Noted in Lake Washington"). Sanitation authorities doubted that the problems stemmed from the increased entry of effluent into the lake. They blamed sunshine, weather, and water conditions for the changes. It turned out, however, that the conditions of the lake that day triggered a new dimension of scientific curiosity in Edmondson's laboratory. Anderson had returned from the lake with a water sample containing an alga never encountered during his doctoral investigations of the lake: the blue-green alga (cyanobacterium) *Oscillatoria rubescens*.

To Edmondson, the appearance of *Oscillatoria* signaled that the lake was deteriorating in classical fashion. The large, deep lakes of Western Europe, particularly Lake Zurich, had been similarly enriched with high-nutrient effluents decades earlier, and water quality had declined. A series of lakes near Madison, Wisconsin, had received treated discharges from that municipality and had deteriorated. Accounts of these cases were in the scientific literature (Hasler, 1947), and Edmondson was struck by the fact that the name *Oscillatoria* appeared in each account in connection with the earliest stages of decline. For Lake Washington, previous data were available for comparison from 1933 (Scheffer and Robinson, 1939) and from 1949-1950 (Comita and Anderson, 1959). The doctoral studies of Anderson and Comita and the earlier investigation from 1933 provided a baseline by which to judge changes (Edmondson *et al.*, 1956). The main change in the watershed had been the increased load of nutrients from secondary-treatment waste discharge. Edmondson shared his observations in the October 13, 1955, *University of Washington Daily* ("Edmondson Announces Pollution May Ruin Lake"), recounting the appearance of *Oscillatoria* and its likely meaning. He defined his own interest as observing and analyzing the transitional nature of the lake and adding to the research done in Germany and Switzerland.

By 1956, the stage was set for developments that would bring civic leaders and scientists together. The scientists took the first step. Edmondson and two University of Washington engineering faculty members, R. O. Sylvester and R. H. Bogan, published a popular-science article in the university's journal *The Trend in Engineering*, "A New Critical Phase of the Lake Washington Pollution Problem" (Sylvester *et al.*, 1956). The article told the history of sewage treatment for the area, described the problems posed by nutrient enrichment, and proposed three procedures for solution: comprehensive regional administration and planning, complete elimination of sewage discharge into Lake Washington, and research on the relationships among temperature, nutrients, and algal growth. It provided a concise layman's explanation of nutrient enrichment and its effects, including the appearance of *O. rubescens*, and it focused particularly on enrichment with the mineral nutrient phosphorus and the difficulty of removing it from sewage. The *Seattle Times* publicized the article with the headline (April 18, 1956), "Lesson of Switzerland Lakes Brought Home to Seattle Area." By October, Edmondson and a new postdoctoral associate, J. Shapiro, had received funds from the National Institutes of Health to study water chemistry and photosynthesis by algae in Lake Washington. In December, Edmondson wrote a letter to James Ellis that marked his first involvement in the public action. James Ellis had been appointed chairman of Seattle's newly established Metropolitan Problems Advisory Committee by Mayor Gordon Clinton, and Edmondson wanted to ensure that Ellis and the committee understood that even well-treated sewage contained enough nutrients to stimulate the growth of plants in the lake. Lake Washington was already showing signs of the same series of changes toward deterioration as had been observed elsewhere.

After his initial letter had produced a cordial and positive response from Ellis, Edmondson sent him, on February 13, 1957, a nine-page summary of the effect of drainage and effluent entry into Lake Washington. The letter was phrased as a question-and-answer document and included references to the professional literature. Edmondson listed answers to 15 questions that he thought Ellis might be asked in connection with his work on the advisory committee—such questions as: How has Lake Washington changed? What will happen if fertilization continues? Why not poison the algae? Edmondson included a rudimentary nutrient budget for the lake constructed from available data, and he developed his case that the mass of algae present varied in strict proportion to the amounts of fertilizing nutrients added to the water. The letter included a mass of limnological cause-and-effect reasoning phrased in jargonless, objective tones. Ellis responded enthusiastically within the week, requesting copies of the letter for distribution to interested groups.

The initiative passed back to the political arena for the remainder of 1957. The immediate obstacle was the absence of provisions whereby municipalities could combine some of their government functions in comprehensive regional matters. Moreover, the notion faced opposition from some, on the grounds that it smacked of "big government." The next step required action at the state level. That aim was met when the Washington state legislature passed a bill permitting the establishment of a metropolitan government ("Metro") with specified functions (Ch. 217, Laws of 1957). The floor manager in the House had been Daniel J. Evans, a first-term representative, former King County engineer, and future governor. The act permitted the formation of a metropolitan government with any or all of six functions: water supply, sewage and garbage disposal, transportation, comprehensive planning, and park administration. Establishment of a Metro would require passage of a public referendum.

The first effort to win public acceptance for spending money to clean up the lake occurred in March 1958, when a proposal to establish a Metro charged with sewage disposal, transportation, and comprehensive planning was placed on the ballot. The proposal won 54.4% of the vote, but was defeated through a complicated system of weighting votes separately in Seattle and the rest of King County. Many people outside Seattle believed that the plan was an effort to tax them for the expenses of the city. Ellis and his committee revised the scope of their plan, targeting only water pollution control and reflecting the urgency posed by the deteriorating state of Lake Washington. A revised proposal, with the single function of sewage disposal, was approved on September 9, 1958, winning 58% of the vote in Seattle and 67% in the rest of the county.

Lake Washington obviously had become a focus of regional concern among the numerous communities that populated its shores. The water and beaches served for recreation, and the lake itself was deemed of aesthetic value. A genuine sense of pride and responsibility is evident in the political arguments that surrounded the issue. Citizens were asked to undertake, at the expense of about $2 per month for each household served, a public project of sewage diversion that was at the time the most costly pollution control effort in the nation. The plan called for construction of a massive trunk sewer to divert all effluent from around the lake, to treat it, and to discharge it at great depth in Puget Sound. Tidal flushing guaranteed that objectionable quantities of nutrients would not accumulate in the estuary.

Edmondson played no part in the partisan politics, but his scientific knowledge and judgment were a deciding asset in the Metro campaign. By supplying facts and generally making himself available to answer questions from the mass media or private citizens, he provided the authority that backed the movement with facts and logic. Ellis praised Edmondson's stand years afterward for providing the facts needed to quiet the critics. When he spoke, Ellis reported, "he made us feel that the Lord God was standing right behind us on this one" (Chasan, 1971, p. 11). Privately and professionally, Edmondson reported the Lake Washington case as an experiment in lake fertilization. His scientific publications during this period traced the departure of chemical and biological conditions from the historical conditions and attempted to discern the general quantitative relationships between nutrient additions and primary productivity in lakes.

Edmondson had been able to predict in his letter to Ellis of February 1957 a serious and rapid decline in water quality. He wrote: "Within a few years we can expect to have serious scum and odor nuisances. . . . Judging by the speed with which the process has gone in other lakes, I would expect distinct trouble here within five years, although isolated occurrences might come earlier." The important elements of his predictions that gave heart to the proponents of the diversion plan were that Edmondson thought that the lake had not yet been irreversibly damaged and that diversion would lead to a decrease in the abundance of blue-green algae. These were the plants that clouded the fertilized water, rafted to shore, and decomposed or otherwise fouled the lake. His predictions were based on fundamental principles of mass balance, stoichiometry, and an opinion that, to a large degree, changes in lake conditions are reversible when factors forcing the changes are reversed.

The specific basis for quantitative predictions was a conceptual and graphic model that related changes in lake properties from known initial conditions to changes in nutrients (Edmondson, 1979). The model assumed a limited return of nutrients from sediment deposits on the basis of chemical conditions in the lake and work done decades earlier in Germany and England (Mortimer, 1941, 1942; Ohle, 1934). Finally, it required knowledge of the lake's water budget, to permit calculation of a rate of dilution. From these basic facts and hypotheses, it was possible to project not only the speed of deterioration, but also the rate of recovery with different diversion schemes. Years later, the same ideas were used by other limnologists to construct mathematical models of lake conditions in response to nutrient income and hydrology (Piontelli and Tonolli, 1964; Vollenweider, 1969, 1975, 1976); and the principles remain guiding tenets of modern lake management (Chapra and Reckhow, 1983; Reckhow, 1979). From a strictly scientific viewpoint, the exercise encouraged thought about nutrient budgets and helped to integrate studies of lakes with their watersheds. Equally important, the scientific studies helped to elaborate the quantitative links between nutrients and productivity. Many limnologists of the early twentieth century had been trained in the shadow of Forbes's (1925) philosophical essay "The Lake as a Microcosm" and had confined their investigations within lakeshore boundaries. Forbes himself took a much broader view of lakes and their watersheds than the title suggests, but most limnologists at the time began and ended their studies at the shoreline. Edmondson, however, had visited Wisconsin as a graduate student while C. N. Sawyer was laboring to construct the first budgets of fertilization for the lakes at Madison (Sawyer, 1947). The perspective he gained was holistic and comprehensive. The new view might better be termed "the lake in an ecosystem."

Groundbreaking ceremonies for the new project were held in July 1961. Meanwhile, the lake deteriorated according to predictions. On July 3, 1962, the *Seattle Post-Intelligencer* reported "Lake Washington Brown—That's Algae, Not Mud and It'll Be There For the Next 10 Years." Visibility in lake water had declined from 4 m in 1950 to less than 1 m by 1962. The first diversions were slated for the next year, and, on the basis of the timetable for later diversions, Edmondson estimated that the lake would revert to its condition of 1949 by about 5 years after completion of the project. By October 5, 1963, the *Post-Intelligencer* had dubbed Lake Washington "Lake Stinko," with nuisance conditions at their peak just before effluent diversion.

The rest of the public record is a series of congratulatory editorials and progress reports in city and suburban newspapers. One by one, waste treatment plants around the lake had their effluent diverted. The first diversion was in 1963, and the last was in 1968. The trend of deterioration stopped in 1964; conditions that summer were no worse than in 1963. By 1965, it was apparent that water transparency, algal abundance, and phosphate concentrations were improving. On November 19, 1965, Edmondson predicted in his address to Sigma Xi, the scientific research society, that the lake would return to its pre-1950s condition within 6 years and that it would be possible to see the bottom as deep as 6 m. In the previous summer, he had made the same prediction to K. Wuhrmann, a skeptical colleague from Zurich. At meetings of the International Association for Theoretical and Applied Limnology, Wuhrmann had argued that sediment release of phosphorus would extend the recovery for decades. The scientists disagreed about one of the principal hypotheses that Edmondson had used for his quantitative predictions.

Phosphorus arrives at the sediments in the form of detrital material that decomposes more slowly than it becomes buried. Thus, the water only a few millimeters below the surface would contain great reservoirs of phosphate ions that owed their presence to the rich conditions of eutrophication. If the ions diffused through porous and unconsolidated sediment back into the water, they could renew productivity from that "internal" source. Edmondson reasoned that the oxidation-reduction potential of surficial sediments guaranteed that iron would exist in its ferric ($+3$) oxidized form. In that state, it could form insoluble ferric-hydroxyl-phosphates of indeterminate stoichiometry, and the sediments would become an "iron trap" for the phosphorus. This was what Mortimer had shown in microcosm experiments with mud from the English Lake District. Rates of decomposition in Lake Washington were not high enough to exhaust the oxygen content of the deep water during stratification each summer. As long as the large hypolimnion remained oxic, redox potentials would favor ferric iron, and the "iron trap" could halt the upward diffusive flux of phosphate. Wuhrmann doubted that principles governing ion speciation and fluxes inside model tanks could be freely extrapolated to whole lakes. Their friendly wager that summer and Wuhrmann's delivery of one bottle of Scotch during the 1971 congress in Leningrad highlight the intellectual excitement and new understanding that the Lake Washington experiment afforded to professional limnologists.

Edmondson's professional publications during the period reported the progress of physical, chemical, and biological changes in the lake (Edmondson, 1961, 1966, 1968, 1969a,b, 1970, 1972a,b). Transparency and algal abundance responded very quickly to the nutrient diversions. Species composition proved somewhat more intransigent. Even though biomass was reduced, *Oscillatoria* persisted into the early 1970s, making occasional appearances each summer. Finally, it too was gone. Trophic equilibrium in response to altered nutrient loading was complete in 1975 (Edmondson, 1977a,b; Edmondson and Lehman, 1981). Concentrations of phosphorus were reduced nearly to equilibrium and were similar from year to year. Chlorophyll concentrations and algal biomass were dramatically reduced, in parallel with the nutrient changes. Transparency had increased, and all filamentous blue-green algae, including *Oscillatoria*, had been eliminated.

The experiment was complete, and the scientific community had learned new lessons about dynamic processes in lakes. The general public enjoyed its own measure of international praise, as witness articles in *Harpers* (Clark, 1967), *Smithsonian* (Chasan, 1971), and *Audubon* (Kenworthy, 1971). Metro became one of the few noncities to win an "All-American City" award.

The public record ends here, but new sources of scientific curiosity grew from continuing investigations. After the few years of constancy, water transparency suddenly increased to a point never before recorded in the lake. Visibility was 12 m at times. Accompanying the change was a further, drastic reduction in algal abundance. This time, there had been little or no change in watershed relations; the limnological conditions were changing despite relatively constant nutrient loading.

What *had* changed were the herbivores (Edmondson and Litt, 1982). Throughout the doctoral studies of Comita and throughout the episode of enrichment and diversion, Lake Washington had been dominated by co-pepods, a group of microscopic crustacean plankton. The changing nutrient supply to the lake exerted controls on the biota from the base of the food chain, because nutrients are essential for plant growth. By the late 1970s, however, Lake Washington was at times dominated by cladocerans, particularly by members of the genus *Daphnia*. Controls on algal abundance had shifted to much higher in the food chain. The cladocerans are able to reproduce faster than the copepods, and their success reduced the algae by sheer numbers and grazing pressure.

Why had the zooplankton community changed? *Daphnia* had not dominated the lake even in 1933. Did it have anything to do with changes set in motion by the enrichment episode? The answers to this new puzzle are being debated now, because similar shifts have been recorded in Lake Tahoe, in Lake Michigan, and in ponds of central Europe. It is known that the answer lies in deciphering the balance of forces that affect birth and death rates among the potentially dominant populations. The success of predatory invertebrates and planktivorous fish is involved in present hypotheses. Edmondson and Litt (1982) proposed that the changes might be traced to the decline of an important predator on *Daphnia*. Selective predation is known to be a major force in determining species composition in zooplankton communities (Hrbacek, 1962; Hrbacek *et al.*, 1961). *Neomysis mercedis*, which was very abundant during the 1950s and early 1960s, suddenly declined in the mid-1960s with the rise of the longfin smelt *Spirinchus thaleichthys*. *Neomysis* strongly selects *Daphnia* over other planktonic crustaceans (Murtaugh, 1981), and *Spirinchus* feeds largely on *Neomysis* (Eggers *et al.*, 1978). Released from this predation, *Daphnia* nonetheless took 10 years to dominate the lake plankton. Delays inherent in life histories or colonization times were insufficient to explain the gap.

The reason for the delay seems to have been the continued presence of *Oscillatoria*. Long individual trichomes of *O. rubescens* and a few other filamentous species clog the feeding mechanism of *Daphnia* and force the animal to eject entire boli of food and to engage in elaborate grooming behavior (Infante and Abella, 1985). Copepods like *Diaptomus* do not seem to exhibit similar evidence of interference and can thrive in the presence of the trichomes, possibly because of hydromechanical differences in food capture. Thus, it was not until manipulations of the nutrient base excluded *Oscillatoria* from the lake that *Daphnia* could assert its dominance among the zooplankton.

Changes at many trophic levels are thus relevant to the scientific side of the case study of Lake Washington. They illustrate the ease with which the solving of environmental "problems" can blend with ecological investigation. Retrospective analyses of the public record make a good lesson in civics, but thoughtful progress in science is cheated if investigations do not uncover new challenges and point to new paths of inquiry, as does the study of Lake Washington and its biological community.

REFERENCES

Algae increase noted in Lake Washington. Seattle Times. August 11, 1955.

Anderson, G. C. 1954. A Limnological Study of the Seasonal Variation of Phytoplankton Populations. Ph.D. thesis, University of Washington, Seattle.

Chapra, S. C., and K. H. Reckhow. 1983. Engineering Approaches for Lake Management. Vol. 2. Mechanistic Modeling. Ann Arbor Sciences, Ann Arbor, Mich.

Chasan, D. J. 1971. The Seattle area wouldn't allow the death of its lake. Smithsonian 2(4):6-13.

Clark, E. 1967. How Seattle is beating water pollution: Metro's project. Harpers 234:91-95 (June).

Comita, G. W. 1953. A Limnological Study of Planktonic Copepod Populations. Ph.D. thesis, University of Washington, Seattle.

Comita, G. W., and G. C. Anderson. 1959. The seasonal development of a population of *Diaptomus ashlandi* Marsh, and related phytoplankton cycles in Lake Washington. Limnol. Oceanogr. 4:37-52.

Edmondson announces pollution may ruin lake. University of Washington Daily. October 13, 1955.

Edmondson, W. T. 1961. Changes in Lake Washington following an increase in the nutrient income. Verh. Int. Verein. Limnol. 14:167-175.

Edmondson, W. T. 1966. Changes in the oxygen deficit of Lake Washington. Verh. Int. Verein. Limnol. 16:153-158.

Edmondson, W. T. 1968. Water-quality management and lake eutrophication: The Lake Washington case. Pp. 139-178 in T. H. Campbell and R. O. Sylvester, eds. Water Resources Management and Public Policy. University of Washington Press, Seattle.

Edmondson, W. T. 1969a. Cultural eutrophication with special reference to Lake Washington. Mitt. Int. Verein. Limnol. 17:19-32.

Edmondson, W. T. 1969b. Eutrophication in North America. Pp. 124-149 in Eutrophication: Causes, Consequences, Correctives. Proceedings of a Symposium. National Academy of Sciences, Washington, D.C.

Edmondson, W. T. 1970. Phosphorus, nitrogen and algae in Lake Washington after diversion of sewage. Science 169:690-691.

Edmondson, W. T. 1972a. The present condition of Lake Washington. Verh. Int. Verein. Limnol. 18:284-291.

Edmondson, W. T. 1972b. Nutrients and phytoplankton in Lake Washington. Am. Soc. Limnol. Oceanogr. Spec. Symp. 1:172-193.

Edmondson, W. T. 1977a. Recovery of Lake Washington from eutrophication. Pp. 102-109 in J. Cairns, Jr., K. L. Dickson, and E. E. Herricks, eds. Recovery and Restoration of Damaged Ecosystems. University Press of Virginia, Charlottesville.

Edmondson, W. T. 1977b. Trophic Equilibrium of Lake Washington. EPA-600/3-77-087. Environmental Research Laboratory, U.S. Environmental Protection Agency, Corvallis, Oreg.

Edmondson, W. T. 1979. Lake Washington and the predictability of limnological events. Arch. Hydrobiol. Beih. 13:234-241.

Edmondson, W. T., and J. T. Lehman. 1981. The effect of changes in the nutrient income on the condition of Lake Washington. Limnol. Oceanogr. 26:1-29.

Edmondson, W. T., and A. H. Litt. 1982. *Daphnia* in Lake Washington. Limnol. Oceanogr. 27:272-293.

Edmondson, W. T., G. C. Anderson, and D. R. Peterson. 1956. Artificial eutrophication of Lake Washington. Limnol. Oceanogr. 1:47-53.

Eggers, D. M., *et al.* 1978. The Lake Washington ecosystem: The perspective from the fish community production and forage base. J. Fish. Res. Bd. Can. 35:1553-1571.

Forbes, S. A. 1925. The lake as a microcosm. [Reprinted.] Bull. Ill. Nat. Hist. Surv. 15:537-550.

Hasler, A. D. 1947. Eutrophication of lakes by domestic drainage. Ecology 28:383-395.

Hrbacek, J. 1962. Species composition and the amount of the zooplankton in relation to the fish stock. Rozpr. Cesk. Akad. Ved. Rada Mat. Prir. Ved. 10:1-116.

Hrbacek, J., M. Dvorakova, M. Korinek, and L. Prochazkova. 1961. Demonstration of the effect of fish stock on the species composition of zooplankton and the intensity of metabolism of the whole plankton association. Verh. Int. Verein. Limnol. 14:192-195.

Infante, A., and S. E. B. Abella. 1985. Inhibition of *Daphnia* by *Oscillatoria* in Lake Washington. Limnol. Oceanogr. 30:1046-1052.

Kenworthy, E. W. 1971. How Seattle cleaned up. Audubon 73:105-106.

Lake's play use periled by pollution. Seattle Times. July 11, 1955.

Lake Stinko. Seattle Post-Intelligencer. October 5, 1963.

Lake Washington brown—That's algae, not mud. . . . Seattle Post-Intelligencer. July 3, 1962.

Lesson of Switzerland lakes brought home to Seattle area. Seattle Times. April 18, 1956.

Metro area citizens should be proud of achievement. Seattle Times. September 19, 1965.

Mortimer, C. H. 1941. The exchange of dissolved substances between mud and water in lakes. Parts 1 and 2. J. Ecol. 29:280-329.

Mortimer, C. H. 1942. The exchange of dissolved substances between mud and water in lakes. Parts 3 and 4. J. Ecol. 30:147-201.

Murtaugh, P. A. 1981. Selective predation by *Neomysis mercedis* in Lake Washington. Limnol. Oceanogr. 26:445-453.

Ohle, W. 1934. Chemische und physikalische Untersuchungen norddeutscher Seen. Arch. Hydrobiol. 26:386-464, 584-658.

Peterson, D. R. 1955. An Investigation of Pollution Effects in Lake Washington (1952-1953). Washington Pollution Control Commission Tech. Bull. 18, Seattle, Wash.

Piontelli, R., and V. Tonolli. 1964. Residence time of lake water in relation to enrichment, with special reference to Lago Maggiore. Mem. Ist. Ital. Idrobiol. 17:247-266. [in Italian]

Reckhow, K. H. 1979. Empirical lake models for phosphorus: Development, applications, limitations and uncertainty. Pp. 193-221 in D. Scavia and A. Robertson, eds. Perspectives on Lake Ecosystem Modeling. Ann Arbor Sciences, Ann Arbor, Mich.

Sawyer, C. N. 1947. Fertilization of lakes by agricultural and urban drainage. J. N. Engl. Water Works Assoc. 61:109-127.

Scheffer, V. B., and R. J. Robinson. 1939. A limnological study of Lake Washington. Ecol. Monogr. 9:95-143.

Sylvester, R. O., W. T. Edmondson, and R. H. Bogan. 1956. A new critical phase of the Lake Washington pollution problem. Trend in Engineering 8(2):8-14.

Vollenweider, R. A. 1969. Möglichkeiten und Grenzen elementärer Modelle der Stoffbilanz von Seen. Arch. Hydrobiol. 66:1-36.

Vollenweider, R. A. 1975. Input-output models with special reference to the phosphorus loading concept in limnology. Schweiz. Z. Hydrol. 37:53-84.

Vollenweider, R. A. 1976. Advances for defining critical loading levels for phosphorus in lake eutrophication. Mem. Ist. Ital. Idrobiol. 33:53-83.

Committee Comment

Several features of environmental problem-solving are illustrated by the Lake Washington example. The project itself was regarded as an experiment, and scientists were able to test their hypotheses during the study. Analogs existed in the scientific literature, so investigators were able to reason partly from first principles and partly by reference to other examples. At one point, Edmondson read through an article by A. D. Hasler (1947) that reviewed the history of cultural eutrophication in Europe and North America. He underlined "*Oscillatoria*" each time the word appeared in the text and discovered that the organism was a nearly ubiquitous indicator of eutrophication. Similarly, the relationship between *Neomysis* and *Daphnia* was suggested in part by experiences in Lake Tahoe when *Mysis relicta* was introduced as a forage food for fish (Richards *et al.*, 1975).

To understand the course of events in Lake Washington, one must draw on most of the sources of knowledge identified here. The initial events were treated as an experiment in lake fertilization that could improve our understanding of the ways that nutrient inputs control the biological character of lakes. Changes in the plankton community after the enrichment experiment required analyses of biological events at the population and community levels. Plankton community structure was seen to be governed by a variety of species interactions, including predation and interference. Superficially, the lake appeared to exhibit alternative stable states with regard to plankton composition. In fact, the transition from one state to another was a logical consequence of changing fields of predators and algae.

Most important, the scientists were able to identify a few factors and processes that were acting with special force among the myriad present. Edmondson and Litt (1982) wrote: "We assume that the processes controlling the population are simultaneously affected by many factors and that changing any one can affect the population. At some times, one factor may dominate the others quantitatively." Scientific judgment was needed to identify the stoichiometries and relations among nutrient income, hydrology, and algal production. Similar reasoning helped to establish the likely causes for species alterations. No amount of descriptive field study alone would establish causality firmly enough to permit quantitative management decisions. Judgments had to be made about how algal production would respond to alterations in nutrient supplies and about the importance of nutrient returns from the sediments. That required knowledge of more than the biota alone. It was necessary to regard the organisms as being integrated with their physical and chemical surroundings.

Spatial relations and the vertical differentiation that arises from thermal stratification were important, too. Lake Washington is a warm, monomictic lake; therefore, it circulates all winter long. Winter is also the time of greatest fluvial discharge, and on the average one-third of the lake's volume is renewed each year. Most of the water drains from the Cascade Range and is very low in dissolved salts or nutrients of any kind. Each winter, the lake is thus diluted by water of low nutrient content. Edmondson could argue securely that, if fertilizing discharges from municipalities ceased, the accumulated nutrients and algal biomass could be flushed from the basin within a few years. Because Lake Washington possessed a large hypolimnion with more than adequate reserves of oxygen to last through summer stratification, most of the phosphorus locked in the sediments would stay there.

Furthermore, in the case of Lake Washington, scientific judgment backed by logic and data was separated from emotional statements. The decision to raise public funds and divert the effluent was political, not scientific; indeed, sound and important discoveries probably would have accompanied a study of continued deterioration of the lake.

The case of Lake Washington is exemplary, not because the forecasts were so accurate, but because events were documented and reported in comprehensible fashion. The documents reveal a remarkable synergism of scientific and public awareness about an environmental issue. The retrospective account makes the scientific issues sound perhaps more cut and dried than they were at the time. It might seem that the only suspense was related to the public's willingness to spend money to improve water quality.

In the 1960s, however, debates about the causes of lake eutrophication were common, and the debates eventually spawned an inquiry by the National Research Council (1969). The Research Council's report directed attention to phosphorus, but the evidence came in part from results seen in Lake Washington. Vocal scientific lobbies had argued that carbon and nitrogen could be limiting elements in many aquatic habitats and that phosphorus control would therefore be insufficient to halt eutrophication. Many of the conflicting opinions were published in the proceedings of a special symposium of the American Society of Limnology and Oceanography (Likens, 1972). It was in the early stages of this debate that the Lake Washington experiment was conceived and executed.

Despite the obvious success in curbing pollution of the lake, the experiment could not by itself prove that phosphorus was the culprit, even though predictions had been based on that assumption. The action of removing waste treatment effluent from a lake lacks the rigor of a conventional laboratory experiment, in that many factors are manipulated simultaneously. Opponents could argue that improvement arose because some unmeasured trace metal or unknown growth factor was removed with the effluent and that phosphorus control elsewhere might be costly and irrelevant. Indeed, many investigators had discovered that, when lake phytoplankton was enclosed in bottles and subjected to single-nutrient additions, carbon, nitrogen, or trace metals could often stimulate their metabolism and growth (Likens, 1972). This very type of observation is the basis of present views about nitrogen limitation in the oceans (Ryther and Dunstan, 1971).

The principal difference between bottle bioassays and Lake Washington, however, lies not in methodological detail, but in the scale of the manipulation. In the case of Lake Washington, an entire ecosystem was manipulated. At lake-wide scales, exchange processes at air-water and sediment-water interfaces become important, and responses can be followed over long periods. Lake Washington became one of the pioneer "whole-lake" experiments. Within a few years, Canadian limnologists had established an Experimental Lakes Area in northwestern Ontario (Johnson and Vallentyne, 1971) and were setting out to test nutrient controls of eutrophication more rigorously than could ever be possible in Lake Washington. Experimental lakes were purposely fertilized with nitrogen, phosphorus, and carbon, singly and in combinations. The results showed beyond doubt that phosphorus was the master controlling nutrient, as far as eutrophication was concerned (Schindler, 1977). When lakes were fertilized with phosphate alone, algal growth reached bloom proportions, because inorganic carbon entered the lake water from the atmosphere and continuously replaced the carbon used by the algae during photosynthesis. Similarly, species composition in these lakes became more strongly represented by nitrogen-fixing blue-green algae (cyanobacteria), which formed the ultimate reservoir for the nutrient. Phosphorus, however, has no gaseous atmospheric phase, so its rate of supply to a lake basin sets an absolute limit on standing crops. With only bottle bioassays or small-scale experiments of short duration, the ultimate consequences of a manipulation could not be forecast. Bottles are closed to gas exchange with the atmosphere, and the experiments are too brief to permit species assemblages to change. In short, no study short of a whole-lake manipulation could have provided an adequate analogy to this experiment. That is why the lessons from Lake Zurich and the lakes in Madison, Wisconsin, were so valuable in the early, predictive stages of the project.

As a historical footnote, the success story of Lake Washington might have heartened those in Switzerland who were trying to clean up Lake Zurich. Cultural eutrophication in Lake Zurich had been accelerating since 1896 (Thomas, 1969). Three-stage waste treatment plants around the lake with chemical precipitation processes for the removal of the phosphate were introduced. The first started operating in 1967; since then, all Zurich treatment plants have had precipitation installations to eliminate phosphate (Dietlicher, 1974). The improvements coincided with reductions in phosphate concentrations in the lake and improvements in water quality.

References

Dietlicher, K. 1974. The Water Quality of the Lakes of Zurich and "Walensee." Zurich Waterworks, Zurich, Switz.

Edmondson, W. T., and A. H. Litt. 1982. *Daphnia* in Lake Washington. Limnol. Oceanogr. 27:272-293.

Hasler, A. D. 1947. Eutrophication of lakes by domestic drainage. Ecology 28:383-395.

Johnson, W. E., and J. R. Vallentyne. 1971. Rationale, background, and development of experimental lake studies in northwestern Ontario. J. Fish Res. Bd. Can. 28:123-128.

Likens, G. E., ed. 1972. Nutrients and eutrophication: The limiting nutrients controversy. Am. Soc. Limnol. Oceanogr. Spec. Symp. 1:1-328.

National Research Council. 1969. Eutrophication: Causes, Consequences, Correctives. Proceedings of a Symposium. National Academy of Sciences, Washington, D.C.

Richards, R. C., C. R. Goldman, T. C. Frantz, and R. Wickwire. 1975. Where have all the *Daphnia* gone? The decline of a major cladoceran in Lake Tahoe, California-Nevada. Verh. Int. Verein. Limnol. 19:835-842.

Ryther, J. H., and W. M. Dunstan. 1971. Nitrogen, phosphorus, and eutrophication in the coastal marine environment. Science 171:1008-1013.

Schindler, D. W. 1977. Evolution of phosphorus limitation in lakes. Science 195:260-262.

Thomas, E. A. 1969. Kulturbeeinflusste chemische und biologische Veränderungen des Zürichsees im Verlaufe von 70 Jahren. Mitt. Int. Verein. Limnol. 17:226-239.

21

Raising the Level of a Subarctic Lake

A common form of environmental intervention is damming a river to produce an artificial lake or raising the level of an existing lake to create a larger one. The environmental problems arising from these manipulations are more complex than those resulting from pollution of lakes, because the manipulations establish new flow patterns, inundate existing shorelines, initiate erosional processes that create new shorelines, and change the morphologic characteristics of the lakes. For these reasons, prediction of the consequences of altering the dimensions of lakes is more difficult and more uncertain than prediction of the responses of lakes to pollutants and their elimination.

Southern Indian Lake, in northern Manitoba, was raised 3 m in 1976, and rivers were diverted so that the flow of water through the lake was reduced by about 75%. The case study reveals that, despite careful planning of the project, many of the predictions were seriously wrong. The limnologists who carried out the preproject studies later analyzed the disparities between their predictions and the outcomes. As a result, future studies for similar projects will be based on better models.

Case Study

JOHN T. LEHMAN, Division of Biological Sciences, University of Michigan, Ann Arbor

Southern Indian Lake (57° N, 99° W), in northern Manitoba, is a riverine basin along the Churchill River that drains northward to Hudson Bay. In 1976, the natural lake outlet was dammed and the lake level raised 3 m, so that the Churchill could be diverted southward across a drainage divide. Of the natural river flow of almost 1,000 m³/second, 75% was diverted into the Nelson River and no longer flows through the main lake. The diversion scheme permitted the combined flow of drainage from over 1.4 million square kilometers of the northern Great Plains to flow through a single series of dams and hydroelectric generating stations. The project had been planned in principle since the 1950s, and the lake had been the object of careful impact assessments. Extensive and detailed knowledge of the later ecological events has come through the efforts of scientists at the Freshwater Institute in Winnipeg. Institute scientists were charged in 1976 with a long-term study to evaluate the predictive capabilities of impact assessments and to generate new capabilities with increased quantitative precision.

The preimpoundment assessment work was done by professional limnologists who used existing modeling techniques and a literature-based paradigm regarding reservoir dynamics. The project was treated as a large-scale experiment to test the hypotheses of the impact assessments and to modify the paradigm. Many predictions were realized, and others differed from expectation only in quantitative detail. Most biological responses above the primary trophic level, however, were predicted either incorrectly or not at all. Assessments based on analogy and scaling were often misleading. The most striking unpredicted change involved the commercial whitefish fishery and the local economy that depended on it.

Southern Indian Lake supported the largest commercial fishery in northern Manitoba. About 85% of the catch was *Coregonus clupeaformis*, the lake whitefish. During the 3 decades before impoundment, the annual catch averaged 334 metric tons, almost exclusively high-quality fish of "export" grade (Bodaly *et al.*, 1984b). The dendritic basin of Southern Indian Lake consists of several subbasins that are separated by islands and channel constrictions. The high quality of the fishery was maintained by selective fishing in the most profitable basins. Elsewhere, stocks were dominated by darker fish with high incidences of muscle cysts of the parasitic cestode *Triaenophorus crassus*.

318

Southern Indian Lake is in a region underlain by Precambrian shield bedrock. The overlying sedimentary deposits are glaciolacustrine or glaciofluvial, tracing either to the glacial Lake Agassiz or to outwash during the retreat of the last continental glaciation. Before the 3-m impoundment, 76% of the shoreline was exposed bedrock. Mean annual temperature in the region is $-5°C$, so permafrost is widespread in the glacial deposits. Vegetation is characteristic of boreal forest or taiga (Newbury *et al.*, 1984).

On the basis of federal and provincial feasibility studies, Manitoba Hydro in the 1960s proposed raising Southern Indian Lake by 10 m. This high-level impoundment was selected to optimize the generation of electricity, without regard for environmental issues or a thorough study of the cost-effectiveness of hydrologic storage in the lake. In response to public concern, the utility contracted with Underwood-McLellan and Associates Ltd., a consulting engineering firm, to produce a predevelopment impact assessment of several diversion scheme options. Published in 1970, that study caused Manitoba Hydro to opt instead for a less costly low-level (3-m) impoundment, on the grounds that it would have proportionally smaller effects on watershed and shoreline. A federal-provincial study board was later commissioned to investigate the effects of the new plan. The Lake Winnipeg, Churchill and Nelson Rivers Study Board undertook its work concurrently with construction activities and issued its final report in April 1975, shortly before the diversion. The two studies differed considerably in their objectives and data bases. The first involved no substantial field work and was aimed at determining costs and benefits of different diversion schemes. The second was initiated after the configuration and construction schedule of the project were fixed; scientists were charged with providing reliable baseline data, predicting future conditions, and then measuring them.

The federal-provincial study was comprehensive. It included studies of hydrology, soils, fisheries, limnology, wildlife, geology, recreation, forestry, archaeology, navigation, and socioeconomic factors. University scientists, private consultants, and government laboratories were involved in different aspects of the investigations. The fisheries-limnology component was summarized by Hecky and Ayles (1974), who described the existing conditions of the basin in concise, quantitative terms and then presented a series of predictions.

The limnologists recognized that the diversion scheme would alter flow patterns in the lake and that some regions would be flushed more quickly than others. In their opinion, the altered flow would have a greater impact on water chemistry than would inundation itself, because Churchill River water greatly influenced the main basins of the lake under natural conditions. If normal flowthrough were reduced by 80%, which was the proposed long-term average, local drainage could dominate conditions in a large area of the lake.

Projections were based in part on conceptual and numerical models that related nutrient loading and water exchange with algal abundance and productivity. These were new, sophisticated versions of the approach that Edmondson had used to evaluate nutrient loading in Lake Washington (see Chapter 20 and, e.g., Dillon and Rigler, 1974, and Vollenweider, 1975). The models had sound empirical support and widespread acceptance as a means to predict algal abundance and trophic state. Predictions were for steady-state conditions that would follow transient inundation effects. Reduced inflow of Churchill River water would lead to diminished turbidity and nutrients from fluvial sediment loading in the long run, but water residence time would increase. Overall annual algal production was expected to decline by 33-50% in the main basins. The diversion was expected to have a positive effect in one basin along the diversion route, which previously had received little Churchill water directly.

Inundation was expected to affect water quality, but not as drastically in the long term as diversion. The conventional reservoir paradigm predicted that nutrient release from soils and vegetation would increase post-impoundment production at all levels in the food chain. The overburden of frozen glaciolacustrine clays around Southern Indian Lake, however, had great potential for erosion. Increased turbidity would negate benefits of increases in nutrients, according to surveys of light penetration and primary productivity. The inundation was thus expected to have no net favorable effects and to be detrimental in littoral regions.

The limnologists expected higher trophic levels generally to track changes in the algae, although not necessarily immediately or in strict proportion. Declines in planktonic and zoobenthic production in main basins were expected to diminish fish production gradually. Suitable spawning sites for whitefish in the inundated areas were of concern, as was the chance that diversion could disorient migrating fish. Hecky and Ayles (1974) foresaw difficulties for the commercial fishery that would cause it to decline more than overall biological production would decline. The traditional fishery was concentrated in the largest basin, which was certain to experience the greatest deprivation from diversion. As production potential fell, fishermen would see their yields per unit of fishing effort decline. No evidence of overfishing yet existed, but many marginal operators would probably leave the fishery if faced with a need to increase their effort substantially. The scientists expected an overall decrease of 13% in commercial yield. Lakewide average biological production was expected to decline by 10%. The declines were forecast on the basis of expected redistributions of nutrient income from the Churchill River. They applied to the long-term, steady-state outcome of the manipulation. Greater uncertainty was attached to short-term responses, and statements on that matter were largely qualitative.

Early in the study, the federal-provincial study board had identified the need for long-term ecological monitoring of the impact areas. Hecky and his colleagues at the Freshwater Institute initiated their own case study of the Southern Indian Lake reservoir in 1974 with the preimpoundment work. The operational regimen has been roughly that considered by the impact study, so predictions can legitimately be compared with results. The lake provides a good opportunity to evaluate scientific judgments retrospectively. The technical results of the decade-long effort were published in 17 papers in Volume 41 of the *Canadian Journal of Fisheries and Aquatic Sciences* (1984).

Quantitative predictions had been generally lacking in the impact assessments of 1970 and 1975. The 10% reduction in biological production offered by Hecky and Ayles was cited repeatedly in the 1975 summary report as one hard prediction (Lake Winnipeg, Churchill and Nelson Rivers Study Board, 1975). The lack of postimpoundment case studies in northern Canada and the absence of verified models for important processes precluded other precise estimates. The limnologists recommended research into process models as one way to improve predictions. They had been forced to rely, as had those who conducted the 1970 study, on results from reservoirs in temperate and tropical regions. Excellent studies had been done in Siberia, but those reservoirs were deep, steep-sided impoundments, and their physical characteristics differed from those of Southern Indian Lake. The Siberian reservoirs, for instance, are deep enough to develop thermal stratification with reduced oxygen in the bottom water during the summer, but Hecky and Ayles expected no such thermal stratification or deoxygenation in Southern Indian Lake. The potential for extensive shoreline erosion caused their predictions to differ from conventional wisdom regarding productivity and fish yield in new reservoirs.

In retrospect, the predevelopment predictions were extremely good with regard to nutrients and algal production. No thermal stratification developed, and shoreline erosion was extensive (Hecky, 1984; Hecky and McCullough, 1984; Newbury and McCullough, 1984. The increased nutrients from erosion and decay of vegetation increased phosphorus concentrations as expected, but turbidity increased also. Physiological studies revealed that phosphorus deficiency among the algae had been replaced by light limitation in turbid regions of the lake, and no significant increase in productivity occurred (Hecky and Guildford, 1984; Planas and Hecky, 1984). Regions with high transparency showed increased productivity. Few changes were predicted incorrectly. The notable exception was in fish recruitment. Both impact assessments had predicted that northern pike (*Esox lucius*) would benefit from the increase in habitat associated with inundation and that the sport fishery would improve. Some short-term spawning problems were anticipated for walleye (*Stizostedion vitreum vitreum*) and the important whitefish, but the fish were expected to exploit new spawning grounds and to recover. *Esox* in fact experienced no increase in survival or growth, and the commercial whitefish fishery collapsed (Bodaly and Lesack, 1984; Bodaly *et al.*, 1984b).

Loss of the viable commercial fishery in Southern Indian Lake was a blow to the local economy, and it had been unexpected. The reasons for the change are complex, but they were anticipated in part by Hecky and Ayles (1974). The catch per unit effort indeed declined on the traditional fishing grounds, to about half the preflooding value. Total catches were maintained by major increases in total fishing effort. The fishermen also began to exploit regions that had previously been avoided because the fish were of lower marketability, being darker and having higher rates of cestode infestation. The lower-quality fish made up 12-72% of the summer catch in the years after impoundment, whereas they were insignificant components of the original fishery. In 1982, the lake was changed from "export" to "continental" classification, and the commercial value of the catch plummeted. Total catch eventually fell to one-third of its preimpoundment size as many operators abandoned the fishery (Bodaly *et al.*, 1984b; Wagner, 1984).

The catch declines apparently stemmed from migrations of fish away from traditional fishing grounds. Genetic markers measured among whitefish stocks before and after impoundment indicate that stocks became redistributed when normal flow patterns were altered. There is an indication of net emigration of whitefish from Southern Indian Lake. Compensation payments to commercial fishermen by Manitoba Hydro subsidized their efforts from 1977 to 1982. Without the payments, production costs would have exceeded fishery revenues. In 1982, Manitoba Hydro provided a one-time cash settlement of Can$2.5 million for all future losses (Bodaly *et al.*, 1984b).

The fishery problems were not confined to collapse of the commercial enterprise. Changes had been triggered by physical manipulations that proved more far-reaching and persistent than anyone expected. The shoreline of Southern Indian Lake had been bedrock-controlled; 76% of the shore was exposed granite or gneiss. After flooding, only 14% of the new shore consisted of bedrock. Erosion had been expected by everyone, but not of the scale and duration produced when lakewater melted the permafrost. The shoreline retreated at up to 12 m/year. Lakewater melted and undercut the backshore zone, and that resulted in massive faulting of the overhanging shoreline. Newbury and McCullough (1984) predicted that it would take at least 35 years for the shoreline to be eroded to preflooding conditions and that high rates of sediment input from inundation would persist for decades.

The continuing flooding of terrestrial areas posed a new threat to fisheries, this time presenting a health hazard for domestic consumers. The flooded soils released mercury, which became concentrated particularly in the piscivorous northern pike and walleye (Bodaly *et al.*, 1984a). Increased mercury concentrations are a common consequence of reservoir creation, but the phenomenon has been recognized only within the last decade (Abernathy and Cumbie, 1977; Cox *et al.*, 1979; Kent and Johnson, 1979; Potter *et al.*, 1975). Soils need not be rich in mercury, and indeed the Southern Indian Lake source materials have a low or average mercury content (Bodaly *et al.*, 1984a). Bacterial methylation of inorganic mercury salts mobilizes the element, and it then associates with particles. Muscle mercury concentrations in the walleye and northern pike now exceed the Canadian marketing standards (0.5 ppm) and will probably remain increased for years.

A few other unexpected changes occurred. Mean lake temperatures decreased, owing to increased mean depth, river diversion, greater surface reflectance, and backscattering of incident solar radiation from the turbid water (Hecky, 1984). The zooplankton community changed, too (Patalas and Salki, 1984). The large-bodied crustacean *Mysis relicta*, formerly rare, became common. Large-bodied calanoid copepods like *Limnocalanus macrurus* increased. Cladocerans and small copepods decreased. Patalas and Salki ascribed the changes primarily to lower water temperatures and increased water depth, which enlarged the habitat of generally deep-dwelling species. They regarded an overall decrease in the abundance and biomass of crustacean zooplankton to be a consequence of the diversion-related temperature decrease. Lower temperatures slow growth and development rates of both eggs and juveniles. The patterns suggest, moreover, that predation pressure from planktivorous fish decreased and that predatory invertebrates became an important force in the case of the plankton. The decline in predation by fish could reflect a decrease in stocks or unsuccessful foraging due to turbidity.

The changes in the zooplankton community in Southern Indian Lake parallel some of the changes that occurred in Lake Tahoe when *Mysis* was introduced (Goldman *et al.*, 1979; Richards *et al.*, 1975). It seems reasonable to ascribe the changes in zooplankton, including *Limnocalanus* and *Mysis*, to declines in fish stocks and foraging success. The findings are united with recent events in Lake Washington by the major theme in zooplankton community ecology of the last 2 decades, which emerged in the seminal works of Hrbacek (Hrbacek, 1962; Hrbacek *et al.*, 1961) and Brooks and Dodson (1965): that the introduction of planktivorous fish causes wholesale alterations of species composition and size distribution in zooplankton communities. That zooplankton communities are shaped by the size-selective predatory behavior of fish is well established (Galbraith, 1967; Hall *et al.*, 1976; Lynch, 1979; O'Brien, 1979; Wells, 1970; Zaret, 1980). The proximate mechanism is prey conspicuousness (Zaret, 1972; Zaret and Kerfoot, 1975; Zaret and Suffern, 1976). Much of the recent excitement in zooplankton community ecology has come from investigations of community structure in the absence of fish. Brooks and Dodson had used allometric estimates of metabolism and feeding to support their "size-efficiency hypothesis"—i.e., large animals simply outcompete small ones through economies of scale. Others contested that claim and argued with convincing empirical support that predation by invertebrates could eliminate small species (Dodson, 1974; Kerfoot, 1977; Zaret, 1975). Unlike vertebrates, these predators rely on mechanoreception to detect their prey and manipulate the prey to consume it. This type of predation by invertebrates, mostly insect larvae and predatory crustacean zooplankton, has been linked to much morphological variation—including cyclomorphosis—within the taxa that they prey on (Halbach, 1971; Jacobs, 1965; Zaret, 1972).

Large invertebrate predators can thrive in the absence of fish, feeding on small-bodied animals and the juvenile stages of larger taxa. Large-bodied zooplankton have an advantage, particularly if they can produce large eggs and boost their offspring past the vulnerable juvenile period. When planktivorous fish abound, the invertebrate predators are held in check; so, too, are large herbivores. The small-bodied animals are released from predation. They might come to dominate the assemblage, even though the fish might eventually be forced to turn to them as prey (Brooks, 1968). This explanation probably figures in the events in Southern Indian Lake, although Patalas and Salki (1984) regarded the direct physiological effects of temperature to be dominant.

Hecky and colleagues undertook in 1984 their own retrospective analysis of environmental impact prediction and assessment based on their experience with Southern Indian Lake (Hecky *et al.*, 1984). Scientists who had formulated the original predictions and then used the experiment to advance understanding provided their scientific insights into the limitations and capabilities of the prediction process. They conceded, for instance, that the 1970 "office" study was nearly as effective in forecasting impacts as was the 1975 study after a year of field work. Improved precision in the 1975 study resulted in part from having a better definition of the project. Hecky and colleagues concluded that both assessments were of only marginal overall utility, because significant impacts had not been predicted and because even the correct predictions had often been qualitative. Qualitative statements could not enter the quantitative cost-benefit analyses that influence major resource development.

The field study had proved invaluable, nonetheless, primarily as the means to document the baseline conditions in the lake. Without such data, any followup study might have been meaningless, and compensation payments might have been more difficult to arrange. The field study, in essence, provided insurance against unpredicted adverse environmental impacts (Hecky *et al.*, 1984).

Three-meter impoundment was selected because of the high cost of accessory ''saddle'' dams that are needed to contain high-level impoundment and because the 1970 study had suggested that the effects of the originally planned 10-m impoundment might be too severe. The authors of the 1970 study had assumed that effects were proportional to the magnitude of manipulation. A 3-m change was assumed to have smaller effect than a 10-m change. In fact, impoundment levels greater than 3 m would have produced effects only slightly worse than actually occurred, because severe shoreline erosion began as soon as the natural range of water levels was exceeded and the impounded water entered the previously frozen backshore zone.

The Freshwater Institute scientists bemoaned the lack of suitable analog studies and the limitations of current reservoir paradigms. Conventional wisdom and expectations were based on experience with deep riverine basins with limited wind fetches. Indigenous biological processes had been emphasized nearly to the exclusion of physical forces. The most likely analog candidates either lacked permafrost features or had little fine-grained erodable material on the shores. In the absence of examples or experience, it was difficult for the scientists to judge the importance of these novel characteristics. However, they developed empirical relations between energy and erosion and can now use shoreline maps and meteorological records to predict lake volumes after erosion (Newbury and McCullough, 1984). Southern Indian Lake has changed the understanding of reservoir dynamics.

Responses at higher trophic levels showed the greatest failures of prediction. Hecky *et al.* (1984) rejected the notion that the responses are intrinsically unpredictable. They argued instead that possible responses, unless parts of existing paradigms, tend to be overlooked or considered only rarely. Hecky *et al.* seemed to blame themselves for not anticipating the problems with fishery quality and with mercury contamination, inasmuch as elements of the logical puzzle were already in their grasp. They suggested that they were on surest ground when predicting energy flow, biomass, and general trophic relations, but that species-level predictions are elusive. Their most severe criticism was that they learned that current impact assessment was incomplete and unacceptable. They could also have mentioned that their studies opened the door to new understanding of reservoir processes and functions.

REFERENCES

Abernathy, A. R., and P. M. Cumbie. 1977. Mercury accumulation by largemouth bass (*Micropterus salmoides*) in recently impounded reservoirs. Bull. Environ. Contam. Toxicol. 17:595-602.

Bodaly, R. A., and L. F. W. Lesack. 1984. Response of a boreal northern pike (*Esox lucius*) population to lake impoundment: Wupaw Bay, Southern Indian Lake, Manitoba. Can. J. Fish. Aquat. Sci. 41: 706-714.

Bodaly, R. A., R. E. Hecky, and R. J. P. Fudge. 1984a. Increases in fish mercury levels in lakes flooded by the Churchill River diversion, northern Manitoba. Can. J. Fish. Aquat. Sci. 41:682-691.

Bodaly, R. A., T. W. D. Johnson, R. J. P. Fudge, and J. W. Clayton. 1984b. Collapse of the lake whitefish (*Coregonus clupeaformis*) fishery in Southern Indian Lake, Manitoba, following lake impoundment and river diversion. Can. J. Fish. Aquat. Sci. 41:692-700.

Brooks, J. L. 1968. The effects of prey size selection by lake planktivores. Syst. Zool. 17:272-291.

Brooks, J. L., and S. I. Dodson. 1965. Predation, body size, and the composition of the plankton. Science 150:28-35.

Cox, J. A., J. Carnahan, J. DiNunzio, J. McCoy, and J. Meister. 1979. Source of mercury in fish in new impoundments. Bull. Environ. Contam. Toxicol. 23:779-783.

Dillon, P. J., and F. H. Rigler. 1974. A test of a simple nutrient budget model predicting the phosphorus concentration in lakewater. J. Fish. Res. Bd. Can. 31:1771-1778.

Dodson, S. I. 1974. Adaptive change in plankton morphology in response to size-selective predation: A new hypothesis of cyclomorphosis. Limnol. Oceanogr. 19:721-729.

Fudge, R. J. P., and R. A. Bodaly. 1984. Postimpoundment winter sedimentation and survival of lake whitefish (*Coregonus clupeaformis*) eggs in Southern Indian Lake, Manitoba. Can. J. Fish. Aquat. Sci. 41:701-705.

Galbraith, M. G., Jr. 1967. Size-selective predation on *Daphnia* by rainbow trout and yellow perch. Trans. Am. Fish. Soc. 96:1-10.

Goldman, C. R., M. D. Morgan, S. T. Threlkeld, and N. Angell. 1979. A population dynamics analysis of the cladoceran disappearance from Lake Tahoe, California-Nevada. Limnol. Oceanogr. 24:289-297.

Halbach, U. 1971. Zum Adaptivwert der zyklomorphen Dornenbildung von *Brachionus calyciflorus* Pallas (Rotatoria). Oecologia 6:267-288.

Hall, D. J., C. W. Burns, and P. H. Crowley. 1976. The size-efficiency hypothesis and the size structure of zooplankton communities. Annu. Rev. Ecol. Syst. 7:177-208.

Hecky, R. E. 1984. Thermal and optical characteristics of Southern Indian Lake before, during, and after impoundment and Churchill River diversion. Can. J. Fish. Aquat. Sci. 41:579-590.

Hecky, R. E., and H. A. Ayles. 1974. Summary of fisheries-limnology investigations on Southern Indian Lake. Lake Winnipeg, Churchill and Nelson Rivers Study Board, 1971-1975. Technical Report Appendix 5, Volume 1A. Environment Canada Fisheries Service, Winnipeg, Man.

Hecky, R. E., and S. J. Guildford. 1984. Primary productivity of Southern Indian Lake before, during, and after impoundment and Churchill River Diversion. Can. J. Fish. Aquat. Sci. 41:591-604.

Hecky, R. E., and G. K. McCullough. 1984. Effect of impoundment and diversion on the sediment budget and nearshore sedimentation of Southern Indian Lake. Can. J. Fish. Aquat. Sci. 41:567-578.

Hecky, R. E., R. W. Newbury, R. A. Bodaly, K. Patalas, and D. M. Rosenberg. 1984. Environmental impact prediction and assessment: The Southern Indian Lake experience. Can. J. Fish. Aquat. Sci. 41:720-732.

Hrbacek, J. 1962. Species composition and the amount of the zooplankton in relation to the fish stock. Rozpr. Cesk. Akad. Ved Rada Mat. Prir. Ved. 10:1-116.

Hrbacek, J., M. Dvorakova, M. Korinek, and L. Prochazkova. 1961. Demonstration of the effect of fish stock on the species composition of zooplankton and the intensity of metabolism of the whole plankton association. Verh. Int. Verein. Limnol. 14:192-195.

Jacobs, J. 1965. Significance of morphology and physiology of *Daphnia* for its survival in predator-prey experiments. Naturwissenschaften 52:141.

Kent, J. C., and D. W. Johnson. 1979. Mercury, arsenic and cadmium in fish, water, and sediment of American Falls Reservoir, Idaho, 1974. Pestic. Monit. J. 13:35-40.

Kerfoot, W. C. 1977. Implications of copepod predation. Limnol. Oceanogr. 22:316-325.

Lake Winnipeg, Churchill and Nelson Rivers Study Board. 1975. Summary Report: Canada-Manitoba Lake Winnipeg, Churchill and Nelson Rivers Study. Allied Printing, Winnipeg, Man.

Lynch, M. 1979. Predation, competition, and zooplankton community structure: An experimental study. Limnol. Oceanogr. 24:253-272.

Newbury, R. W., and G. K. McCullough. 1984. Shoreline erosion and restabilization in the Southern Indian Lake reservoir. Can. J. Fish. Aquat. Sci. 41:558-566.

Newbury, R. W., G. K. McCullough, and R. E. Hecky. 1984. The Southern Indian Lake impoundment and Churchill River diversion. Can. J. Fish. Aquat. Sci. 41:548-557.

O'Brien, W. J. 1979. The predator-prey interaction of planktivorous fish and zooplankton. Am. Sci. 67:572-581.

Patalas, K., and A. Salki. 1984. Effects of impoundment and diversion on the crustacean plankton of Southern Indian Lake. Can. J. Fish. Aquat. Sci. 41:613-637.

Planas, D., and R. E. Hecky. 1984. Comparison of phosphorus turnover times in northern Manitoba reservoirs with lakes of the Experimental Lakes Area. Can. J. Fish. Aquat. Sci. 41:605-612.

Potter, L., D. Kidd, and D. Standiford. 1975. Mercury levels in Lake Powell: Bioamplification of mercury in man-made desert reservoir. Environ. Sci. Tech. 9:41-46.

Richards, R. C., C. R. Goldman, T. C. Frantz, and R. Wickwire. 1975. Where have all the *Daphnia* gone? The decline of a major cladoceran in Lake Tahoe, California-Nevada. Verh. Int. Verein. Limnol. 19:835-842.

Vollenweider, R. A. 1975. Input-output models with special reference to the phosphorus loading concept in limnology. Schweiz. Z. Hydrol. 37:53-84.

Wagner, M. W. 1984. Postimpoundment change in financial performance of the Southern Indian Lake commercial fishery. Can. J. Fish. Aquat. Sci. 41:715-719.

Wells, L. 1970. Effects of alewife predation on zooplankton populations in Lake Michigan. Limnol. Oceanogr. 14:556-565.

Zaret, T. M. 1972. Predators, invisible prey, and the nature of polymorphism in the Cladocera (Class Crustacea). Limnol. Oceanogr. 17:171-184.

Zaret, T. M. 1975. Strategies for existence of zooplankton prey in homogeneous environments. Verh. Int. Verein. Limnol. 19: 1484-1489.

Zaret, T. M. 1980. Predation and freshwater communities. Yale University Press, New Haven, Conn.

Zaret, T. M., and W. C. Kerfoot. 1975. Fish predation on *Bosmina longirostris*: Body-size selection versus visibility selection. Ecology 56:232-237.

Zaret, T. M., and J. S. Suffern. 1976. Vertical migration in zooplankton as a predator avoidance mechanism. Limnol. Oceanogr. 21:804-813.

Committee Comment

Like the Lake Washington work (Chapter 20), the investigation of Southern Indian Lake called on scientists to use nearly all the sources of ecological knowledge identified in this case study. Freshwater Institute scientists were explicit at the outset about using impoundment and diversion as an experiment in reservoir processes. They coordinated and focused scientific and technical talent on a scale rarely achieved in such ventures. Productivity of fish and plankton communities was a dominant theme of the biological investigations. The observational program was designed to study spatial differences in the dendritic basin and to identify regions that were influenced, either positively or negatively, by the physical manipulation. The specific incorporation of genetic markers into the study of whitefish populations (Bodaly *et al.*, 1984), for instance, made it possible to document the redistribution and migration of stocks.

Productivity and biomass were the measures of choice for all trophic levels, because they reduce biological entities and rates to a common currency. For a study that encompasses an entire ecosystem, such a common means of presentation of data is necessary. Nonetheless, events in Southern Indian Lake are traceable to the life histories and physiological and behavioral characteristics of individual interacting populations. Increased turbidity and lowered lake temperatures triggered biological responses in ways not explainable strictly by thermodynamics or mass flux alone. Temperature affects the rates of metabolic and developmental processes, and inorganic particles in suspension attenuate light that would otherwise be available for photosynthesis. Turbidity, moreover, obscures the vision of predators that hunt by sight and reduces the risk to their preferred prey. The experiment in Southern Indian Lake apparently introduced manipulations at both the base and the apex of the food web. Primary producers were affected by the altered light and by changes in nutrient input. Whereas selection might have operated for species adept at nutrient acquisition before impoundment, the postimpoundment community clearly was governed by light limitation (Hecky and Guildford, 1984). One would expect changes in species composition, and the phytoplankton community showed such changes (H. Kling, personal communication, cited by Patalas and Salki, 1984).

The increase in large-bodied crustaceans suggests that predatory pressure on the species had relaxed. Deeper, more turbid water provided a refuge from the fish. The predatory behavior of these large zooplankters, in turn, could have contributed to the decline in cladocerans and small copepods reported by Patalas and Salki (1984). Effects of the manipulation thus became focused on the small herbivores, beset on the one side by abundant carnivorous zooplankton and on the other by reduced abundance of phytoplankton owing to the new light regime. The cladocerans declined by 75% and small cyclopoid copepods by 50% (Patalas and Salki, 1984).

In contrast with the alterations of the plankton communities, little or no change occurred among the profundal macrobenthos (Wiens and Rosenberg, 1984). The benthic organisms seemed to be influenced principally by the input of organic material from the Churchill River or from shoreline erosion and thus to be divorced from biological interactions in the turbid waters. The scientific papers include no discussion of meroplanktonic taxa like *Chaoborus*, which can enter the plankton each night and are known to be influenced by planktivorous fish (e.g., Northcote *et al.*, 1978). If fish predation were substantially relaxed, these species would likely be affected.

The professional publications produced as a result of the Southern Indian Lake study are valuable contributions to current knowledge about reservoir processes. The environmental manipulation was an experiment that had never been tried previously. Most of the results could be guessed only qualitatively, if at all. From a strictly scientific point of view, the project was exploited very profitably, in that new principles were uncovered and present paradigms have been enriched. But the knowledge was gained at a cost: the native peoples of the Southern Indian Lake region who suffered the loss of livelihood and threats to the quality of their food did not share in the scientific adventure. Hecky and colleagues seemed to regret most their inability to forecast the fisheries problems—the eventual compensation program was somewhat arbitrary and inequitable. Had they predicted the decline in fishery quality and the hazards associated with mercury, the forecast need not have halted the experiment, but might have led to a well-planned agreement for adequate and just compensation. Such a plan could very easily have entered the cost-benefit analysis of the project.

References

Bodaly, R.A., T. W. D. Johnson, R. J. P. Fudge, and J. W. Clayton. 1984. Collapse of the lake whitefish (*Coregonus clupeaformis*) fishery in Southern Indian Lake, Manitoba, following lake impoundment and river diversion. Can. J. Fish. Aquat. Sci. 41:692-700.

Hecky, R. E., and S. J. Guildford. 1984. Primary productivity of Southern Indian Lake before, during, and after impoundment and Churchill River Diversion. Can. J. Fish. Aquat. Sci. 41:591-604.

Northcote, T. G., C. J. Walters, and J. M. B. Hume. 1978. Initial impacts of experimental fish introductions on the macrozooplankton of small oligotrophic lakes. Proc. Int. Assoc. Theor. Appl. Limnol. 20:2003-2012.

Patalas, K., and A. Salki. 1984. Effects of impoundment and diversion on the crustacean plankton of Southern Indian Lake. Can. J. Fish. Aquat. Sci. 41:613-637.

Wiens, A. P., and D. M. Rosenberg. 1984. Effect of impoundment and river diversion on profundal macrobenthos of Southern Indian Lake, Manitoba. Can. J. Fish. Aquat. Sci. 41:638-648.

22

Ecological Effects of
Nuclear Radiation

Particular kinds of environmental perturbation are essentially replicated in many places. Because no two sites are identical, detailed prediction of effects requires knowledge of the ecosystem in question. Much can be learned, however, by carrying out generic studies designed to discover results of general applicability to many conditions. Studies supported by the U.S. Atomic Energy Commission to determine the effects of radiation on living organisms and how radionuclides move through natural environments have been the most extensive attempts to use a generic approach to obtain information required for making major policy decisions. This case study summarizes and analyzes these studies and their contributions both to the solution of problems at which they were directed and to ecological theory generally.

Case Study

CARL F. JORDAN, Institute of Ecology, University of Georgia, Athens, Georgia

INTRODUCTION

Ionizing radiation resulting from production of radionuclides by bombs and reactors was the first pollutant given major national and international attention. It became an environmental concern soon after the first test of a nuclear weapon, which occurred on July 16, 1945, at Trinity, New Mexico. Starting almost immediately after the test and continuing for years thereafter, field surveys were conducted at Trinity to discover the extent and degree of environmental contamination by radionuclides, persistence of radionuclides, and effects of radiation on organisms and ecosystems (Larson, 1963). The studies showed that it would be extremely important to understand the effect of this pollutant on organisms and how it moves through the environment. It was thought that dispersion of radioisotopes in the environment as a result of fallout, reactor development, waste disposal, nuclear war, and technological projects could pose serious environmental problems (Wolfe, 1963).

As nuclear energy was developed, for both peaceful and military uses, programs were established to evaluate the environmental effects of human-produced radiation. Many of the programs were at laboratories that became parts of the complex supported by the U.S. Atomic Energy Commission (AEC), such as those at Argonne, Illinois; Brookhaven, New York; Hanford, Washington; Idaho Engineering Laboratory; Los Alamos, New Mexico; Livermore, California; the Nevada test site; Oak Ridge, Tennessee; and Savannah River, South Carolina (Whicker and Schultz, 1982). Other programs were established at universities or in conjunction with state agencies.

The effects of ionizing radiation and radionuclide movement in the environment were also studied in other countries. A series of symposia sponsored by the International Atomic Energy Agency dealt with the uses of radionuclides in various disciplines, such as hydrology and plant nutrition, as well as with environmental contamination by radioactive materials. The series of studies sponsored by the Environmental Sciences Branch of the Division of Biology and Medicine of AEC (later the Energy Research and Development Administration and now the Department of Energy) probably constituted the greatest concentrated effort ever expended to understand the environmental impact of a pollutant. (Lists of proceedings, books, and bibliographies of primary references were compiled by Klement and Schultz in 1980 and Whicker and Schultz in 1982.)

The studies of environmental effects of nuclear radiation covered most aspects of ecology. Some dealt with life histories to determine at which stage in its life cycle an organism was most sensitive to ionizing radiation. Studies of population dynamics and population interactions were crucial in understanding the dynamics of radionuclides in food chains. Studies sponsored by AEC were carried out in various habitats to determine the effects of radiation on community structure and community pattern. Determining changes in nutrient cycling and productivity of ecosystems also was a major goal of many studies.

This case study illustrates a generic approach to evaluating environmental pollutants. Ecological theory often cannot be used to make accurate predictions about individual cases. Valuable predictions for specific cases often are based on local field experience, not on formal theory. Furthermore, predictions are usually difficult to apply beyond the bounds of a specific case. In a generic approach, the effects of a pollutant on a large number of different organisms in different environments are studied, thereby providing a framework for predicting effects on new organisms or environments.

THE ENVIRONMENTAL PROBLEMS

Ionizing radiation is radiation with sufficient energy for its interactions with matter to produce an ejected electron and a positively charged ion (Whicker and Schultz, 1982). In large numbers, such interactions in the cells of living organisms can cause genetic and physiological damage and death. Low levels of ionizing radiation from cosmic rays and radionuclides in the earth's crust have always been present. Life has evolved in an environment of low background radiation.

Some of the first environmental studies concerned the radioactivity discharged in the mid-1940s from reactors at Hanford, Washington, into the Columbia River (Whicker and Schultz, 1982). Others concerned the magnitude and duration of radioactivity at weapon test sites (Hines, 1962; Koranda, 1965, 1969; Larson, 1963). Studies at nuclear test sites were important, not only because of the environmental dangers at the sites themselves, but also because results could be used to predict conditions at sites affected by nuclear war.

An important early stimulus for studies of radionuclides in the environment was observation of the fate of radioactive fallout from atmospheric tests of nuclear weapons. One particularly disturbing case involved the movement of cesium-137 in the lichen-reindeer-human food chain in tundra ecosystems (Liden and Gustafsson, 1967; Nevstrueva et al., 1967). Cesium-137, a relatively strong emitter of gamma radiation, has a relatively long half-life (30 years) and metabolic effects similar to those of potassium. It is adsorbed on the surface of lichens, which are abundant in the tundra. Lichens are an important food of reindeer, and the nuclide became concentrated in tissue of reindeer and of the Finnish Lapps and Alaskan eskimos, who depend heavily on reindeer for meat (Hanson et al., 1967; Lindell and Magi, 1967; Miettinen and Hasanen, 1967). Another potentially hazardous combination of nuclide and food chain that was identified early involved iodine-131, which, if deposited on pasture grass, quickly moves through dairy cattle to milk (Barth and Seal, 1967; Bergstrom, 1967) and becomes concentrated in the thyroid (Turner and Jennrich, 1967). Although there was never any conclusive evidence of damage to humans, the potential for danger was recognized.

Observation of potential hazards of fallout gave rise to systematic studies of radionuclide concentrations in several species, such as deer (Schultz and Longhurst, 1963), and to studies of the environmental factors important in radionuclide accumulation (Davis et al., 1963). An important result of these analyses (see Auerbach, 1965, for review) was a series of international symposia (Whicker and Schultz, 1982) in which the problem of radioactive contamination and accumulation in food chains was highlighted and brought to international attention.

In 1957, AEC established the Plowshare Program to investigate and develop peaceful uses for nuclear explosions (Auerbach, 1971a; Kelly, 1966). Studies were carried out to obtain food-chain and transport data needed for calculating radiation doses to human populations and to assess the impact on the local environment. One of the first was designed to predict the environmental impact of the use of nuclear explosives to excavate a harbor in the Cape Thompson region of Alaska (Wolfe, 1966). Another focused on the feasibility of using thermonuclear devices to create a new transisthmian canal in Panama (Atlantic-Pacific Interoceanic Canal Study Commission, 1970; Martin, 1969). Although the studies did not conclusively predict damage to human health as a result of using nuclear explosives in these regions, neither of the proposed excavations ever took place (for reasons never made public).

As nuclear technology advanced to the point where nuclear energy could be used to generate electricity, studies began to address the ecological problems in siting nuclear power plants, particularly the problem of radioactive discharge into the environment, both accidental and as a result of normal operations (Auerbach, 1971b; Schultz and Whicker, 1980).

Underground detonations were also used experimentally to stimulate gas flow in geological formations of low permeability (Alldredge *et al.*, 1976).

APPROACHES TO THE PROBLEM OF RADIOACTIVITY IN THE ENVIRONMENT

After the early observations of environmental radioactivity caused by nuclear testing, experimental studies were begun to evaluate the problem. These studies had two basic aspects. One was the movement of radionuclides in the environment after accidental or deliberate release from a nuclear device or power plant (Comar, 1965); radioactive tracers were often used to determine the pathway of each potentially dangerous nuclide and the rate of movement along that pathway. The second aspect was the effect of ionizing radiation on organisms in the environment; various animal populations (French, 1965; Turner, 1975), plant communities (Whicker and Fraley, 1974), aquatic organisms (Blaylock and Trabalka, 1978), and other ecosystem components (Platt, 1965) were irradiated experimentally.

Movement of Radionuclides in the Environment

Before the advent of radioecology, studies of food chains and of whole ecosystems had scarcely been initiated. An important contribution of the studies of radionuclide dynamics was to show that species in ecosystems were connected with each other and how particular species depended on the flows and cycles of nutrients and energy among all the other species in the ecosystem. Evidence of feedback in ecosystems also emerged from those studies; e.g., they showed that the rate of return of a nuclide to an individual or species can depend on other species and on environmental factors.

Perhaps the most important idea used in the efforts to understand radionuclide movement in food chains was the specific-activity concept discussed by Kaye and Nelson (1968). Specific activity is defined as the ratio of radioactive atoms to total atoms of the same element. By using the stable-element distribution in environmental samples as a chemical analog for the radionuclide, we can predict the dispersal of radionuclides through environmental pathways, if we know the stable-element chemistry of organisms constituting the links in the biological pathways of food chains and the ratio of the radionuclide to its stable element at the source of entry of the radionuclide into the food chain (Reichle *et al.*, 1970).

A modification of the specific-activity approach uses the ratio of the potentially dangerous radionuclides to analogous nutrient elements. Several radionuclides produced in important quantities during nuclear reactions are chemically similar to nutrient elements important in animal diets. For example, strontium-90 is similar to calcium and is accumulated in bones, and cesium-137 and potassium are metabolized similarly. Known pathways and concentrations of calcium and potassium in food chains can be used to predict concentrations of the analogous radionuclides, after corrections are made for discrimination or concentration factors (Comar and Lengemann, 1967). Discrimination and concentration are affected by the atomic weight of the isotope, as well as feeding habits and other food-chain characteristics (Whicker and Schultz, 1982).

The need to make predictions about the fates and concentrations of radionuclides in ecosystems where experimental tests were not feasible gave rise to the development of systems analysis techniques in ecology (Kaye and Ball, 1969; Shugart and O'Neill, 1979). These techniques used models of the flux and turnover of radionuclides in ecosystems. When data were not available, assumptions were based on studies in other ecosystems, on known metabolism of stable-element analogs in the species of interest, or on other appropriate physical, chemical, and biological models. Once all important ecosystem turnovers and transfers were formulated, equations were solved simultaneously to predict radionuclide dynamics through an entire ecosystem.

One of the efforts to predict radionuclide movement in an ecosystem was an analysis of the fate of radioactivity, if thermonuclear devices were used to excavate a new canal across Central America (Kaye and Ball, 1969). The predictive model was based on stable-element data collected at the proposed site and at similar sites and on results of laboratory analyses. Because the canal was never excavated, the model was never tested. But Jordan et al. (1973) used a similar approach to predict the environmental residence half-time of strontium-90 after its release into a tropical rain forest. They then used atmospheric fallout measured before the atmospheric-test ban in 1963 as model input and predicted concentrations of the nuclide in the forest through the end of the century. Measurements in 1974 (Jordan and Kline, 1976) showed that actual concentrations in the forest were higher than those predicted, because of atmospheric tests after 1963. The strontium-90 study showed that the environmental half-time of the isotope in this system was about 20 years and that loss was predominantly by physical decay.

Predictions of movement of radionuclides in food chains also were based on laboratory data on the biological turnover of radionuclides in organisms and on the factors—such as intake, assimilation, metabolism, and excretion—that affect turnover (Reichle et al., 1970). Other tracer studies showed the rate of movement of radionuclides between living and non-living portions of ecosystems (Auerbach, 1965).

An important early study that showed how changes in ecosystem structure affected radionuclide dynamics used a series of microcosms in combination with mathematical modeling (Patten and Witkamp, 1967). Leaf litter tagged with cesium-134 was introduced into microecosystems composed of different combinations of soil, microflora, millipedes, and aqueous leachate. Field studies were much more difficult, because of the health hazard of radioactivity. However, the section of Oak Ridge National Laboratory that is now the Environmental Sciences Division carried out a study in which a whole stand of trees was inoculated with cesium-137 in 1962 and the movement of the nuclide in the ecosystem was followed for a number of years thereafter (Francis and Tamura, 1971). An unexpected finding of that study was that much of the cesium did not move up through the leaves, but rather moved down into the roots and then into the soil when roots died and decomposed.

Radiation Effects

In addition to predicting the rate of nuclide movement through food chains and the amounts of radioactivity reaching valued species, it was necessary to know what effect a given amount of radioactivity would have on the valued species.

The behavior of individuals of a species is obviously important in the effect of radiation release. For example, burrowing animals are shielded from radiation (Buchsbaum, 1958). Other factors that influence radiation sensitivity in complex ways are body size, temperature, rate of reproduction, and life span. It was initially predicted that the most important biological factor affecting sensitivity to radiation would be the volume of chromosomes, but the first tests suggested that interphase chromosomal volume was a better predictor of both species sensitivity (Sparrow et al., 1968) and pattern of community response (Woodwell and Whittaker, 1968). Another important hypothesis regarding radiation sensitivity (Henshaw, 1963) was that there is a threshold of radiation tolerance, which may be different for each species. For radiation exposure below this threshold, damage does not exceed that caused by natural background radiation.

Studies of the effects of ionizing radiation on ecosystems were carried out in an oak-pine forest at Brookhaven, New York (Woodwell and Rebuck, 1967), a tropical rain forest (Odum et al., 1970), a northern hardwood forest (Murphy et al., 1977), southern pine-hardwood forests in Georgia (Cotter and McGinnis, 1965) and in Tennessee (Witherspoon, 1965), a pine forest (McCormick, 1969), and a shortgrass prairie (Fraley and Whicker, 1971). There were many other studies on the effects on populations and species (Appendix). In these studies, a shielded source of radiation was used, so that scientists could enter the irradiated areas. At the completion of an experiment, the source was removed. Because the source of radioactivity was thereby contained, there was no residual contamination, which was a problem in tracer studies.

Results showed that radiation sensitivity of plants was correlated to some extent with interphase chromosomal volume, but there were frequent exceptions (Koo and deIrizarry, 1970; Woodwell and Whittaker, 1968). A much more useful generalization from the radiation studies is that sensitivity of plants depends on the ratio of photosynthetic tissue to total tissue (Woodwell, 1967, 1970). The most sensitive plants are trees, which have a relatively low ratio of photosynthetic mass (leaves) to nonphoto-synthetic mass (stem and root). Shrubs are less sensitive than trees, and herbs and grasses are less sensitive than shrubs. Plants like algae, in which much of the tissue is photosynthetic, are highly resistant. Among trees, pine trees—which produce long-lived leaves—are more sensitive than deciduous hardwoods, in which replacement of leaves represents a smaller drain on energy reserves. Rhizomatous species, such as sedges, a large proportion of whose biomass is shielded by the soil, usually are relatively resistant. A generalization that applied to both plants and animals is that radiation sensitivity is correlated with size: the largest species are usually the most sensitive (Woodwell, 1967), as they are to stress in general (Woodwell, 1970).

Two irradiation experiments contrasted the effects of long exposure (Woodwell, 1967) and short exposure (Odum, 1970). In the site exposed for a short time, sprouting of trees played an important part in ecosystem recovery (Jordan, 1969). In a site chronically irradiated, root carbohydrate reserves were exhausted and sprouting was not important. Disturbed areas at that site were colonized by forbs and grasses with seeds that are widely dispersed and that germinate rapidly (Woodwell, 1967).

CONCLUSION

The AEC-sponsored studies of radiation in the environment resulted in two major conclusions. First, some but not all radionuclides released into the environment are concentrated as they are passed through food chains, and, if concentration factors are high, relatively low releases of radio-activity can pose a danger. Because every ecosystem and every food chain is different, potential danger depends in part on characteristics of the particular ecosystem and food chain and in part on the radionuclide. Concentration of radionuclides in food chains is a generic characteristic with the potential to occur in any ecosystem; for purposes of environmental safety, it must be predicted specifically for the ecosystem of interest.

Second, although radiosensitivity is sometimes correlated with interphase chromosomal volume, a more practical index of radiation sensitivity is simply the size of an organism and its life span. Large organisms are almost always more sensitive to radiation than small organisms. Long-lived species usually suffer more from radiation exposure than short-lived species.

In evaluating the effect of a particular radionuclide in a particular environment, both the food-chain accumulation factors and the sensitivity of the organisms in the food chain to the predicted dose must be known. These factors are now known for some organisms in many types of ecosystem, but must be evaluated in light of the specific conditions at particular sites.

REFERENCES

Alldredge, A. W., F. W. Whicker, and W. C. Hanson. 1976. Some environmental impacts associated with project Rio Blanco. Pp. 65-73 in C. E. Cushing, ed. Radioecology and Energy Resources. Proceedings of the Fourth National Symposium on Radioecology. Dowden, Hutchinson, and Ross, Stroudsburg, Pa.

Atlantic-Pacific Interoceanic Canal Study Commission. 1970. Interoceanic Canal Studies; Annexes 1-5; Vols. 1-6. Atlantic-Pacific Interoceanic Canal Study Commission, Washington, D.C.

Auerbach, S. I. 1965. Radionuclide cycling: Current status and future needs. Health Phys. 11:1355-1361.

Auerbach, S. I. 1971a. Contributions of radioecology to AEC mission programs. Pp. 3-8 in D. J. Nelson, ed. Radionuclides in Ecosystems. Proceedings of the Third National Symposium on Radioecology. U.S. Atomic Energy Commission, Washington, D.C.

Auerbach, S. I. 1971b. Ecological considerations in siting nuclear power plants: The long-term biotic effects problem. Nucl. Safety 12:25-34.

Barth, D. S., and M. S. Seal. 1967. Radioiodine transport through the ecosystem, air-forage-cow-milk using a synthetic dry aerosol. Pp. 151-158 in B. Aberg and F. P. Hungate, eds. Radioecological Concentration Processes. Proceedings of an International Symposium. Pergamon Press, Oxford, Eng.

Bergstrom, S. O. W. 1967. Transport of fallout ^{131}I into milk. Pp. 159-174 in B. Aberg and F. P. Hungate, eds. Radioecological Concentration Processes. Proceedings of an International Symposium. Pergamon Press, Oxford, Eng.

Blaylock, B. G., and J. R. Trabalka. 1978. Evaluating the effects of ionizing radiation on aquatic organisms. Adv. Radiat. Biol. 7:103-152.

Buchsbaum, R. 1958. Species response to radiation: Radioecology. Pp. 124-141 in W. D. Claus, ed. Radiation Biology and Medicine. Addison Wesley, Reading, Mass.

Comar, C. L. 1965. Movement of fallout radionuclides through the biosphere and man. Annu. Rev. Nucl. Sci. 15:175-206.

Comar, C. L., and F. W. Lengemann. 1967. General principles of the distribution and movement of artificial fallout through biosphere to man. Pp. 1-18 in B. Aberg and F.

P. Hungate, eds. Radioecological Concentration Processes. Proceedings of an International Symposium. Pergamon Press, Oxford, Eng.

Cotter, D. J., and J. T. McGinnis. 1965. Recovery of hardwood stands 3-5 years following acute irradiation. Health Phys. 11:1663-1673.

Davis, J. J., W. C. Hanson, and D. G. Watson. 1963. Some effects of environmental factors upon accumulation of worldwide fallout in natural populations. Pp. 35-38 in V. Schultz and A. W. Klement, eds. Radioecology. Proceedings of the First National Symposium on Radioecology. Reinhold, New York.

Fraley, L., and F. W. Whicker. 1971. Response of a native shortgrass plant stand to ionizing radiation. Pp. 999-1006 in D. J. Nelson, ed. Radionuclides in Ecosystems. Proceedings of the Third National Symposium on Radioecology. U.S. Atomic Energy Commission, Washington, D.C.

Francis, C. W., and T. Tamura. 1971. Cesium-137 soil inventory of a tagged *Liriodendron* forest, 1962 and 1969. Pp. 140-149 in D. J. Nelson, ed. Radionuclides in Ecosystems. Proceedings of the Third National Symposium on Radioecology. U.S. Atomic Energy Commission, Washington, D.C.

French, N. R. 1965. Radiation and animal populations: Problems, progress, and projections. Health Phys. 11:1157-1568.

Hanson, W. C., D. G. Watson, and R. W. Perkins. 1967. Concentration and retention of radionuclides in Alaskan Arctic ecosystems. Pp. 233-245 in B. Aberg and F. P. Hungate, eds. Radioecological Concentration Processes. Proceedings of an International Symposium. Pergamon Press, Oxford, Eng.

Henshaw, P. S. 1963. Radiation effects and peaceful uses of atomic energy in the animal sciences: Radiation and biologic capability. Pp. 13-17 in V. Schultz and A. W. Klement, eds. Radioecology. Proceedings of the First National Symposium. Reinhold, New York.

Hines, N. O. 1962. Proving Ground: An Account of the Radiobiological Studies in the Pacific, 1946-1961. University of Washington Press, Seattle.

Jordan, C. F. 1969. Recovery of a tropical rain forest after gamma irradiation. Pp. 88-98 in D. J. Nelson and F. C. Evans, eds. Symposium on Radioecology. Proceedings of the Second National Symposium. CONF 670-503. U.S. Department of Commerce, Springfield, Va.

Jordan, C. F., and J. R. Kline. 1976. Strontium-90 in a tropical rain forest: 12th-year validation of a 32-year prediction. Health Phys. 30:199-201.

Jordan, C. F., J. R. Kline, and D. S. Sasser. 1973. A simple model of strontium and manganese dynamics in a tropical rain forest. Health Phys. 24:477-489.

Kaye, S. V., and S. J. Ball. 1969. Systems analysis of a coupled compartment model for radionuclide transfer in a tropical environment. Pp. 731-739 in D. J. Nelson and F. C. Evans, eds. Symposium on Radioecology. Proceedings of the Second National Symposium. CONF 670-503. U.S. Department of Commerce, Springfield, Va.

Kaye, S. V., and D. J. Nelson. 1968. Analysis of specific-activity concept as related to environmental concentration of radionuclides. Nucl. Safety 9:53-58.

Kelly, J. S. 1966. Foreword. Pp. iii-iv in N. J. Wilimovsky and J. N. Wolfe, eds. Environment of the Cape Thompson Region, Alaska. U.S. Atomic Energy Commission, Washington, D.C.

Klement, A. W., and V. Schultz. 1980. Terrestrial and Freshwater Radioecology. A Selected Bibliography. Dowden, Hutchinson and Ross, Stroudsburg, Pa.

Koo, F. K. S., and E. R. deIrizarry. 1970. Nuclear volume and radiosensitivity of plant species at El Verde. Pp. G-15—G-20 in H. T. Odum and R. F. Pigeon, eds. A Tropical Rain Forest. U.S. Atomic Energy Commission, Washington, D.C.

Koranda, J. J. 1965. Preliminary studies of the persistence of tritium and ^{14}C in the Pacific Proving Ground. Health Phys. 11:1445-1457.

Koranda, J. J. 1969. Residual tritium at Sedan Crater. Pp. 696-708 in D. J. Nelson and F. C. Evans, eds. Symposium on Radioecology. Proceedings of the Second National Symposium. CONF 670-503. U.S. Department of Commerce, Springfield, Va.

Larson, K. H. 1963. Continental close-in fallout: Its history, measurement, and characteristics. Pp. 19-25 in V. Schultz and A. W. Klement, eds. Radioecology. Proceedings of the First National Symposium. Reinhold, New York.

Liden, K., and M. Gustafsson. 1967. Relationships and seasonal variation of ^{137}Cs in lichen, reindeer, and man in northern Sweden 1961-1965. Pp. 193-208 in B. Aberg and F. P. Hungate, eds. Radioecological Concentration Processes. Proceedings of an International Symposium. Pergamon Press, Oxford, Eng.

Lindell, B., and A. Magi. 1967. Observed levels of ^{137}Cs in Swedish reindeer meat. Pp. 217-219 in B. Aberg and F. P. Hungate, eds. Radioecological Concentration Processes. Proceedings of an International Symposium. Pergamon Press, Oxford, Eng.

Martin, W. E. 1969. Radioecology and the feasibility of nuclear canal excavation. Pp. 9-22 in D. J. Nelson and F. C. Evans, eds. Symposium on Radioecology. Proceedings of the Second National Symposium. CONF 670-503. U.S. Department of Commerce, Springfield, Va.

McCormick, J. F. 1969. Effects of ionizing radiation on a pine forest. Pp. 78-87 in D. J. Nelson and F. C. Evans, eds. Symposium on Radioecology. Proceedings of the Second National Symposium. CONF 670-503. U.S. Department of Commerce, Springfield, Va.

Miettinen, J. K., and E. Hasanen. 1967. [137]Cs in Finnish Lapps and other Finns in 1962-6. Pp. 221-231 in B. Aberg and F. P. Hungate, eds. Radioecological Concentration Processes. Proceedings of an International Symposium. Pergamon Press, Oxford, Eng.

Murphy, P. G., R. R. Sharitz, and A. J. Murphy. 1977. Response of a forest ecotone to ionizing radiation. Pp. 43-48 in J. Zavitkovski, ed. The Enterprise, Wisconsin, Radiation Forest. USERDA TID-26113-p2. U.S. Energy Research and Development Administration, Washington, D.C.

Nevstrueva, M. A., P. V. Ramzaev, A. A. Moiseer, M. S. Ibatullin, and L. A. Teplykh. 1967. The nature of ^{137}Cs and ^{90}Sr transport over the lichen-reindeer-man food chain. Pp. 209-215 in B. Aberg and F. P. Hungate, eds. Radioecological Concentration Processes. Proceedings of an International Symposium. Pergamon Press, Oxford, Eng.

Odum, H. T. 1970. Summary. An emerging view of the ecological system at El Verde. Pp. I-191—I-281 in H. T. Odum and R. F. Pigeon, eds. A Tropical Rain Forest. U.S. Atomic Energy Commission, Washington, D.C.

Odum, H. T., P. Murphy, G. Drewry, F. McCormick, C. Schinan, E. Morales, and J. A. McIntyre. 1970. Effects of gamma radiation on the forest at El Verde. Pp. D-3—D-75 in H. T. Odum and R. F. Pigeon, eds. A Tropical Rain Forest. U.S. Atomic Energy Commission, Washington, D.C.

Parzyck, D. C., J. P. Witherspoon, and J. E. Till. 1976. Validation of environmental transport models in the CUEX methodology. Pp. 194-198 in C. E. Cushing, ed. Radioecology and Energy Resources. Proceedings of the Fourth National Symposium on Radioecology. Dowden, Hutchinson, and Ross, Stroudsburg, Pa.

Patten, B. C., and M. Witkamp. 1967. Systems analysis of [134]cesium kinetics in terrestrial microcosms. Ecology 48:813-824.

Platt, R. B. 1965. Radiation effects on plant populations and communities: Research status and potential. Health Phys. 11:1601-1606.

Reichle, D. E., P. B. Dunaway, and D. J. Nelson. 1970. Turnover and concentration of radionuclides in food chains. Nucl. Safety 11:43-55.

342 SELECTED CASE STUDIES

Schultz, V., and W. M. Longhurst. 1963. Accumulation of strontium-90 in yearling Columbian black-tailed deer, 1950-1960. Pp. 73-76 in V. Schultz and A. W. Clement, eds. Radioecology. Proceedings of the First National Symposium. Reinhold, New York.

Schultz, V., and F. W. Whicker. 1980. Nuclear fuel cycle, ionizing radiation, and effects on biota of the natural environment. CRC Crit. Rev. Environ. Control 10:225-268.

Shugart, H. H., and R. V. O'Neill. 1979. Introduction. Pp. 1-6 in H. H. Shugart and R. V. O'Neill, eds. Systems Ecology. Benchmark Papers in Ecology. 9. Dowden, Hutchinson, and Ross, Stroudsburg, Pa.

Sparrow, A. H., A. F. Rogers, and S. S. Schwemmer. 1968. Radiosensitivity studies with woody plants. I. Radiat. Bot. 8:149-186.

Turner, F. B. 1975. Effects of continuous irradiation on animal populations. Adv. Radiat. Biol. 5:83-144.

Turner, F. B., and R. I. Jennrich. 1967. The concentration of ^{131}I in the thyroids of herbivores and a theoretical consideration of the expected frequency distribution of thyroidal ^{131}I in a large consumer population. Pp. 175-182 in B. Aberg and F. P. Hungate, eds. Radioecological Concentration Processes. Proceedings of an International Symposium. Pergamon Press, Oxford, Eng.

Whicker, F. W., and L. Fraley. 1974. Effects of ionizing radiation on terrestrial plant communities. Adv. Radiat. Biol. 4:317-366.

Whicker, F. W., and V. Schultz. 1982. Radioecology: Nuclear Energy and the Environment. Vols. I and II. CRC Press, Boca Raton, Fla.

Witherspoon, J. P. 1965. Radiation damage to forest surrounding an unshielded fast reactor. Health Phys. 11:1637-1642.

Wolfe, J. N. 1963. Impact of atomic energy on the environment and environmental science. Pp. 1-2 in V. Schultz and A. W. Klement, eds. Radioecology. Proceedings of the First National Symposium. Reinhold, New York.

Wolfe, J. N. 1966. Committee on environmental studies for Project Chariot, Plowshare Program. Pp. ix-x in Environment of the Cape Thompson Region, Alaska. U.S. Atomic Energy Commission, Washington, D.C.

Woodwell, G. M. 1967. Radiation and the pattern of nature. Science 156:461-470.

Woodwell, G. M. 1970. Effects of pollution on the structure and physiology of ecosystems. Science 168:429-433.

Woodwell, G. M., and A. L. Rebuck. 1967. Effects of chronic gamma radiation on the structure and diversity of an oak-pine forest. Ecol. Monogr. 37:53-69.

Woodwell, G. M., and R. H. Whittaker. 1968. Effects of chronic gamma irradiation on plant communities. Q. Rev. Biol. 43:42-55.

APPENDIX: Some Sources of Information on Radioecology

Aberg, B., and F. P. Hungate, eds. 1967. Radioecological Concentration Processes. Proceedings of an International Symposium. Pergamon Press, Oxford, Eng.

Cushing, C. E., ed. 1976. Radioecology and Energy Resources. Proceedings of the Fourth National Symposium on Radioecology. Dowden, Hutchinson, and Ross, Stroudsburg, Pa.

Hanson, W. C., ed. 1980. Transuranic elements in the environment. U.S. DOE Rep. DOE/TIC-22800. U.S. Department of Energy, Washington, D.C.

Hungate, F. P., ed. 1965. Hanford Symposium on Radiation and Terrestrial Ecosystems. Health Phys. 11:1255-1675.

Klement, A. W., and V. Schultz. 1980. Freshwater and Terrestrial Radioecology: A Selected Bibliography. Dowden, Hutchinson, and Ross, Stroudsburg, Pa.

Nelson, D. J., ed. 1971. Radionuclides in Ecosystems. Proceedings of the Third National Symposium on Radioecology. U.S. Atomic Energy Commission, Washington, D.C.

Nelson, D. J., and F. C. Evans, eds. 1969. Symposium on Radioecology. Proceedings of the Second National Symposium. CONF 670-503. U.S. Department of Commerce, Springfield, Va.

Odum, H. T., and R. F. Pigeon, eds. 1970. A Tropical Rain Forest. A Study of Irradiation and Ecology at El Verde, Puerto Rico. U.S. Atomic Energy Commission, Washington, D.C.

Schultz, V., and A. W. Klement, eds. 1963. Radioecology. Proceedings of the First National Symposium on Radioecology. Reinhold, New York.

Thompson, R. C., and W. J. Blair, eds. 1972. Hanford Symposium on the Biological Implications of the Transuranium Elements. Health Phys. 22:533-957.

Whicker, F. W., and V. Schultz. 1982. Radioecology: Nuclear Energy and the Environment. Vols. I and II. CRC Press, Boca Raton, Fla.

Committee Comment

Environmental problem-solving is hindered by differences in insight derived from general and specific approaches. General ecological theory usually makes only crude predictions about specific conditions or impacts at a specific site. Useful predictions for a specific case often are based on local field experience, rather than on formal theory, and it is difficult to determine their applicability beyond the bounds of the specific case.

The AEC studies used both generic and specific approaches. Studies were carried out on a wide variety of ecosystems in an effort to form generalizations about radioactivity in the environment. The relative sensitivity of plants as a function of the ratio of photosynthetic tissue to total tissue (photosynthesis:respiration ratio) is an example of a generalization that emerged from these comparative studies. Studies on conditions at a specific site or with a specific nuclide were useful in predicting effects of the same nuclide under similar conditions. For example, the Arctic studies suggested that scavenging might be important wherever lichens were dominant members of the community.

This case study suggests the value of approaching many types of environmental perturbations both generically and specifically. The generic approach predicts what, in general, to expect from a perturbation, regardless of where it occurs; the specific approach addresses the question of whether the specific case differs in any important way from the general case.

The many studies in radioecology have resulted in one of the most successful evaluations of the impact of an environmental hazard. One reason is that they were adequately funded. Most environmental evaluation must be carried out with less support than was available for the radio-ecological work. The results of these studies show that thorough under-standing can be achieved if enough time and money are allocated.

Another notable characteristic of the AEC studies was the separation of studies of radionuclide movement in the environment from studies of the effects of ionizing radiation. Had these two aspects of the radiation problem not been separated, much less progress would have been made. Experiments in which exposure to radiation was great enough to reveal dose-response relationships would have been impossible with radionu-clides released into the environment. Conversely, laboratory studies cannot reveal how radionuclides behave in nature.

Studies sponsored by AEC, particularly those of nuclide movement through the environment, made an important contribution to the emergence of "ecosystem ecology" (Odum, 1965; Odum and Golley, 1963), in which a major focus is the flow of energy and elements through a unit of land-scape. This work supplements and extends studies oriented toward the ecology of individuals, populations, species, and communities.

This examination of the history of AEC-sponsored environmental ra-diation studies suggests that their contribution to ecological knowledge has been as important as, or perhaps even more important than, the con-tribution of ecological knowledge to the design and interpretation of the AEC studies. During the four decades of radioecological studies, the constant interplay between experimental results and general theory has proved fruitful to the basic science of ecology and the applied field of radioecology.

References

Odum, E. P. 1965. Feedback between radiation ecology and general ecology. Health Phys. 11:1257-1262.

Odum, E. P., and F. B. Golley. 1963. Radioactive tracers as an aid to the measurement of energy flow at the population level in nature. Pp. 403-410 in V. Schultz and A. W. Klement, eds. Radioecology. Proceedings of the First National Symposium. Reinhold, New York.

23

Ecological Effects of
Forest Clearcutting

Management of the harvesting of renewable biological resources requires knowledge of the stocks of those resources and of the rates at which they recover after harvesting. Gaining this knowledge is especially difficult when the intervals between cropping are long and rates of recovery are low, as they are in the case of timber harvesting. Not only are trees long-lived, but differences between sites make it difficult to extrapolate results from one location to another. This case study reviews the ways in which these problems have been approached and assesses the current state of prediction of the amounts and significance of nutrient losses from forests after clearcutting.

Case Study

CARL F. JORDAN, Institute of Ecology, University of Georgia, Athens, Georgia

INTRODUCTION

A major goal of the forestry profession is to sustain the production of wood in forests. In the early days of the profession, much attention centered on prevention of fire and outbreaks of insect pests as means of sustaining high productivity. These factors still cause extensive losses of trees, but now that many forests are managed as agricultural crops, other problems have increased in importance. One of the common intensive management techniques in modern forestry is clearcut logging. Clearcut logging is often more economical than other forest management techniques, but it has the potential to create serious environmental problems out of effects that would be negligible if less intensive techniques were used. Examples include the elimination of habitat of endangered species and increases in soil erosion. This case study reviews research that has evaluated the impact of forest clearcutting on nutrient stocks and site productivity.

The first studies of nutrient loss due to clearcutting were concerned primarily with only parts of the nutrient budget of an ecosystem, for example, leaching losses. Improved understanding of ecosystem functioning made it apparent that useful interpretation of nutrient losses due to clearcutting required an analysis of both the nutrient stocks in the ecosystem and the nutrient inputs and losses. To evaluate the effect of nutrient loss during clearcutting on future productivity of a site, it is necessary to know the quantities not only of nutrient losses due to leaching and erosion after clearcutting, but also of losses due to tree removal, volatilization, and fixation in the soil. Also important are the rate of nutrient replacement through atmospheric input, rock weathering, and fertilization and the size of nutrient stocks in the ecosystem that can buffer short-term fluctuations in input and output.

A nutrient-budget approach to forest management is analogous to budget management in business. Successful business management demands an understanding of the entire operating budget, including revenues, expenditures, and financial balance. This need in business management is so obvious as to be taken for granted. Yet, in forest management, it has become clear only recently that successful management for sustained yield requires knowledge of nutrient inputs, nutrient losses, and nutrient stocks.

346

These quantities are often difficult to measure, but complete budgets must be constructed, if the impact of clearcutting on nutrient loss is to be evaluated adequately. This case study examines how such understanding evolved.

BACKGROUND

We do not know when the first agriculturalist observed that ash, litter, and animal carcasses improved crop growth. Undoubtedly, the knowledge that the addition of particular materials to the soil increases productivity arose independently in many regions. However, only in the nineteenth century, after chemical elements had been identified, did it become understood that nutrients constituted the common factor in soil amendments that maintained agricultural productivity (Brady, 1974).

The first scientist to make a systematic study of nutrient circulation in forests in relation to growth of trees was Ebermayer. The objective of his study, "Complete Treatise of Forest Litter," was to explain the adverse effect on forest quality of the litter removal then common in middle European forests (Tamm, 1979). The litter was used in cow stables, and some of the plant nutrients in it eventually reached arable fields in dung, thus contributing to food production at a time when commercial fertilizer was not available. However, forest growth decreased, particularly in already poor sites, and Ebermayer attributed this to the export of plant nutrients (Tamm, 1979).

Interest in the role of nutrients in forest growth continued throughout the early part of the twentieth century, but only within the last several decades have scientists begun to quantify total forest-ecosystem nutrient budgets and changes in budgets due to various logging practices. One reason for interest in nutrient loss resulting from the clearcut method of forest harvesting is the obvious increase in soil erosion, nutrient leaching, and consequent stream eutrophication after such operations (Tamm *et al.*, 1974). Another, which is examined here, is the effect of nutrient loss on the ability of the soil to supply enough nutrients to produce another stand of trees.

Observation of soil erosion and increased eutrophication of drainage streams after clearcutting left little doubt that nutrient loss was occurring, but did not indicate whether it was great enough to affect productivity of the site. Because tree growth responds to fertilization in at least some cases, nutrient loss is potentially harmful. But nutrients lost through leaching and erosion might be quickly replaced by weathering and other processes. No long-term studies have been designed specifically to test the effects of nutrient leaching on site productivity. Studies at Hubbard Brook,

New Hampshire (Bormann and Likens, 1979), showed that a northern hardwood forest that had been clearcut and then treated with herbicide experienced substantial amounts of nutrient leaching, but recovery of the forest through natural succession did not appear to be noticeably inhibited by lack of nutrients.

Clearcutting has rarely resulted in nutrient loss so serious that trees could not grow on the denuded site. Occasionally, on very steep slopes in regions of heavy rainfall, such as some areas in western British Columbia, clearcutting has been followed by soil erosion severe enough to expose bedrock, and sites have consequently been permanently deforested (J. P. Kimmins, personal communication). In most cases, however, at least some tree growth has occurred after clearcutting, so the environmental question is not whether trees can grow after clearcutting, but how fast they can grow. The rate is important to industries or agencies concerned with timber as a crop, because trees that grow rapidly yield a greater profit than trees that grow slowly. Rate of forest recovery also is important for nonmarket values of forests, such as the reduction of soil erosion on watersheds, scrubbing of polluted air, support of fish and game, and provision of habitat for some rare species (Farnworth et al., 1981).

APPROACHES TO EVALUATING LOSSES

Leaching and Erosion

Nutrient leaching losses often increase as a result of clearcutting. After removal of trees, evapotranspiration on the site decreases, so the amount of water percolating through the soil increases; the result is an increase in leaching potential. A decrease in nutrient recycling is also important in increasing leaching. In undisturbed forests, root uptake of nutrients and their incorporation in biomass reduces nutrient loss. Nutrients from decomposing litter and soil organic matter on a clearcut are not taken up by trees, but are exchanged on the clay surfaces of mineral soil, where they are susceptible to loss through leaching and erosion. These ecological ideas are commonly accepted and are discussed in many texts on ecology, agronomy, soil conservation, and forestry. However, emphasis often is on the effect of the nutrients on lake and stream eutrophication and fish productivity, rather than on future productive capacity of the site losing the nutrients.

Recent studies have revealed some of the mechanisms by which nutrients are leached. After clearcutting, the activity of nitrifying bacteria in the soil increases (Likens et al., 1969; Vitousek et al., 1979), owing to increased temperatures, decreased competition for ammonium from tree roots, decreased allelopathic inhibition of nitrifying bacteria, and other factors (Reiners, 1981; Swift *et al.*, 1979). As a result, ammonium from the mineralization of organic matter is oxidized to nitrate, and nitrate anions and nutrient cations that have been exchanged for hydrogen ions on soil surfaces are rapidly leached.

An important effect of forest disturbance, especially in mountainous terrain, is soil erosion. Soil erosion results from exposure of mineral soil to the direct impact of raindrops. The impact breaks soil aggregates, causing pores and channels in the soil to be filled, the soil surface to become less permeable, and the surficial runoff to be greater than when soil surfaces are covered with litter. Although soil erosion often accompanies clearcut logging, recent studies have shown that it is construction of access roads, not the clearcutting itself, that causes most of the erosion (Douglass and Swift, 1977); removal of logs by cable appears to lead to less erosion.

A general approach to determining nutrient loss during and after clearcutting of forests and during site preparation for a new stand has been to measure leaching and erosional losses from clearcut areas directly and to compare losses with those in undisturbed areas that serve as a control. Measurements have usually been made where drainage from a watershed flows over a weir resting on bedrock, so that subsurface drainage is insignificant. Continuous monitoring of water flow and nutrient content permits measurement of total nutrient loss. Clearcut watersheds are compared with control watersheds or watersheds subjected to other treatments, such as conversion to grassland. In some cases where discrete watersheds were not available, losses have been studied by means of lysimeters (soil water collectors). Losses have also been measured by comparing soil and ecosystem nutrient stocks before and after cutting or between cut and control plots.

Many studies have shown increases in rate of nutrient loss during forest disturbance. For example, the studies of the watersheds at Coweeta Hydrologic Laboratory in North Carolina showed higher rates of nitrate and sediment loss in recently cut catchments than in control catchments (Monk 1975; Swank and Douglass, 1975, 1977; Webster and Patten, 1979). Studies of nutrient dynamics in conifer forests of the Pacific Northwest also showed an increase in loss rates after clearcut harvesting and slash burning (Feller and Kimmins, 1984; Gessel and Cole, 1965; Miller and Newton, 1983).

An important study of nutrient loss after clearcutting was carried out in northern hardwoods at the Hubbard Brook watershed site in New Hampshire (Bormann and Likens, 1970; Bormann *et al.*, 1968; Likens *et al.*, 1970). Bormann *et al.* (1968) concluded that clear-cutting tends to deplete the nutrients of a forest ecosystem by (i) reducing transpiration and so increasing the amount of water passing through the system; (ii) simultaneously reducing root surfaces able to remove nutrients from the leaching waters; (iii) removal of nutrients in forest products; (iv) adding to the organic substrate available for immediate mineralization; and (v) in some instances, producing a microclimate more favorable to rapid mineralization.

This conclusion was controversial, because clearcutting at the Hubbard Brook site had been followed by herbicide treatment to prevent regrowth of vegetation. Aubertin and Patric (1974) wrote that "there is a substantial difference between the Hubbard Brook treatment and conventional clear-cutting. Conventional clearcutting also features complete forest cutting; but all saleable wood is harvested and rapid forest regeneration is encouraged." In an experimental hardwood watershed in West Virginia, Aubertin and Patric measured negligible nutrient losses after clearcutting that was carried out to resemble conventional clearcut logging techniques. Other clearcutting studies in the Hubbard Brook region showed that nutrient losses are less if herbicides are not used after clearcutting, but can still be important because the soils in the region are shallow (Pierce *et al.*, 1972); harvesting removes a larger proportion of the total nutrient pool there than in areas with deeper soils.

Because of the importance of nutrient uptake by vegetation, alternating contour strip cuts with undisturbed forest on a mountainside should decrease nutrient leaching. Experiments with strip cuts showed that this technique can reduce nutrient leaching in northern forests (Hornbeck *et al.*, 1975). The effect should be even greater in the humid tropics, where the potential for leaching is extremely high (Jordan, 1982). Once vegetation becomes re-established in the clearcut strips, the uncut strips can be harvested.

Biomass Removal

Nutrient losses due to leaching are brief, and often small, compared with those due to tree removal (Cole and Bigger, undated; Hornbeck and Kropelin, 1982; Kimmins, 1977; Sollins and McCorison, 1981; Swank and Waide, 1980). As evidence accumulated that nutrient losses due to biomass removal during clearcut operations were often much greater than losses due to leaching and erosion, scientists shifted their attention to studies of nutrient stocks in forests and soils and to losses from ecosystems due to biomass harvest. Results from various sites and management strategies were presented in several major symposia (Ballard and Gessel, 1983; Leaf, 1979). Generalizing about the studies is difficult, because each site had its own combination of soil type, soil nutrient stocks, biomass nutrient stocks, management technique, rates of nutrient input via rainfall, nitrogen fixation, fertilization, and nutrient losses through leaching and denitrification. Generalizations valid for sites of a specific type in a given region do not apply to sites of other types. Each nutrient also appears to behave differently. Only by measuring the stocks and dynamics of all the critical nutrients at each site can one estimate accurately the impact of nutrient removal on productivity of the site. Even then, estimation might be insufficient, because trees can exhibit uptake beyond immediate needs, and this can be misleading in the prediction of requirements.

One general finding is that the extent of nutrient loss depends on which parts of the trees are removed from the site. Leaves have higher concentrations of nutrients than do stems, so stripping logs of leaves before the logs are removed from the site decreases nutrient losses. For evergreens, branches with leaves are cut off before the tree bole is removed. In the case of deciduous species, an alternative is to harvest during the winter. However, because some nutrients are withdrawn from leaves into stems before the leaves are shed, nutrient losses would be smaller if trees were cut while in full leaf and mechanically defoliated. Harvesting of whole trees, including roots, removes larger proportions of nutrients.

Nutrients Remaining in Soil

Simply measuring the nutrients lost because of clearcutting and site preparation does not indicate the effect of the nutrient loss in site productivity. Estimates of the quantities of nutrients remaining in the site and of the rates at which they are replenished are also needed. Scientists attempting to predict the effect of nutrient loss on site productivity must determine nutrient stocks remaining in the soil and the fractions of those stocks available to growing trees (Johnson *et al.*, 1982).

Nutrients are held in the soil in various ways (Brady, 1974). Some cations, such as potassium, can be part of the lattice structure of minerals, where they are relatively unavailable to roots; or they can be exchanged on clay surfaces, where they are readily available. Phosphorus can exist in relatively soluble forms, but, in the presence of low pH and high aluminum concentrations, it becomes bound in compounds that are not readily taken up by plants. Soil can contain large amounts of nitrogen; if the nitrogen is bound in organic matter that is resistant to decomposition, however, nitrogen shortages might be critical. Thus, knowing the total stocks of nutrients in the soil is not sufficient.

For many agricultural crops, reagent-soluble nutrients in soil have been correlated with crop growth, and solubility of nutrients has been used as an index of the availability of nutrients to crop plants. However, such an index usually is not applicable to tree growth. Many tree species are adapted for extracting nutrients from soil when "availability," as measured by solubility, is low. Some of these adaptations are large root biomass, mycorrhizal symbiosis, slow root uptake, and long life, which enable trees to survive during periods of nutrient shortage (Chapin, 1980). The effects of these variables are not easy to measure.

Nutrient Replenishment

Site productivity depends on relative rates of nutrient replenishment and nutrient loss. The atmosphere contains a stock of nutrients adsorbed on the surface of aerosols, such as dust and pollen, that either settle out of the atmosphere gradually as "dry fall" or are washed out by precipitation. Nutrient input from the atmosphere can be substantial. Nitrogen can be contributed to a site by nitrogen-fixing species, which can be present because they occur naturally or through planting of species symbiotic with nitrogen-fixing bacteria. Nutrients can also enter an ecosystem through the weathering of subsoil or parent rock.

Nutrients lost through clearcutting can be replaced by fertilizers, but fertilization of forests is often economically infeasible. For most agricultural crops, fertilization is profitable, because annual sales are high and the time between investment in fertilizers and return of investment in crop sales is short. The economic benefits of fertilizing forests are more difficult to calculate, because the value of harvestable products that accumulate each year is low, because the period between investment and harvest is long, and because forests have other values besides their use for wood.

Budgets and Sensitivity Analysis

Nutrient budgets of ecosystems have proved extremely difficult to measure precisely, because of high variability in both space and time. In addition, such important quantities as the proportions of nutrients available to plants and the rate of mineral weathering are often hard to measure; errors in their measurement could result in large errors in estimates of total ecosystem nutrient budgets. Therefore, these estimates are often viewed skeptically by both scientists and managers. The key question, however, is whether the error is important in relation to the environmental or ecological problem being addressed.

As an example of a simple sensitivity analysis to assess the effect of biomass removal on remaining nutrient stocks and the effect of error on the predictions, consider the following hypothetical case. Assume that the standard deviation around the amount of calcium in the soil is relatively large and that the average value of a set of samples could be different from the true average value by 30%. Is this error important? If calcium were present in the aboveground biomass at 500 kg/ha, all of which would be removed by clearcutting, and present in the soil at 5,000 kg/ha, an error of 30% in the estimate of calcium in the soil would not change the conclusion that clearcutting will not have an important effect on total calcium stock. If calcium were present at 500 kg/ha in the biomass and 100 kg/ha in the soil, an error of 30% would not change the conclusion that clearcutting would have a very important effect on total calcium stock. However, if calcium were present at 500 kg/ha in both the biomass and the soil, it would be important to be able to estimate more accurately the true value of calcium in the soil, because a 30% error could lead to very different management conclusions.

This relatively simple type of analysis has yielded predictions of nutrient depletion due to biomass removal in several regions. For example, calcium depletion due to clearcutting of ridge forests in the southern Appalachians might result in reduced growth of the next crop (West and Mann, 1983), and nitrogen is often a limiting nutrient in the conifer forests of the Pacific Northwest (Peterson and Gessel, 1983).

Simulation models of forest growth that incorporate the effects of nutrients can assist in answering questions about the required degree of accuracy of field measurements and experiments. Only recently have models with increased sophistication been developed. A model like the nitrogen model of Swank and Waide (1980), which predicts yield under various management strategies on the basis of nitrogen dynamics in the treated ecosystems, can be used to assess the effect of an error of a given magnitude on the overall conclusion and can aid in a decision as to whether more accurate measurements are necessary. The latter model permits evaluation of management alternatives and their consequences, such as the effect on forest yield of a change in the length of rotation.

The importance of sensitivity analyses is often not appreciated by ecological and environmental scientists. Perhaps the reason is that sensitivity analyses often contradict what many scientists have been taught, which is that the greatest possible accuracy should always be sought. Sensitivity analyses show us where a large amount of less accurate data might be more valuable than a smaller amount of more accurate data.

CONCLUSION

Studies to evaluate the effect of clearcutting on nutrient depletion and site productivity have progressed a long way. Early studies were concerned with nutrients removed by leaching and erosion after clearcutting. Then it was recognized that nutrient stocks remaining in the ecosystem after logging were important, and studies of those stocks were carried out. Nutrient dynamics and their effects on stocks and productivity were included in the analyses. Recent work has used systems analysis to keep a better account of the multitude of continuously changing factors in the ecosystem.

The proof of the effectiveness of the approach will lie in tests of the predictions generated by the models. From the perspective of applications, the most important prediction is that of wood yield. Because of the time required for such validation, it is too early to appraise the approach. Regardless of how accurate the predictions prove to be, they will be useful to the scientists who formulated the models on which they were based. Discrepancies between predictions and results will point up weaknesses in the models and indicate what studies are necessary to improve the accuracy of the predictions.

REFERENCES

Aubertin, G. M., and J. H. Patric. 1974. Water quality after clearcutting a small watershed in West Virginia. J. Environ. Qual. 3:243-249.

Ballard, R., and S. P. Gessel, eds. 1983. IUFRO Symposium on Forest Site and Continuous Productivity, Seattle, Washington, August 22-28, 1982. PNW-163. U.S. Department of Agriculture Forest Service, Portland, Oreg.

Bormann, F. H., and G. E. Likens. 1970. The nutrient cycles of an ecosystem. Sci. Am. 223(4):92-101.

Bormann, F. H., and G. E. Likens. 1979. Pattern and Process in a Forested Ecosystem. Springer, New York.

Bormann, F. H., G. E. Likens, D. W. Fisher, and R. S. Pierce. 1968. Nutrient loss accelerated by clear-cutting of a forest ecosystem. Science 159:882-884.

Brady, N. C. 1974. The Nature and Properties of Soils. Macmillan, New York.

Chapin, F. S. 1980. The mineral nutrition of wild plants. Annu. Rev. Ecol. Syst. 11:233-260.

Cole, D. W., and C. M. Bigger. Undated. Effect of Harvesting and Residue Removal on Nutrient Losses and Productivity. Fifth Annual Report. University of Washington College of Forest Resources, Seattle.

Douglass, J. E., and L. W. Swift. 1977. Forest service studies of soil and nutrient losses caused by roads, mechanical site preparation, and prescribed burning in the Southeast. Pp. 489-503 in D. L. Correll, ed. Watershed Research in Eastern North America. A Workshop to Compare Results. Chesapeake Bay Center for Environmental Studies, Edgewater, Maryland. Smithsonian Institution, Washington, D.C.

Farnworth, E. G., T. T. Tidrick, C. F. Jordan, and W. M. Smathers. 1981. The value of natural ecosystems: An economic and ecological framework. Environ. Conserv. 8:275-282.

Feller, M. C., and J. P. Kimmins. 1984. Effects of clearcutting and slash burning on streamwater chemistry and watershed nutrient budgets in southwestern British Columbia. Water Resour. Res. 20:29-40.

Gessel, S. P., and D. W. Cole. 1965. Influence of removal of forest cover on movement of water and associated elements through soil. J. Am. Water Works Assoc. 57:1301-1310.

Hornbeck, J. W., and W. Kropelin. 1982. Nutrient removal and leaching from a whole-tree harvest of northern hardwoods. J. Environ. Qual. 11:309-316.

Hornbeck, J. W., G. E. Likens, R. S. Pierce, and F. H., Bormann. 1975. Strip cutting as a means of protecting site and streamflow quality when clearcutting northern hardwoods. Pp. 209-225 in B. Bernier and C. H. Winget, eds. Forest Soils and Forest Land Management. Proceedings of the Fourth North American Forest Soils Conference. Les Presses de l'Universite Laval, Que.

Johnson, D. W., D. C. West, D. E. Todd, and L. K. Mann. 1982. Effects of sawlog vs. whole-tree harvesting on the nitrogen, phosphorus, potassium, and calcium budgets of an upland mixed oak forest. Soil Sci. Soc. Am. J. 46:1304-1309.

Jordan, C. F. 1982. Amazon rain forests. Am. Sci. 70:394-401.

Kimmins, J. P. 1977. Evaluation of the consequences for future tree productivity of the loss of nutrients in whole-tree harvesting. For. Ecol. Manage. 1:169-183.

Leaf, A. L., ed. 1979. Impact of Intensive Harvesting on Forest Nutrient Cycling. Proceedings of a Symposium at Syracuse, New York, August 13-16, 1979. Northeast Forest Experiment Station, Broomall, Pa.

Likens, G. E., F. H. Bormann, and N. M. Johnson. 1969. Nitrification: Importance to nutrient losses from a cutover forested ecosystem. Science 163:1205-1206.

Likens, G. E., F. H. Bormann, N. M. Johnson, D. W. Fisher, and R. S. Pierce. 1970. Effects of forest cutting and herbicide treatment on nutrient budgets in the Hubbard Brook watershed-ecosystem. Ecol. Monogr. 40:23-47.

Miller, J. H., and M. Newton. 1983. Nutrient loss from disturbed forest watersheds in Oregon's coast range. Agro-Ecosystems 8:158-167.

Monk, C. D. 1975. Nutrient losses in particulate form as weir pond sediments from four unit watersheds in the southern Appalachians. Pp. 862-867 in F. G. Howell, J. B. Gentry, and M. H. Smith, eds. Mineral Cycling in Southeastern Ecosystems. ERDA Symposium Series. CONF 740-513. U.S. Energy Research and Development Administration, Washington, D.C.

Peterson, C. E., and S. P. Gessel. 1983. Forest fertilization in the Pacific Northwest: Results of the regional forest nutrition research project. Pp. 365-369 in R. Ballard and S. P. Gessel, eds. IUFRO Symposium on Forest Site and Continuous Productivity, Seattle, Washington, August 22-28, 1982. PNW-163. U.S. Department of Agriculture Forest Service, Portland, Oreg.

Pierce, R. S., C. W. Martin, C. C. Reeves, G. E. Likens, and F. H. Bormann. 1972. Nutrient loss from clearcuttings in New Hampshire. Pp. 285-295 in Watersheds in Transition. American Water Resources Association Symposium. Colorado State University, Ft. Collins, Colo.

Reiners, W. A. 1981. Nitrogen cycling in relation to ecosystem succession. Pp. 507-528 in F. E. Clark and T. Rosswall, eds. Terrestrial Nitrogen Cycles. Proceedings of a Workshop. Ecological Bulletins (Stockholm) 33. Swedish Natural Science Research Council (NFR), Stockholm.

Sollins, P., and F. M. McCorison. 1981. Nitrogen and carbon solution chemistry of an old growth coniferous forest watershed before and after cutting. Water Resour. Res. 17:1409-1418.

Swank, W. T., and J. E. Douglass. 1975. Nutrient flux in undisturbed and manipulated forest ecosystems in the southern Appalachian mountains. Pp. 445-456 in Symposium de Tokyo. Publication 117 de l'Association Internationale des Sciences Hydrologiques, Reading, Eng.

Swank, W. T., and J. E. Douglass. 1977. Nutrient budgets for undisturbed and manipulated hardwood forest ecosystems in the mountains of North Carolina. Pp. 343-364 in D. L. Correll, ed. Watershed Research in Eastern North America. A Workshop to Compare Results, Chesapeake Bay Center for Environmental Studies, Edgewater, Maryland. Smithsonian Institution, Washington, D.C.

Swank, W. T., and J. B. Waide. 1980. Interpretation of nutrient cycling research in a management context: Evaluating potential effects of alternative management strategies on site productivity. Pp. 137-158 in Forests: Fresh Perspectives from Ecosystem Analysis. Proc. 40th Annu. Biol. Colloq. Oregon State University Press, Corvallis.

Swift, M. J., O. W. Heal, and J. M. Anderson. 1979. Decomposition in Terrestrial Ecosystems. Studies in Ecology. Vol. 5. University of California Press, Berkeley.

Tamm, C. O. 1979. Nutrient cycling and productivity of forest ecosystems. Pp. 2-21 in A. L. Leaf, program chairman. Impact of Intensive Harvesting on Forest Nutrient Cycling. Proceedings of a Symposium at Syracuse, New York, August 13-15, 1979. Northeast Forest Experiment Station, Broomall, Pa.

Tamm, C. O., H. Holmen, B. Popovic, and G. Wiklander. 1974. Leaching of plant nutrients from soil as a consequence of forestry operations. Ambio 3:211-221.

Vitousek, P. M., J. R. Gosz, C. C. Grier, J. M. Melillo, W. A. Reiners, and R. L. Todd. 1979. Nitrate losses from undisturbed ecosystems. Science 204:469-474.

Webster, J. R., and B. C. Patten. 1979. Effects of watershed perturbation on stream potassium and calcium dynamics. Ecol. Monogr. 49:51-72.

West, D. C., and K. L. Mann. 1983. Whole-Tree Harvesting: Fourth Year Progress Report for 1982. Nutrient Depletion Estimates, Postharvest Impacts on Nutrient Dynamics and Regeneration. Environmental Sciences Division Publication No. 2184. Oak Ridge National Laboratory, Oak Ridge, Tenn.

Committee Comment

As illustrated by this case study, methods of analyzing nutrient budgets of ecosystems have developed slowly as scientists gradually have recognized the need to perform a complete accounting of nutrient stocks and fluxes if the significance of changes in flux rates induced by clearcutting is to be understood. This represents discovery of methods that were already well known in the business world, and it might be asked whether ecologists would have progressed more rapidly if they had been better versed in budget analyses as practiced in various disciplines. The answer is probably yes, but it is also clear that ecosystems are so different from businesses that those techniques have to be modified for use in studying ecosystem nutrient dynamics. Such features of ecosystems as differences in availability of nutrients due to variation in soil types and in the types of plants growing on them have no close parallels in the business world.

Whereas much progress has been made in measuring total stocks and fluxes of nutrients, site variation is great enough that very few general predictions are possible. There is no reason to believe, in principle, that developing a broad predictive ability is impossible, but it seems evident that the necessary data base is large and that its components might not yet all be identified. That is true even if we simply wish to predict growth rates of trees. If we also wish to predict the dynamics of other constituents of an ecosystem, such as herbivores, carnivores, and detritivores—as well as interactions among plant species—even more extensive data are required. Results of such analyses could be surprising. For example, lower availability of nutrients, in combination with physical stress, such as drought, might render trees much more susceptible to attacks by defoliating insects or fungi (Fearnside and Rankin, 1985). These attacks might decrease yields and change growth forms of the trees, leaving yields of usable wood products much lower than would be predicted simply on the basis of the availability of nutrients to support tree growth.

An important conceptual advance illustrated by this case study is sensitivity analysis. Sensitivity analysis is especially important when systems consist of many interacting factors that are of uncertain influence or that are difficult to measure accurately. Time and budgetary constraints prevent accurate measurement of all factors of interest, and choices need to be made as to which factors should receive the most attention. Predictions generated by models are unequally sensitive to variation in different parameters. Sensitivity analysis is a powerful method for deciding which factors need to be measured most accurately and which ones require only crude estimation. Sensitivity analysis helps to avoid consequential errors— errors that could lead to inappropriate management decisions. The natural inclination of most scientists to measure all quantities as accurately as possible might actually lead to poorer predictive abilities, given the investment of comparable resources, than an approach designed to obtain accurate measurements only for the quantities identified as critical.

The study of ecosystem nutrient dynamics is made difficult not only by the number of interacting factors, but also by the long duration of the most important processes. Development and testing of models might require decades of work that taxes the patience of investigators, funding agencies, and managers who must make decisions, whether or not sufficient data are available for them to predict the consequences of their decisions. This emphasizes the need for cooperation among all who are involved in developing plans for managing and using forest ecosystems.

Reference

Fearnside, P. M., and J. R. Rankin. 1985. Jari revisited: Changes and the outlook for sustainability in Amazonia's largest silvicultural estate. Interciencia 10:121-129.

24

Environmental Effects of DDT

Dichlorodiphenyltrichloroethane, better known as DDT, was a potent insecticide when first used in the late 1930s. It had no obvious side effects and was active against many insect pests. As a result, it was extremely widely used, both in the United States and elsewhere. But, as has since happened with so many other chemicals, DDT had unforeseen effects. Those effects resulted from its persistence—one of its initially attractive features. Because of this persistence, DDT was transported from its initial site of application by both biotic and abiotic factors until almost no part of the earth's surface was free of it. Animals in the oceans, lakes, and rivers, on land, and in the air had detectable amounts of DDT in their tissues. The resulting concern led to detailed monitoring, to increased understanding of how chemicals are transported through the environment and how quickly resistance to pesticides can evolve, and to a new awareness of how even the most careful scoping of a problem does not guarantee freedom from unpleasant surprises.

Case Study

JOHN BUCKLEY, Whitney Point, New York

INTRODUCTION

DDT (dichlorodiphenyltrichloroethane) was first synthesized in 1874, but its properties as an insecticide were not discovered until 1939. By the early 1940s, the United States was producing large amounts of DDT for control of vectorborne diseases, and its use is credited with saving millions of lives from diseases with insect vectors, such as typhus and malaria. Its use in the United States for agricultural purposes expanded rapidly after World War II, reached a peak of 80 million pounds in 1959, and decreased thereafter until it was terminated by cancellation of its registration in 1972. Approximately 1,350 million pounds of DDT were used in the United States during those 30 years. In addition to domestic use, hundreds of millions of pounds were exported.

Public concern over use of DDT became widespread in 1962, when Rachel Carson's *Silent Spring* was published. As a result of this concern, the pros and cons of DDT use were considered by the President's Science Advisory Committee, whose report, *Use of Pesticides*, issued in May 1963, recommended an orderly phasing out of DDT over a short period. As early as 1957, the federal government began restricting its own use of DDT, and, beginning in 1967, many registered uses of DDT were canceled. Finally, on June 14, 1972, after a hearing that generated 9,312 pages of testimony (from 125 experts) and more than 350 documents, the administrator of the Environmental Protection Agency (EPA) announced the cancellation of all remaining uses of DDT on crops. Uses for public health and quarantine purposes were not affected, and export was permitted. After appeals by both sides, the Court of Appeals for the District of Columbia ruled on December 13, 1973, that there was "substantive evidence" in the record to support the ban of DDT.

The cancellation of DDT was so strongly opposed by some agricultural interests that the 1974 report of the Committee on Appropriations of the House of Representatives (Congressional Record, November 1973, H. 9619) stated:

The [Environmental Protection] Agency was also directed to initiate a complete and thorough review, based on scientific evidence, of the decision banning the use of DDT. This review of DDT must take into consideration all of the costs and benefits and the importance of protecting the Nation's supply of food and fiber.

359

The resulting review, *DDT: A Review of Scientific and Economic Aspects of the Decision to Ban Its Use as a Pesticide*, was published by EPA in July 1975. All data reviewed supported the 1972 findings.

ENVIRONMENTAL PROBLEMS

The cancellation of DDT was based on its persistence, transport, biomagnification, and toxic effects and on the absence of benefits of DDT that were not available through less environmentally harmful substances. In the course of research on DDT and its metabolites, many insights into the movement, fate, and effects of pollutants in the environment were developed. DDT was the first pesticide to which rapid and extensive resistance evolved.

The earliest field studies were of DDT applied at 5 lb/acre for spruce budworm control. These studies concentrated on acute effects, such as deaths of birds, fish, and insects and other invertebrates. The acute LD_{50} (the dose lethal to 50% of the test organisms) in laboratory mammals varied from 60 mg/kg in dogs to 800 mg/kg in rats. Bioassays of fish, however, revealed LC_{50} values (concentrations in water lethal to 50% of the test organisms in a specified number of hours) of a few parts per billion (ppb). Captive birds had acute LC_{50}s (concentrations in food lethal to 50% of the test organisms) of 400-1,200 ppm (parts per million) when given in the diet for 5 days (Hill *et al.*, 1975). Because of the extreme toxicity of DDT to fish, attempts were made to avoid bodies of water in the course of aerial spraying. Avoiding such areas completely was not possible, because more than 50% of the DDT applied typically drifts outside the target area, often onto water.

DDT had been in widespread use only a short time before resistance developed in some pest species. The first recognized evidence of resistance to DDT was the failure of house fly control with DDT in southern Italy in 1947 after exceptional success in 1946. Shortly after discovery of resistance to DDT, resistance to other insecticides was observed in the field, but development of resistance was not universally recognized as late as 1956 (Brown and Pal, 1971). Standardized laboratory methods for detection of resistance were worked out under the sponsorship of the World Health Organization. By the time of cancellation of DDT in 1972, more than 200 species of invertebrates were resistant to DDT, cyclodiene, or organophosphorus insecticides; some species were resistant to all three. Resistance in vertebrates is less common, but has been reported to occur in some frogs in Mississippi (Boyd *et al.*, 1963).

Resistance has several ramifications in relation to environmental effects. Indirectly, inability to control resistant insects has led to an increase in entomological research to find alternative methods of control. Generally referred to as IPM (integrated pest management), these methods depend on greater understanding of the life history and ecology of the pest, and they result in much more selective control. Pesticides are used in most applications of IPM, but only sparingly and in combination with other methods, so that there is less contamination of the environment.

In the mid-1950s, massive spraying programs were undertaken to save the American elm by killing the elm bark beetle, which spread the imported Dutch elm disease fungus from tree to tree. Robin mortality was widespread, and robins dying with tremors characteristic of DDT poisoning were commonly seen in treated areas. Dying robins appeared not only at the time of spraying, but also in the following spring before additional spraying. Chemical analyses showed that soils in these areas contained DDT at up to 18 ppm, and earthworms that fed on the leaves in the soils contained DDT at 53-204 ppm. Dead robins that had fed on the worms had up to 3 mg of DDT in their tissues (Barker 1958). Tests for residues in forest soils revealed that DDT disappeared slowly (Woodwell, 1961). Here was unequivocal evidence of the persistence of DDT and of the movement and concentration of DDT through a terrestrial food chain. Also during the mid-1950s, dichlorodiphenyldichloroethane* (DDD, also called tetrachlorodiphenylethane, or TDE) was applied to Clear Lake, California, to control gnats. Initial concentrations of 0.02 ppm in water yielded DDD residues of 10 ppm in plankton, 903 ppm in the fat of plankton-eating fish, 2,690 ppm in the fat of predatory fish, and 2,134 ppm in the fat of grebes that fed on fish (Hunt and Bischoff, 1960).

In the late 1950s, the first indication of worldwide spread of DDT residue was observed. It was discovered (Buckley, 1979)

that shark liver contained detectable amounts of DDT and its metabolites. This discovery was so surprising that it was not believed. The initial presumption was that the sample analyzed had been inadvertently contaminated; but a second sample, taken with precautions to avoid contamination, also contained DDT residues. Because DDT was not applied to the oceans, it appeared that the residue in shark liver had to be a naturally occurring substance, but residue analysis of pre-DDT shark liver oil detected no DDT. Finally, it was assumed that the DDT must have been accidentally dumped into the ocean and that the shark liver contamination was local. Residue analysis showed at least small amounts of DDT widely distributed in sharks and other marine fish, thus requiring us to acknowledge long-range movement of DDT and biological magnification of these residues. Laboratory studies confirmed the bioaccumulation.

Later analyses of tissues from penguins, skuas, and seals from the

---

*DDD was used directly as an insecticide, but also formed by metabolism of DDT.

Antarctic and fur seals from the Pribiloff Islands in the Bering Sea confirmed the occurrence of DDT and its metabolites far from any area where it had been applied. DDT was truly ubiquitous (Anas and Wilson, 1970; Sladen *et al.*, 1966).

The degree of bioconcentration of DDT from the environment into living tissues and of residues from prey to predator was so striking as to be almost unbelievable (U.S. EPA, 1975):

Experimental data showed that DDT can be biologically concentrated by a variety of aquatic organisms at all trophic levels. Phytoplankton, the dominant oceanic vegetation and primary food source for marine animals, concentrates DDT from seawater into its cell membranes. Water-fleas (*Daphnia*), a food source for many freshwater fish species, accumulated 9.0 ppm in tissues after three days exposure to 80 pptr (parts per trillion). This represents a bioconcentration factor of 112,500 times the exposure level. Rainbow trout exposed to 1.0 ppm DDT (wet weight) in food and 10 pptr in water for 84 days contained 2.3 ppm as whole body residues. Exposure to food alone resulted in residues of 1.8 ppm (a concentration factor of 1.8 X) and exposure to water alone yielded residues of 0.72 ppm (a concentration factor of 72,000 X). In fish fed 1 mg/kg DDT/day, 73% of the DDT residues were present 90 days after the fish were transferred to clean food.

DDT is not equally toxic to all species of animals. Control programs for arthropods have sometimes resulted in increased pest damage, because predators of a pest were eliminated by DDT. For example, DDT applications in apple orchards eliminated populations of predaceous ladybird beetles, so that red-mite populations formerly controlled by the ladybird beetles reached outbreak proportions. This particular mite is not susceptible to DDT and was hardly influenced directly by the DDT that killed the beetles (Helle, 1965). In other cases, less susceptible populations have flourished after control programs, presumably because of reduction in competition as a result of elimination or reduction of more susceptible species. Cope (1961) reported that, after treatment of 72,000 acres of the Yellowstone River watershed, the total numbers of invertebrates had recovered within a year, but the species composition was still altered. Plecopterans and ephemopterans were reduced, but trichopterans and dipterans occurred at higher numbers at the end of the year. In some cases, selective toxicity has resulted in reductions of species that are normally used as food by valued predators. Ide (1967), in studies of the effects of DDT applied at 0.5 lb/acre in the forested watershed of the Mirimachi River in New Brunswick, observed that fewer insect species emerged in streams affected by DDT, and that those most severely reduced were the large ones, such as caddis flies, on which salmon mainly feed. Recovery of stream fauna required up to 4 years.

Studies carried out at the Patuxent Wildlife Research Center demonstrated that stored residues of DDT could be mobilized during weight loss, and this resulted in mortality after exposure to DDT had stopped (Van Velzen *et al.*, 1972). These studies provide a basis for understanding the cause of some of the delayed mortality observed in the field.

During the 1960s, more subtle effects of DDT and its metabolites were discovered. The striking eggshell thinning in some predatory birds—such as bald eagles, ospreys, peregrine falcons, and pelicans—was clearly the cause of decreases in their populations. Eggshells of the affected species collected before 1945 were found to be noticeably thicker than those collected in the late 1940s and later (Anderson and Hickey, 1972). It was hypothesized that this eggshell thinning resulted in decreased reproductive success, and field studies of the brown pelican supported the hypothesis (Risebrough *et al.*, 1971). Laboratory studies confirmed that thinning of shells occurred in mallards (Davison and Sell, 1974), American kestrels (Peakall *et al.*, 1973), and ring doves (Haegele and Hudson, 1973) when dichlorodiphenyldichloroethylene (DDE) was incorporated in the diet of these birds. Since the cancellation of DDT in 1972, there has been a marked decrease of DDT residues in brown pelicans and improved reproduction in those pelicans on the Pacific, Gulf, and Atlantic coasts and in ospreys in the eastern United States (U.S. EPA, 1975).

DDT alters the behavior of some fish, and some of the changes are clearly detrimental to the welfare of the species. Atlantic salmon parr, for example, select water temperatures below those to which they are acclimated when exposed to DDT at 10 ppb or less and temperatures higher than acclimation temperature when exposed to DDT at 10-100 ppb. Furthermore, exposed salmon are less active than control fish. Rainbow trout select higher than acclimation temperatures and exhibit high-temperature shock followed by death (Javaid, 1972a).

KEY ISSUES

By 1970, it was evident that DDT residues were everywhere and that reducing the residues would require huge reductions in the use of DDT. Contamination of fish used for human food beyond the contamination limit approved by the Food and Drug Administration (FDA) occurred widely, especially in freshwater fish. Residues also appeared in many meats and dairy products. Surveys of human fat revealed virtually universal contamination by DDT, and residues in human milk gave rise to concern over the desirability of breastfeeding. That DDT residues appeared far from where DDT had been applied and lasted for years or even decades suggested that most uses would have to stop; control of application sites and care in application would not be enough to reduce environmental residues substantially.

The low cost of DDT, its relatively low toxicity to mammals (including humans), and its persistence made its continued use highly desirable for public health vector control programs. Although concern for its toxicity is greater now than when DDT was first used in the United States (FDA workers are said to have attested to its harmlessness by putting a spoonful in their coffee), there is still no clinical or epidemiological evidence of damage to man from approved uses of DDT, despite its demonstrated tumorigenicity in mice.

ECOLOGICAL APPROACHES TO THE ENVIRONMENTAL PROBLEMS ASSOCIATED WITH DDT

Ecologists first became concerned about ecological effects of DDT when it began to be used for control of forest insects, especially when applied from aircraft over hundreds of thousands of acres. The earliest studies consisted of field observations of treated areas. Mortality at the time of treatment or shortly thereafter was the criterion of the effect of treatment. These first field observations revealed some mortality of birds and fish in streams in treated areas and massive mortality of stream insects. For example, it was noted in one study that cutthroat and brook trout stomachs contained 99% crayfish after spraying, compared with none before, presumably because of the almost complete elimination of their insect food (Adams *et al.*, 1949). An experimental area was treated annually for 4 years with DDT, and breeding bird populations were studied. Three species decreased substantially, but no changes were detected in 23 others (Robbins *et al.*, 1951).

Studies during the early 1950s, principally in the laboratory, continued to concentrate on acute effects of DDT. In the late 1950s, the Dutch elm disease control program resulted in insights into food-chain transfer and accumulation of toxic materials in terrestrial environments. Studies of DDT (and other chlorinated organic pesticides) thereafter regularly looked for phenomena associated with food-chain accumulations. The general belief among toxicologists had been that metabolism of toxic substances invariably decreased their toxicity, but it was discovered that some metabolites of DDT, especially DDE, were more persistent and in some cases more active biologically than the original compound. This discovery laid the groundwork for research that eventually explained the decline of a number of bird species through eggshell thinning.

Pesticide-wildlife studies also led to insights into repopulation by songbirds of areas from which birds were removed. A study by Robbins and colleagues demonstrated that, if half or more of the songbirds were removed from a 40-acre tract of bottomland forest in Maryland during the early part of the nesting season, they were replaced within a few days, as judged by counts of singing males before and after the removal (U.S. FWS, 1963).

By the early 1960s, it had become evident that many freshwater sport and commercial fish were becoming so contaminated with DDT and its metabolites (and with other chlorinated hydrocarbons) that they were not suitable for human food. Concentrations were sometimes high enough to impair fish reproduction. For example, lake trout from Lake George, New York, produced eggs containing DDT and metabolites at 2.9 ppm or more; these eggs hatched, but the fry died (Burdick *et al.*, 1964).

USES OF ECOLOGICAL KNOWLEDGE AND UNDERSTANDING

Valued Ecosystem Components

No formal effort was made to identify valued ecosystem components, although it was the effects on valued ecosystem components that eventually resulted in eliminating the use of DDT in the United States. Among these effects were the decrease in songbird populations caused by massive control programs, such as those for spruce budworms in forests and Dutch elm disease in suburban areas; the declines in sport fish, such as lake trout in Lake George, caused by blackfly control programs; the contamination of fish and game desired as human food to a point greater than that permitted in domestic animals used for food; and the decline of avian predators at the top of food chains, particularly aquatic ones.

Importance of the Impacts

Probably of greatest importance were the drastic population decreases over large regions or continents of bird species at the top of their food chains. Some of these decreases—such as those of the osprey, brown pelican, peregrine falcon, and bald eagle—resulted in the extirpation of the species over vast areas of the United States and were grounds for legitimate concern over total extinction of these species.

Boundaries of the Problem

Effects of DDT in the ecosystem were at first perceived to be local and restricted to areas where DDT was applied, but findings from studies of DDT kept expanding the boundaries of the problem. Eventually, it was recognized that DDT residues appeared throughout the world. Studies were conducted, often cooperatively with scientists in other nations, to elucidate the problems wherever they existed, e.g., raptor decline in Europe and residue appearance in the Arctic and Antarctic and in marine environments throughout the Northern Hemisphere. Actions by the United States to resolve the problems, however, were restricted to the control of DDT within the United States. In fact, because DDT manufactured in the United States was superior to others, manufacture for use in international health programs was permitted after the ban on use in the United States took effect. For several years, unfortunately, the manufacturer was less adept at controlling wastes from the manufacturing process than at maintaining purity of the product, so considerable waste DDT and related products continued to be emptied into the Pacific Ocean, with detrimental effects on the brown pelican and other species.

Study Strategy

Because of the nature of the problem, there was no formal study strategy. Studies simply built on earlier studies. The earliest studies were carried out largely by government scientists, especially fish and wildlife biologists. As results were disseminated, more people became interested in the problem, and more academic scientists began to participate. Extensive, informal cooperation developed among scientists in the United States, Canada, and Great Britain. Results of field and laboratory studies were exchanged, and these results were often used in shaping the work of other investigators. In addition, more formal cooperation was arranged through such international agencies as the North Atlantic Treaty Organization, the Organisation for Economic Co-operation and Development (OECD), and various bodies of the United Nations.

Monitoring

Measurements of eggshell thickness were made in many areas of North America and Europe by using eggs of predatory birds in museums and in nests. Thinning of 20% or more was found in many species (see Cooke, 1973, for a review of eggshell thinning). The thinning seems to be caused only by DDE, inasmuch as experiments with other chlorinated hydrocarbons showed thinning only when the other compounds were at nearly toxic concentrations, and it disappeared as soon as signs of toxicity diminished (Haegele and Tucker, 1974). In the years after cancellation of DDT use in the United States and after elimination of industrial discharge by Montrose Chemical Company's DDT manufacturing plant, residues in many aquatic species began to decrease, and field studies of predatory birds showed increasing shell thickness and improved reproductive success (U.S. EPA, 1975).

Extensive programs for monitoring pesticide residues, including those of DDT and its metabolites, began in the United States during the 1960s under the sponsorship of the Federal Committee on Pest Control. Results of these programs—in which sampling and analytical methods were standardized and which covered human food and animal feeds, air, freshwater, estuarine areas, soils, and selected members of the biota—were published in the *Pesticide Monitoring Journal*. Less extensive results were available through a cooperative program sponsored by OECD that provided comparable data from Canada, the United States, and several European countries.

It was in the course of these monitoring studies involving marine species that substances interfering with DDT residue analyses were identified as polychlorinated biphenyls (PCBs). This discovery, by the Swedish chemist Jensen (1966), laid the groundwork for extensive studies of what has turned out to be the most severe nonpesticide chemical contamination problem in the world. Enactment of toxic-substance control legislation in the United States was probably a result of studies on PCBs.

Effects of DDT and its metabolites undoubtedly are aggravated by other hydrocarbon residues. Hardly any wild species on which residue analyses have been completed contain only DDT residues, and laboratory studies have shown that many of the other compounds are at least as toxic as DDT. The interaction of these residues is poorly understood.

SOURCES OF KNOWLEDGE AND UNDERSTANDING

Generally Accepted Ecological Facts

Studies of acute toxicity of DDT to birds and fish were undertaken with bioassay methods devised by pollution control biologists and toxicologists. Data on mammals, at least laboratory rats and mice, were available and, in the absence of specific tests on wild mammals, were used to predict effects on wild mammals. Feeding studies were conducted on pheasants and bobwhite quail, which were available because they were routinely raised for release to the wild. Later, to answer specific questions, studies were carried out on species not routinely kept in captivity. Such studies usually required prior development of appropriate husbandry. The captive birds studied eventually included bald eagles, kestrels, cowbirds, mallards, pigeons, and others. Similarly, trout and goldfish were exposed, and LC_{50}s for 24-48 hours were determined. The aquatic organisms studied included other species of fish and a number of invertebrates. Relatively elaborate exposure systems for aquatic species were devised to overcome the problems of DDT adsorption on surfaces and loss to the air through codistillation.

On the basis of these laboratory-determined estimates of toxicity, naturalists observed field applications of DDT for control of forest insects. In addition to observations of deaths of birds and aquatic organisms, analytical chemistry was used to measure DDT residues in organisms from treated areas. General knowlege of food chains, coupled with residue analyses, led to predictions of effects on organisms at higher trophic levels, which were indeed observed.

An iterative process of field observation, chemical analysis, laboratory bioassay, and field experimentation was followed, and a reasonable understanding of the movement, concentration, and effects of DDT began to emerge. The ecological theory of food chains and cycling of nutrients (substituting "toxicants" for "nutrients") was used. In the case of decreasing populations of predatory birds, a hypothesis of pesticide causation emerged. So, too, did a series of hypotheses on sublethal effects operating through behavioral changes—e.g., increased stress—and selection of undesirable temperatures.

Some results of field studies of effects on bird populations were surprising. In some environments, few dead birds were found, although many should have been, because residue in or on prey insects exceeded the predicted LC_{50} for birds. At the same time, populations of birds did not decrease noticeably. Two kinds of studies were carried out: one to determine disappearance of bird carcasses, the other to determine repopulation. It was hypothesized that only a small fraction of bird carcasses present were found and that, at least in areas of a few to hundreds of acres in relatively homogeneous habitat during the early nesting season, repopulation would take place very rapidly. Studies carried out at Patuxent Wildlife Research Center in Laurel, Maryland, confirmed both hypotheses.

Some 30-90% of the small-bird carcasses spread along a powerline through a wooded area were not found by trained observers 4 days later; some of the carcasses were simply not seen, but many were later determined to have been buried by beetles or carried away by scavengers. Where 44% of the singing males of nine species of songbirds on a 100-acre tract of bottomland forest were removed by mist netting, the removal was estimated at only one-fourth of the actual removal. Banding data showed that an influx of new birds,which took place immediately after the 3-day removal period, made it impossible to measure accurately the number of birds removed (U.S. FWS, 1963).

Theory and General Principles of Ecology

Knowledge of trophic interactions and food webs was an inherent part of the studies of accumulation of DDD at Clear Lake, California, and of the studies of robin mortality after DDT spraying for Dutch elm disease control. Mortality of bald eagles and other predators was shown to be caused by DDT (and other pesticides) accumulated through the food webs of which these predators were a part. Eggshell thinning also was caused by food-chain accumulations of DDE, which was formed from DDT by organisms lower in the food chain. The importance of the eggshell thinning and consequent reduced reproductive success was understandable only in terms of population dynamics. The observations of distribution of DDT in field studies and monitoring led Metcalf (1972) to devise microcosms in which radioactively labeled substances could be followed as they were partitioned into different environmental compartments.

The Value of Experiments

Studies in the 1940s involved treatment of experimental areas with amounts of DDT similar to those used in actual forest insect control programs. It was hoped that the use of control and experimental areas and the study of the areas before and at various periods after treatment would make it possible to detect changes in the nontarget biota, including repopulation if depopulation occurred. These studies were helpful in planning large-scale studies.

When first introduced in the United States, DDT was believed to be harmless to people and all other valued organisms. Therefore, the registration permitted use in concentrations high enough to be effective. Tolerances in food were based on the amount that would be present as a result of the approved uses. In the 1950s, DDT residues were discovered in cows' milk, so applications to animal feed crops were reduced to the point where the residues were not detectable in milk (i.e., to below about 1 ppm at that time).

Expert Judgment

Probably the most outstanding example of the role of expert judgment was that played by trained ornithologists (especially Hickey), who first noted the thinning of eggshells and hypothesized that it was caused by pesticides. In 1965, Hickey had convened an international conference to bring together what was known about decreasing peregrine falcon populations (Hickey, 1969). Hickey suggested that pesticide residues were probably responsible for reduced reproduction, primarily through egg destruction and egg-eating, but there was no mention of shell thinning. In the next several years, the hypothesis of eggshell thinning was formulated. Another pertinent example of expert judgment was provided by Rachel Carson, a marine biologist, who saw and forcefully exposed the effects of widespread pesticide use on the functioning of the natural environment (Carson, 1962).

REFERENCES

Adams, L., M. G. Hanavan, N. W. Hosley, and D. W. Johnston. 1949. The effects on fish, birds, and mammals of DDT used in the control of forest insects in Idaho and Wyoming. J. Wildl. Manage. 13:245-254.

Anas, R. E., and A. J. Wilson. 1970. Residues in fish, wildlife, and estuaries. Organochlorine pesticides in fur seals. Pestic. Monit. J. 3:198-200.

Anderson, D. W., and J. J. Hickey. 1972. Eggshell changes in certain North American birds. Pp. 514-540 in Proceedings of the 15th International Ornithological Congress, Symposium on Chemical Pollutants. E. J. Brill, Leyden, Neth.

Barker, R. J. 1958. Notes on some ecological effects of DDT sprayed on elms. J. Wildl. Manage. 22:269-274.

Boyd, C. E., B. Vinson, and D. E. Ferguson. 1963. Possible DDT resistance in two species of frogs. Copeia 1963:426-429.

Brown, A. W. A., and R. Pal. 1971. Insecticide Resistance in Arthropods. World Health Organization, Geneva.

Buckley, J. L. 1979. Nontarget effects of pesticides in the environment. Pp. 73-81 in Pesticides: Their Contemporary Roles in Agriculture, Health and the Environment. Humana, Clifton, N.J.

Burdick, G. E., E. J. Harris, H. J. Dean, T. M. Walker, J. Skea, and D. Colby. 1964. The accumulation of DDT in lake trout and the effect on reproduction. Trans. Am. Fish. Soc. 93:127-136.

Carson, R. 1962. Silent Spring. Houghton Mifflin, Boston.

Cooke, A. S. 1973. Shell thinning in avian eggs by environmental pollutants. Environ. Pollut. 4:85-152.

Cope, O. B. 1961. Effects of DDT spraying for spruce budworm on fish in the Yellowstone River System. Trans. Am. Fish. Soc. 90:239-251.

Davison, K. L., and J. L. Sell. 1974. DDT thins shells of eggs from mallard ducks maintained on *ad libitum* or control-feeding regimens. Arch. Environ. Contam. Toxicol. 2:222-232.

Haegele, M. A., and R. H. Hudson. 1973. DDE effects on reproduction of ring doves. Environ. Pollut. 4:53-57.

Haegele, M. A., and R. K. Tucker. 1974. Effects of 15 common environmental pollutants on eggshell thickness in mallards and coturnix. Bull. Environ. Contam. Toxicol. 11:98-102.

Helle, W. 1965. Resistance in the Acarina: Mites. Advan. Acarol. 2:71-93.

Hickey, J. J., ed. 1969. Peregrine Falcon Populations: Their Biology and Decline. University of Wisconsin Press, Madison.

Hill, F. F., R. G. Health, J. W. Spann, and D. Williams. 1975. Lethal Dietary Toxicities of Environmental Pollutants to Birds. U.S. FWS Special Scientific Report—Wildlife No. 191. U.S. Fish and Wildlife Service, Washington, D.C.

Hunt, E. C., and A. I. Bischoff. 1960. Inimical effects on wildlife of periodic DDD applications to Clear Lake, Calif. Calif. Fish Game Bull. 46:91-106.

Ide, F. P. 1967. Effects of forest spraying with DDT on aquatic insects in salmon streams in New Brunswick. J. Can. Fish. Res. Bd. 24:769-805.

Javaid, M. Y. 1972a. Effect of DDT on the locomotor activity of Atlantic salmon, *Salmo salar*, in a horizontal temperature gradient. Pak. J. Zool. 4:17-26.

Javaid, M. Y. 1972b. Effect of DDT on temperature selection of some salmonids. Pak. J. Sci. Ind. Res. 15:171-176.

Jensen, S. 1966. A new chemical hazard. New Sci. 32:612.

Metcalf, R. L. 1972. A model ecosystem for the evaluation of pesticide biodegradability and ecological magnification. Outlook Agric. 7:55-59.

Peakall, D. B., J. L. Lincer, R. W. Risebrough, J. B. Pritchard, and W. B. Kinter. 1973. DDE-induced eggshell thinning: Structural and physiological effects in three species. Comp. Gen. Pharmacol. 4:305-313.

Risebrough, R. W., F. C. Sibley, and M. N. Kirven. 1971. Reproduction failure of the brown pelican on Anacapa in 1969. Am. Birds 25:8-9.

Robbins, C. S., P. F. Springer, and C. G. Webster. 1951. Effects of five-year DDT application on breeding bird population. J. Wildl. Manage. 15:213-216.

Sladen, W. J. L., C. M. Menzie, and W. L. Reichel. 1966. DDT residues in Adelie penguins and a crab-eater seal from Antarctica: Ecological implications. Nature 210:670-671.

U.S. EPA (U.S. Environmental Protection Agency). 1975. DDT: A Review of Scientific and Economic Aspects of the Decision to Ban Its Use as a Pesticide. EPA-540/1-75-022. U.S. Environmental Protection Agency, Washington, D.C.

U.S. FWS (U.S. Fish and Wildlife Service). 1963. Pesticide-Wildlife Studies: 1963. A Review of Fish and Wildlife Service Investigations During the Calendar Year 1963. FWS Circular 199. U.S. Fish and Wildlife Service, Washington, D.C.

Van Velzen, A. C., W. B. Stiles, and L. F. Stickel. 1972. Lethal mobilization of DDT by cowbirds. J. Wildl. Manage. 36:733-739.

Woodwell, G. M. 1961. The persistence of DDT in a forest soil. For. Sci. 7:194-196.

Committee Comment

The wide use of DDT after World War II for the control of insects that caused public health problems or were agricultural and forest pests created a global environmental problem. Scientists and regulators were unprepared to deal with this problem. Before 1947, the sole concern of regulatory processes for pesticides was protection of purchasers from fraud. The passage of the Federal Insecticide, Fungicide, and Rodenticide Act in 1947 permitted attention to the possible effects of a pesticide on nontarget organisms other than people and domestic animals; but in practice, if the proponent of a use certified that the product was both effective and safe, registration was granted. The regulatory agencies—the Food and Drug

Administration (FDA) and the U.S. Department of Agriculture (USDA)—required additional information from the proponent; until they were satisfied, they could withhold registration (USDA) or a tolerance (FDA). However, a proponent could obtain a "protest registration," shifting the burden of proof from the proponent to the government, which had to disprove the proponent's claim. Furthermore, during this period, few on the staff of the pesticide registration office were knowledgeable about fish and wildlife. Little concern for ecological problems of any kind was evident in the regulatory agencies.

Scientists were little better prepared to deal with the problems posed by DDT. Toxicologists believed that acute toxic effects of the chemicals were the central issues of concern and that the metabolism of toxic substances invariably resulted in decreased toxicity. The possibility that metabolites of DDT might be as toxic as or even more toxic than DDT itself was not considered initially.

Ecological investigations were similarly narrow. At first, ecologists concentrated their attention on field observations of areas that had been directly treated with DDT. These studies revealed much about acute toxicity of DDT to different organisms and rates of recolonization of treated areas. But ecologists were not prepared for the striking phenomena of movement of DDT far from the sites of its application or for the remarkable bioconcentration of DDT that resulted in reproductive failure of top carnivores. These ideas are now generally accepted, but in the late 1940s and 1950s were not. As a result, the first data indicating that DDT was globally distributed and that concentrations were extremely high in the tissues of organisms high on food chains were not believed.

The history of research to determine the results of the massive use of DDT and of how those results were achieved reveals strikingly how the state of science at any time constrains the questions that are likely to be asked and the range of answers that are considered reasonable. Had concepts of the trophic-dynamic organization of ecological communities been developed before 1950, progress in understanding the roots of the problems caused by DDT might have been much more rapid. Lindemann (1942) had published his pioneering paper on trophic-dynamic ecology, but this perspective had yet to become a part of the thinking patterns of ecologists at the end of World War II. Indeed, as this case study demonstrates, efforts to determine the causes of the effects of DDT were a major stimulus to the development of this field of ecology. Persistence of organic chemicals in their original form, or as toxic metabolites, is now routinely measured as an indicator of likelihood of ecological effects. Bioaccumulation of lipid-soluble chemicals to 10^4 or 10^5 times the concentrations present in the physical environment is now expected as a normal part of food-chain dynamics. Spread of materials far beyond the sites of their application is now considered characteristic of many chemicals in ecological systems.

Toxicology has likewise been influenced by research on the effects of DDT. That metabolites of toxins can be toxic is now commonly accepted. Efforts are no longer directed exclusively or even primarily to determining lethal doses of toxins. Much more attention is now paid to sublethal effects that might affect behavior, survival, and population dynamics, because investigators recognize that severe decreases in animal populations can occur in nature without massive direct mortality of adults. As a result, although it can be argued that one "DDT" was inevitable, it is equally clear that another "DDT" would be inexcusable.

Striking though these advances are, it is clear that the basic problems encountered in dealing with DDT can express themselves in the future. Metabolites, transport, and bioaccumulation are now routinely investigated, but we still know almost nothing about the effects of toxicants in mixtures. Nearly all toxicity studies still deal with chemicals one at a time. In nature, however, organisms encounter potentially toxic chemicals in complex mixtures. How these chemicals interact in the environment and in the bodies of animals to produce different and unexpected effects is still generally unknown. As long as this is true, effects that are unexpected and difficult to believe or explain are likely. The general recognition that complex mixtures of toxicants require concerted efforts means that the speed with which scientists and regulators can respond to the surprises of the future should be greater than was the case with DDT. Nonetheless, the major lesson of the DDT story would be lost if it did not heighten our awareness of the current state of ignorance about important processes that affect the responses of individual organisms, populations, and ecological communities to toxic materials introduced into the environment.

The real basis for banning of DDT was its ecological effects, rather than its effects on human health. Even today, the evidence on human health effects of DDT is inconclusive, whereas the data on adverse ecological effects are vast and convincing. Also, the decision to ban DDT, rather than to restrict its use, was based on the conclusion that there was no way to control its movement once it was released into the environment. The decision was made more palatable politically by the rapid evolution of resistance to DDT among crop pests. It is doubtful that the ban would have been proposed or sustained if DDT had not decreased substantially in its effectiveness as a result of evolved resistance. The agreement to allow continued manufacture of high-grade DDT for export for use in public health vector control programs also made the ban politically more acceptable.

The assessment of the environmental effects of DDT is a good example of the interplay between laboratory analyses, field tests, and conceptual developments in both ecology and toxicology. The constraints imposed by the state of both sciences are clear in retrospect. We hope that our awareness of those limits and of the deficiencies in knowledge will make scientists and regulators more alert to evidence that does not fit into our current picture of how the world works.

Reference

Lindemann, R. L. 1942. The trophic-dynamic concept of ecology. Ecology 23:399-418.

Index

A

Acid rain, 2, 98

Africa, *see* West Africa

Age
 factors in forest management programs,
  280–284
 harvesting and, 36
 size as indicator of, 32–33
 structure and life expectancy variations,
  33–34

Age classes
 of forests, 282–295, 297–298
 of halibut, 141
 influence of numbers of, on stability,
  33–34

Aggregations
 influence on harvestability, 32
 influence on stability of fish populations,
  33

Agricultural techniques to reclaim derelict
  land in Great Britain, 252–254

Algae, as indicators of eutrophication, 49,
  83, 303–304, 306, 308, 310, 312,
  315

Analog studies of environmental problems,
  75–76
 in California red scale control, 180
 in caribou protection, 76, 221
 in Lake Washington eutrophication
  control, 76
 in land reclamation projects, 264–265

Annual surplus production (ASP), in
  assessment of fishery stock
  abundance, 145

Anopheles funestus, 195

Anopheles gambiae s.l., 195, 201, 202

Anopheles gambiae s.s., 202

Anopheles species, 190–204

Anticoagulants, in vampire bat control, 35,
  155–158

Aphytis africanus, 177

Aphytis chrysomphali, 170, 186

Aphytis linganensis, 170, 171–172, 175,
  179, 180, 181, 186, 187, 188

Aphytis melinus, 170, 171–172, 174, 177–
  178, 179, 180, 181, 182, 186, 187,
  188

Aphytis species, in California red scale
  control, 166–167, 170, 171–172,
  174–182, 186–188

Atlantic halibut (*Hippoglossus
  hippoglossus*), 149

Atomic Energy Commission, U.S., studies
  of radiation effects, 79, 112, 331–344

B

Banked specimens, for preservation of
  records, 85–86

Baseline studies, value of, 111, 113

Bats, *see* Vampire bat control

Bean weevil (*Callosobruchus chinensis*), 5

Behavior, of individuals, 26–28

Biological introductions, *see* Introduced
  species

Biological magnification, 6

Biological monitoring, 3, 4, 81–87, 113–
  114. *See also* Monitoring studies

375

effects of size, 68–69, 99
Patuxent Wildlife Research Center, 363, 368
Peak District National Park, Great Britain, 250
Peak Park Planning Board, 253, 267
Pest control, 10, 14, 165–189. *See also* Pesticides; *and specific pests*
Pesticides
  in California red scale control, 167, 169
  evolution of resistance to, 10, 35–36, 73, 96
  in integrated pest management, 361
  in malaria control, 191, 192, 194, 196, 200–202
  timing in applications of, 73
  in vampire bat control, 154–158
  see also DDT; Toxic substances; *and specific pesticides*
pH factors affecting derelict lands in Great Britain, 260, 263
Phosphorus, effects of, in water, 65, 66, 308, 314–315
Pilot-scale studies of environmental problems, 78–79, 111
  in caribou protection, 78, 221
  of DDT effects, 78–79
  in land reclamation projects, 78
  in vampire bat control, 159
Plantation management in forest development, *see* Forest management program in New Brunswick
Plantation model, in forest development program, 288, 289–290
Plants, 5
  and biological monitoring, 83
  competition among, 53–54, 258–259, 263–264
  as invaders, 57
  mutualistic interactions of, 45–46
  natural defenses of, 41–42
  and population interactions, 38–39
  radiation studies in, 338
  role of, in reclamation of derelict lands in Great Britain, 248–273
  species diversity, 51
  and stability boundaries, 7
Plasmodium falciparum, 192, 193, 196, 200, 202
Plasmodium malariae, 192, 196, 202
Plasmodium ovale, 192, 196, 202

Plasmodium species, 190–204
Plasmodium vivax, 192
Plowshare Program, 334
Pollutants, 11
  in atmosphere, 98–99
  biological effects of, 11, 63–64
  and biological monitoring, 82–85
  cumulative addition of, 95, 96, 98, 100, 102
  effects on Lake Washington, 302–315
  and eutrophication in Lake Washington, 301–315
  in food chain, 63, 64, 94
  in Lake Erie, 63–64
  monitoring for effects of, 81–87, 113–114
  monitoring of, 82–86
  from nuclear radiation, 331–344
  see also DDT; Pesticides; Radionuclides; Toxic substances
Polychlorinated biphenyls (PCBs), 63, 114, 367
Polygynous mating systems, 34
Population biology, 23–37. *See also* Populations
Population fragmentation, 6
Population interactions, 38–46
  competitive, 42–45
  indirect effects of, 46, 47–48, 88
  mutualistic, 6, 45–46
  patterns of succession, 59
  predator-prey, 6, 40–42, 309–310, 324, 329
Populations
  age structure of, 33–34
  aggregations of, 32
  behavior of, 26–28
  biology of, 23–37
  carrying capacity and, 28
  control of, 27–28
  density-dependent factors in, 5, 28–29, 30
  dispersion patterns and migratory behavior in, 31–32
  dynamics of, 24–25, 28–35, 47–48
  effects of habitat changes upon, 26
  in forest development, 280–284
  genetic and evolutionary changes in, 24–25, 35–37
  growth rates in, 32–33
  harvesting practices affecting, 95

role of ecological knowledge in, 256–
261, 267–268
seminatural grasslands in, 254–256
soil seed banks in, 259, 273
sources of ecological information in,
261–266
successions in, 260–261
sward diversification in, 253–259, 261,
263
theoretical concepts in, 264
yield and relative growth rate in, 254,
257
Recolonization, 72, 99, 364–365, 368,
372. *See also* Reclamation of derelict
lands in Great Britain
Recommendations, 15–17
Red mites, nonsusceptibility to DDT, 362
Red scale control, *see* California red scale
control
Regulation of populations, 28–29
Reindeer, cesium-137 concentrations in,
334
Relative growth rate (RGR), 254, 257, 258
Renewable-resource management, 9–10,
14, 137–150, 275–300, 345–357
Reproductive potential (*r*), 28
harvesting species with low *r*, 30
Reproductive rate
change in, as indicator of stress, 83
and colonizing ability, 71
of mosquitoes, 197, 198, 201
of vampire bats, 154, 164
Resilience, of ecological communities, 56–
57
Resistance to pesticides, evolution of, 10,
35–36, 73, 96
Restoration projects, 12, 248–274
Roe deer (*Capreolus capreolus*), 33
Root rot, 169

S

Saguaro cactus (*Carnegiea gigantea*), 34
Salmon, 25, 31, 34, 36, 362
San Joaquin Valley, control of California
red scale in, 168, 177, 186
Scales, temporal and spatial, 3, 8–9, 16–
17, 68–74, 96, 98–100, 110
Scoping, in environmental problem-
solving, 107–109

Sea birds, and biological monitoring, 84–
85
Seasonal changes, and natural variability,
89
Seattle, Washington, *see* Lake Washington
eutrophication control
Seed banks, soil, 259, 273
Seedling yellows (*Tristeza*), 169
Sensitivity analysis
of clearcut logging effects, 352–353,
357
and management plans, 92
of nuclear radiation effects, 337–338
Sewage discharge into Lake Washington,
73, 95, 102, 302–303
Sex biases, 34–35
Sex ratios, 34–35, 36, 37
Size
as indicator of age, 32–33
of habitats, 32
harvesting and, 36
Social behavior, and extinction of small
populations, 70
Social interactions, 27–28, 70
Sockeye salmon (*Oncorhynchus nerka*),
34
Soil erosion, *see* Erosion
Soil seed banks, 259, 273
Southern Indian Lake project, 13, 14, 106,
111, 114, 317–330
analog studies of, 76
basis of predictions in, 112–113
changes in zooplankton community,
323–324, 329
committee comment on, 328–329
compared to Siberian reservoirs, 321
effects on whitefish, 26, 318, 320, 322,
329
experimental nature of, 318
flow diversion in, 72
mercury release in, 96, 323, 329
models used in, 320
nutrient fluxes in, 64–65, 320–321
predevelopment studies in, 319, 321
retrospective analysis of, 76, 321–326
shoreline erosion in, 320, 321, 322–323
temperature changes in, 323, 328
turbidity of water in, 65, 96, 320, 328–
329
Spatial factors, 3, 8–9, 16–17, 54, 68–74,
96, 99, 110